SECOND EDITION

ASSET MANAGEMENT EXCELLENCE

Optimizing Equipment Life-Cycle Decisions

SECOND EDITION

ASSET MANAGEMENT EXCELLENCE

Optimizing Equipment Life-Cycle Decisions

Edited by

John D. Campbell
Andrew K. S. Jardine
Joel McGlynn

CRC Press
Taylor & Francis Group
Boca Raton London New York

CRC Press is an imprint of the
Taylor & Francis Group, an **informa** business

CRC Press
Taylor & Francis Group
6000 Broken Sound Parkway NW, Suite 300
Boca Raton, FL 33487-2742

First issued in paperback 2019

© 2011 by Taylor and Francis Group, LLC
CRC Press is an imprint of Taylor & Francis Group, an Informa business

No claim to original U.S. Government works

ISBN-13: 978-0-8493-0300-5 (hbk)
ISBN-13: 978-0-367-86437-8 (pbk)

Library of Congress Cataloging-in-Publication Data

Asset management excellence : optimizing equipment life-cycle decisions / editors, John D. Campbell, Andrew K. S. Jardine, Joel McGlynn. -- 2nd ed.
  p. cm.
Previous ed. published under title: Maintenance excellence, 2001.
Includes bibliographical references and index.
ISBN 978-0-8493-0300-5 (hardcover : alk. paper)
  1. Production engineering. 2. Product life cycle. 3. Production management. I. Campbell, John D., 1946- II. Jardine, A. K. S. (Andrew Kennedy Skilling) III. McGlynn, Joel. IV. Maintenance excellence.

TS176.M336 2010
658.2'02--dc22   2010008254

Visit the Taylor & Francis Web site at
http://www.taylorandfrancis.com

and the CRC Press Web site at
http://www.crcpress.com

Contents

SECTION I   Maintenance Management Fundamentals

SECTION II   Managing Equipment Reliability

SECTION III   Optimizing Maintenance Decisions

SECTION IV   Achieving Maintenance Excellence

Preface

This is the second edition of the *Maintenance Excellence* book, which is now taking on the title of *Asset Management Excellence* as a result of the ever-changing nature of the business. In the time that has passed since the first edition, suffice it to say that shifts have occurred. The original authors and contributors of content provided sound information and principles related to working toward maintenance excellence at that time. For this edition, new authors and contributors have revisited the content and have updated and added information based on changes in thinking and the introduction of and improvement in technologies since the first edition.

It has been the opinion of many maintenance and asset management personnel in multiple industries that at the root of the discipline, "maintenance is maintenance." This has been true for many years, from the era of paper-based work-order systems through the evolution to computerized software, the Internet, and wireless technologies. The root principles are the same: personnel with tools (electronic or manual) address the needs of maintaining assets. The application of root principles—as well as the way enterprises are perceiving maintenance organizations today—is changing.

Maintenance and asset management organizations have some of the same pressures today as in the past, such as asset availability and reliability and regulatory requirements. Prevalent areas that have driven major transformations in recent times are globalization and consolidation and technology changes. These elements reflect changes in thinking. They challenge asset management and maintenance professionals to be more efficient in what they do at various levels. Globalization and consolidation have been particularly instrumental in the changes in maintenance standards, approaches, and the use of technology to become more efficient and cost-effective. For example, emerging wireless and radio-frequency identification (RFID) technologies are being heavily leveraged. RFID allows the status of certain components to be "read" without taking apart an assembly to physically inspect the component. Wireless technology allows maintenance personnel to have direct access to information in the field and to send information from the field. Through RFID, assets can provide information about themselves (e.g., expiration dates) or even "talk" to other assets with wireless technology. Some industries are using RFID technology to tell maintenance personnel the area in which the assets are physically located. In addition, organizations are using geographic information system (GIS) software and tools to visually display the location information and spatially enable their asset management and maintenance organizations. Now organizations have the ability to know where their assets are and to understand relationships between how assets have behaved over time and how assets relate to the changing world around them. The current edition of this book reflects some of these changes, trends, and concepts.

In recent years, an evolution of many of the tools, technologies, and thought processes has occurred. Many of these elements have matured and have allowed the deeper maintenance processes to be rethought. For example, there are trends in the mix of asset and service management principles. There are also trends in adjusting core solutions to provide strong industry-specific solutions. Solution providers have implemented consolidations to make decisions more focused. Increased regulatory pressures have forced many organizations to standardize processes and procedures to become more efficient and also to simply stay in business.

Many organization consolidations have taken place from both geographical and global perspectives. Enterprise leaders at all levels have realized that recognizing maintenance contributions to the organization or enterprise can have a major impact on various aspects of the business. Understanding the areas that need change or improvement to achieve maintenance excellence is challenging at times, if not overwhelming.

The focus on measurement programs, both strategic and operational, is being revisited, developed, or reinvented—in some cases to reflect the changes and challenges that have taken place over time. Many mature maintenance organizations at the enterprise and field level are starting to recognize how to use measurement programs as valuable "tools" to fix larger enterprise problems as well as to work the day-to-day local operational efficiencies desired in the field.

As time has passed, patterns and trends have emerged around the world in asset management. Consolidations have sometimes forced benchmarking some organizations and enterprises in order to take advantage of leading practices from other groups or companies to make them the best practices of the new and larger entity. Consolidations have helped mature the maintenance organizations through the use of enterprise standards in the areas of data, policies and procedures, software applications, and new technologies.

This second edition is a product of change and consolidation. Owing to acquisitions, some of the leading asset management thought leaders and consultants of IBM's global business services asset management solutions organization from around the world have contributed, updated, and added to the concepts and principles in this book. Many have seen and lived the changes in maintenance and asset management and understand the evolution of change. They have been on the forefront of providing services in order to move organizations and enterprises to the next generation of asset management and maintenance excellence.

It is not necessary to have a large organization or enterprise to use this book properly. Even the smallest maintenance departments can benefit. The information provided is all-encompassing. The authors and contributors recognize that there are different levels of maturity from one group to another. What is needed by one may be well established in another. It is necessary to understand which leading practice principles and concepts are the correct ones for one group's needs and to adopt them as its best practices. This edition includes the leading concepts and trends as well as new information on emerging areas and technologies.

It is recognized that maintenance entities in general have various and diverse needs in several key areas. Trends in using frameworks, or models, help organize and prioritize areas of focus, which is one of the concepts shared in this book. At the highest level of identifying needs to achieve maintenance excellence, there are major differences in asset classes and how they are maintained. Additionally, organization maturity and operational principles, as well as lower-level concepts for maintaining assets through their entire life cycle, should be considered.

The aim of this second edition is to provide a combination of practical and deep information. The content, theories, and methodologies can be used by maintenance entities of varying size for their own benefit or to generate thought for personal, group, or academic rigor.

Once armed with the information contained herein, a maintenance enterprise, organization, team, or individual can then take stock of the areas they personally need to address and organize for success. They can use models and frameworks to set a road map and priorities for improvement. They can apply what they have learned to evolve into something "new" or move themselves to the next generation of maintenance excellence.

Acknowledgments

We would like to acknowledge all of the first edition contributors and, in particular, the first edition editors, John D. Campbell, who has since passed on, and Andrew K. S. Jardine. Both have added much that is still an important part of the core content.

Many thanks to our colleagues who are part of IBM Global Services Asset Management Solutions, the largest asset management consulting organization in the world. They have come from varying industries, consulting companies, and locations around the world to add a new and updated view based on the changes that have taken place since the first edition. Their dedication to the discipline of asset management and vision of reaching maintenance excellence have made this second edition possible.

Organizing, reviewing, and adding content require effort from various individuals at many different levels. There is a mix of knowledge and personal dedication that can't be accurately measured but that is recognized by all involved with the original book as well as with this edition. We will not name each one, since we feel everyone has had an equal share in the efforts from reviewing and revising to consulting and adding new content. Each can be found in the Contributors section.

We would also like to thank our maintenance, engineering, and industry professionals, as well as our business partners, whom many of the editors and contributors learned from, communicated with, or consulted to discuss or confirm concepts and information in this book. The collaboration and academic discussion validated existing ideas and concepts or have driven new ones.

It is also recognized that many leading software providers and suppliers of technologies are an indirect source of input to many of the contributors' thoughts and input. These types of companies and their personnel consistently try to improve or develop their products with leading practice, innovation, and easy-to-use functionality. Exposure to—and sometimes interaction with—these companies has provided new ways of thinking and additional thoughts and concepts that have found their way into various content portions in both editions of the book.

Finally, we must express our most sincere appreciation to our clients, who have provided us with the opportunity to understand their needs over time. It is they who have made sweeping changes by proving us experiences and challenges faced during start-ups, mergers, acquisitions, and consolidations. They have provided a global environment in which we have been observers of sector and industry changes as well as participants and partners in driving new forefronts. They have sometimes taken themselves, and many of us, on cutting-edge journeys that have provided us with experiences that make us subject-matter experts in various fields. Without them and what they have provided, this book and this latest edition would be nothing more than an academic exercise.

We would like to extend our gratitude to Don Fenhagen. The *Asset Management Excellence* book development has been a multi-year process and involved the careful coordination of IBM asset management experts in geographically dispersed regions of the world. Don has tirelessly collaborated with these IBM asset management experts to ensure that we get the best possible contributions and portray a global view of best practices in asset management.

Don is a leader in the IBM Asset Management practice and has led consulting engagements in the public and private sectors. Don has built and leads the IBM Spatial Services area, which is focused on Spatially Enabling Asset Management solutions in the asset management space. Don is experienced in a wide variety of leading asset management systems and provides leadership and expertise in asset management enabling technologies including RFID, GPS, and GIS technologies. In addition to leading the IBM Spatial Service Practice, Don also leads the IBM GBS Smarter Water Delivery Team, which is part of the IBM Smarter Planet Solutions delivery campaign.

Joel McGlynn and Andrew Jardine

Editors

John D. Campbell was one of a very few individuals in Canada who put Canadian expertise in physical asset management on the world map. He authored the widely used book *Uptime: Strategies for Excellence in Maintenance Management,* which was published in 1995 and by early 2006 had sold over 16,000 copies. He coauthored *Planning and Control of Maintenance Systems: Modelling and Analysis* and coedited *Maintenance Excellence.* In addition to being an author and leading thinker in maintenance management, Campbell was a sought-after public speaker, appearing once on television in the United States. He was a recognized subject matter expert and advisor in maintenance management to several of the world's most widely recognized organizations ranging from Disneyland to NASA, or "Mickey Mouse to Rocket Science," as he used to joke. He developed a very strong and internationally recognized maintenance management consulting practice in Canada, consisting of about 40 consultants. In recognition of that capability and expertise, PricewaterhouseCoopers appointed Campbell as its Global Leader for the PwC Physical Asset Management practice, consisting of about 150 consultants worldwide. With his sponsorship and support, the University of Toronto's Physical Asset Management certificate program was born in 1999 and continues to this day, franchised for delivery worldwide.

Campbell was an incredibly busy person, but always made time to promote excellence in plant engineering and maintenance activities through his participation in numerous related voluntary events. He received, posthumously in 2006, the Sergio Guy Memorial Award from the Plant Engineering and Maintenance Association of Canada in recognition of his outstanding contributions to the maintenance profession.

In 2000, Campbell began discussions with his practice director about potential updates to *Uptime.* Although he soon began a long battle with cancer, he helped put in place the thinking and necessary encouragement for that 2nd edition to be written and published. John passed away in November 2002, but his work continued: The 2nd edition of *Uptime* was published in 2006. His colleagues from PwC formed both a perpetual scholarship and award at the University of Toronto for leading students in the university's maintenance management course. Through those awards, Campbell continues to contribute to both excellence in asset management and support to students in need.

Andrew K.S. Jardine is director of the Centre for Maintenance Optimization and Reliability Engineering (C-MORE) at the University of Toronto, where the focus is on real-world research in engineering asset management in the areas of condition-based maintenance, spares management, protective devices, maintenance and repair contracts, and failure finding intervals for protective devices. Details can be found at http://www.mie.utoronto.ca/cmore. Dr. Jardine is also professor emeritus in the Department of Mechanical and Industrial Engineering at the University of

Toronto. Dr. Jardine wrote *Maintenance, Replacement, and Reliability*, first published in 1973 and now in its sixth printing. He is coeditor with J.D. Campbell of the 2001 book *Maintenance Excellence: Optimizing Equipment Life Cycle Decisions*. His most recent book *Maintenance, Replacement and Reliablity: Theory and Applications*, co-authored with Dr. A.H.C. Tsang, was published by CRC Press in 2006. In addition to writing, researching, and teaching, Dr. Jardine has, for years, been in high demand as an independent consultant to corporations and governments the world over in matters related to the optimum use of their physical assets. Dr. Jardine has garnered an impressive array of awards, honors, and tributes. He was the Eminent Speaker to the Maintenance Engineering Society of Australia, as well as the first recipient of the Sergio Guy Memorial Award from the Plan Engineering and Maintenance Association of Canada, in recognition of his outstanding contribution to the maintenance profession. In 2008, Dr. Jardine received the Award for the Best Paper, which was presented in the category of Academic Developments and sponsored by the Salvetti Foundation, at the bi-annual conference of the European Federation of National Maintenance Societies, in Brussels. He is listed in the Canadian *Who's Who*.

Joel McGlynn is an IBM vice president and the enterprise asset management global leader for IBM Business Consulting Services within IBM Global Services. He was responsible for creating IBM's asset management practice in 1995 and now manages over 1,000 consultants on many different asset management engagements. He has built IBM's asset management practice into the largest focused asset management consultancies in the world. He has accomplished this by creating an asset management framework and a holistic approach to managing assets that allow organizations to better optimize and align their asset investment to support their mission and strategy. He focuses on managing assets using a portfolio view across the asset's total life cycle, including strategy and planning, evaluation and design, acquisition or building, operation maintenance, modification, and disposition. He also categorizes asset classes based on similar management attributes such as financial, real property, personal property, fleet, infrastructure, and continuous or linear assets. In recent years McGlynn recognized the emergence of smarter assets and the convergence of operational and IT asset classes. He is now heavily involved with IBM's smarter initiatives including IBM's Smart Buildings, Advanced Water Management and the Intelligent Utility Network offerings. He has over 20 years of experience in assisting clients to improve maintenance and manufacturing operations, inventory management, supply chain management, and procurement processes. McGlynn also assists clients in designing and implementing technology solutions to support these processes. McGlynn has emerged as one of the foremost thought leaders in the field of asset management. He is certified in APICS and both CPIM and CIRM.

Contributors

Joseph Ashun is a registered professional engineer in Ontario, Canada. Ashun is currently the global manager for systems and data with Barrick Gold Corporation in Toronto, Canada. Ashun graduated with a bachelor's of engineering in marine mechanical engineering from the Alexandria Maritime Academy in Alexandria. He is also a licensed professional merchant marine engineer officer. Ashun has worked for several years as a senior marine engineer officer and in maintenance management with Marietta Corporation, a leading Amenities Manufacturing Company. He has worked as a physical asset management consultant for about 10 years with Coopers & Lybrand, PricewaterhouseCoopers, and IBM. Ashun had the privilege of working with the late John Dixon Campbell in Coopers & Lybrand and later, PricewaterhouseCoopers.

Don Barry is an associate partner and leader of IBM Canada's Supply Chain Operations and Enterprise Asset Management Consulting Practice within IBM's Global Business Services Group. He has over 32 years in maintenance service delivery and support systems and application development including 3 years in field service management and 15 years directly in business process development and maintenance parts supply chain management. Barry is lead instructor of the University of Toronto Physical Asset Management Certification training, held twice a year at the University of Toronto and also taught at the University of West Indies. He has contributed to many IBM Thought Leadership articles. His consulting clients have included leading companies in computer technologies, airlines, train manufacturing, mining, oil and gas, CPG, power generation, and power transmission and distribution. In September 2007, Barry received the Federated Press Lifetime Achievement Award in Plant and Production Maintenance.

Thomas J. Callaghan graduated from Ryerson Polytechnical Institute in 1971 with a degree in power engineering technology. He was employed by Inco Ltd. from 1971 to 2008. He worked for 26 years as a superintendent in both maintenance and operating roles responsible for mining, refineries, construction, power, divisional shops, and central maintenance. He provided expertise to several external interest groups and regulatory authorities, including Ministry of Labour (Mining Electrical Regulations), Ontario Electrical Safety Authority, Canadian Standards Association (Technical Committee on Use of Electricity Underground), University of Toronto Professional Development Program (Industry Advisory for Physical Asset Management Program), Cambrian College (chair of Skytech Advisory Board), and Canadian Welding Bureau (vice chair of Authorizing National Body Committee).

Andrew Carey is an associate partner in IBM's global Business Services' Asset Management Group. Carey has subject matter expert level experience in facilities, capital projects and estates management procurement and processes. Within

IBM global Business Services he is the Thought Leader on Real Estate & Facilities Management within EMEA. Carey is particularly skilled at managing change in organizations via large, complex and demanding projects. He has led major organization change projects in complex clients (in both the public and private sectors) from inception to launch. Carey has European and global strategic sourcing consultancy experience with multi-national "blue chip" clients. Carey holds both a BSc (hons) degree in mechanical engineering and an MBA and is a Chartered Member (CEng) of the Institution of Mechanical Engineers.

Patrice Catoir holds a bachelor's degree in industrial engineering with a specialization in computer science from the École polytechnique de Montréal (1989) and is a certified member of the Ordre des ingénieurs du Québec (Quebec Order of Engineers). Catoir is a highly accomplished professional with solid national and international experience in a broad variety of industries. He has spent the last half of his 19-year career working for a number of global companies in Canada, the United States, England, France, Sweden, and Indonesia. Specializing in maintenance, operations, supply chain, and IT, he is currently in charge of capital planning and energy management for Molson, Canada.

Don Fenhagen is a service area leader in the IBM Global Business Service Asset Management Practice. Fenhagen is a leader in the Asset Management space and has led consulting engagements in the public and private sectors. He currently runs the IBM Spatial Services practice, which works to link traditional asset management systems, people, and processes with Geographic Information Systems. Fenhagen is experienced in a wide variety of leading asset management systems and provides leadership and expertise in asset management enabling technologies including RFID, GPS, and GIS technologies. His consulting engagements include asset management customers in the United States federal government, state and local municipalities, water utilities, DOTs, computer technologies, airlines, RFID technologies providers, and railroads. Fenhagen is the overall program manager and core editor for the second edition of the *Asset Management Excellence* book. Fenhagen has led the coordination and editing of *Asset Management Excellence*, which involves a worldwide group of specialized subject matter experts. He holds a bachelor's in international business from the College of Charleston and a MBA from Georgetown University.

Ron Green has over 25 years of experience applying information technology to many industry verticals, including state and local government, federal government, and the Department of Defense, healthcare, pharmaceuticals, and retail and distribution. Green has performed varied roles, including solutions consulting project management, systems integration, strategic planning, and facilitating organizational change. He is experienced in enterprise asset management, product life-cycle management, enterprise resource planning, supply chain management, e-marketplaces, collaboration, customer relationship management, warehouse and distribution system, financial management systems and supporting technologies. Green has a bachelor's of business administration from the College of Charleston and a master's of business administration from Savannah State University.

Brian Helstrom is a senior consulting architect with IBM in the Enterprise Integration Practice with expertise in the integration of asset management and maintenance management solutions. Dr. Helstrom is an information systems professional. He graduated from Brock University with a B.Sc. in computer science, from the University of Guelph with a master's degree in human–computer interaction, and from Parkwood University with a doctorate in information systems. Dr. Helstrom is also a master certified IT architect, master certified IT specialist, ITIL certified, certified information systems professional, and is an IBM certified consultant in technology solutions. He spent 8 years in manufacturing maintenance, working as a certified tradesman and supervisor as well as assisting industrial engineers. Over the past 10 years he has been implementing CMM solutions to support both technology asset life-cycle and physical asset life-cycle functions across multiple industries.

Frank "Chip" Knowlton, an associate partner within IBM's Global Business Services Business Unit, leverages his 25 years of industrial operations environment experience to perform process improvement/business process reengineering, team facilitation, project management, financial analysis, and maintenance/materials management services for a variety of public and private sector clients. As a leader within IBM's Public Sector Supply Chain Management (SCM) Practice, Knowlton is responsible for the operational, solution design and deployment, resource development and assignment, and delivery excellence aspects of a variety of SCM and enterprise asset management engagements. These engagements encompass the Total Life-Cycle Asset Management key processes (strategy, plan, design, procure, operate, maintain, modify, dispose) for asset types such as facilities, fleet, equipment, continuous, and IT. Knowlton is a graduate of the United States Merchant Marine Academy with a BS in marine engineering.

Jeffrey Kurkowski is an asset management professional with over 30 years of experience in the asset management discipline. He has direct IT experience, including 8 years in the information technology and programming fields for maintenance and engineering organizations. He has consulted for clients in the multiple asset classes in both asset intensive and service industries worldwide. Kurkowski has personal work experience in field maintenance, parts inventory, work flow process, heavy and light industry, and retail- and service-oriented areas. He is one of the founding principals of IBM's worldwide Asset Management Solutions consulting group, the largest in the world. He also manages IBM's Asset Management Center of Excellence. Today, he works with many Fortune 500 corporation executives, consults as a subject matter expert to clients and project teams, and is an external speaker on the topic of asset management to various groups around the world.

Eric Olson has more than 30 years of experience in the North American upstream petroleum sector. Now an associate partner in IBM Global Business Services, Chemicals and Petroleum practice, Dr. Olson previously had executive, managerial, and operational responsibilities in one of Canada's largest integrated oil companies. He also guided supplier relations and inventory management as vice president of a leading oilfield supply organization. His areas of expertise include operations management, petroleum engineering, asset management, and supply chain management. He holds a Ph.D. (geology, McMaster University) and a B. Eng. (geological, University of Saskatchewan), and is a member of the Alberta Association of Professional Engineers, Geologists and Geophysicists, and the Society of Petroleum Engineers.

Thomas Port is a senior consultant with IBM Business Consulting Services' global asset management practice, based in London, Ontario. Port holds a Bachelor's of Administration degree from Lakehead University in Thunder Bay, Ontario, and is a Certified Maximo Consultant. He has more than 20 years experience consulting management and employees in several major companies on maintenance performance and system improvements, specializing in strategic asset management assessments, planning and implementation, work order systems, preventive maintenance, maintenance planning and scheduling, process re-engineering, computerized maintenance system implementation and training, performance measurement and reporting and maintenance wrench time studies. Port has consulted with companies in mining, aerospace, manufacturing, forest products, utilities, and oil and gas. He has worked for a number of global companies in Canada, United Kingdom, United States, Caribbean, South Africa, and Australia.

Joe Potter is an associate partner in IBM's Global Business Services' Asset Management Group. He has subject matter expert level experience in asset management systems across all asset types and disciplines. Within IBM Global Business Services, he is a Thought Leader on asset management systems, particularly as they apply to real estate & facilities management situations. Potter is particularly skilled at helping organizations devise and implement their asset management systems strategy, typically as part of a larger business transformation agenda. He has led systems programs in complex asset intensive clients (both public and private sector) from strategy to deployment. Potter holds both a B.Sc. (Hons) degree in environmental sciences and an M.B.A. He is a member of the Institute of Asset Management.

Siegfried F. Sanders is a managing consultant with IBM's Global Business Services group, based in Raleigh, North Carolina. Sanders is a licensed professional engineer. He graduated from North Carolina State University with a bachelor's degree in mechanical engineering and a master's degree in business administration, and is a certified energy manager. Specializing in maintenance and materials management, he has over 25 years of experience in the assessment and implementation of strategy, management, and systems for maintenance, materials, and physical asset life-cycle functions. He has authored several journal publications, the latest which is "Amino Acid Production Plant Chooses Ejectors for Vacuum Drying" in *Process Heating* (June 2002). Sanders has also presented topics relating to maintenance and energy management to various national and international organizations.

Doug Stretton, P.Eng., is a director of PWC's Enterprise Asset Management practice and has 22 years of experience in mining, steel, utilities, oil and gas, and manufacturing. He is a recognized expert with specialized knowledge in maintenance and supply chain processes, maintenance software, and total productive maintenance. Stretton has worked in a variety of roles, from consultant to maintenance manager to manager of global maintenance systems.

Murray Wiseman is a founding director of Optimal Maintenance Decisions, Inc. and president of LivingReliability, Inc., where he developed the Living Reliability Centered Maintenance (LRCM) methodology. He manages LRCM and Condition Based Maintenance (CBM) optimization projects in the mining, energy, and transportation sectors. He specializes in continuous maintenance process improvement through the reconciliation of RCM knowledge with recorded failure events. Wiseman is the senior reliability consultant at the Naval Engineering Test Establishment in Montreal, Canada, and lead LRCM project manager at the world's largest integrated coal mining operation, Cerrejon Coal, in Colombia. In these roles he brings a new perspective to maintenance information management with the purpose of achieving stated reliability goals.

1 Asset Management Excellence

Don Barry

CONTENTS

1.1 INTRODUCTION

Asset management excellence is many things, done well. It's when a plant performs up to its design standards and equipment operates smoothly when needed. It's maintenance costs tracking on budget, with reasonable capital investment. It's high service levels and fast inventory turnover. It's motivated, competent trades.

Most of all, asset management excellence is the balance of performance, risk, and cost to achieve an optimal solution. This is complicated because much of what happens in an industrial environment is by chance. Our goal is exceptional performance. That isn't made any easier by the random nature of what we're often dealing with.

Maintenance management, though, has evolved tremendously during the past century. As Coetzee[1] said, it wasn't even contemplated by early equipment designers. Since then, Parks[2] has described how maintenance evolved initially from uncomplicated and robust equipment to planned obsolescence. From Kelly[3] we learned the progression from preventive and planned maintenance after World War II to industrial engineering, industrial psychology, reliability, and operations research in the 1960s and 1970s to condition monitoring, computerization, and life-cycle management in the 1980s. Moubray[4] described how our expectations have changed with the generations and how safety, environmental, and product quality issues now are as important as reliability. Campbell[5] showed us the way equipment changes have advanced maintenance practices, with predominant tactics changing from run to failure to prevention and then to prediction and eventually reliance. It's the reliance tactic that is the focus of this book.

Many organizations had developed a systematic approach to maintenance planning and control in the late 1970s and early 1980s, only to abandon it as we shifted to a global marketplace and major economic harmonization. Today, we must start over. We must first rebuild basic capabilities in maintenance management before we can prove the value of reliability management and maintenance optimization. We also must prepare for the technological advances that are rapidly coming available in concert with our maintenance strategy maturity.

This second edition introduces or reaffirms some key concepts we believe to be fundamental in how a leading organization should look at asset management:

- Five distinct yet congruous asset classes are developing (Figure 1.1):
  - Real estate and facilities
  - Plant and production
  - Mobile assets
  - Infrastructure
  - Information technology

  More and more assets are being developed with electronic intelligence and Internet protocol (IP) addresses that can allow them to communicate via the Internet. To acknowledge this fact, some solutions have evolved to recognize that the maintenance of one asset class is perhaps 80% the same process as another.
- The maintenance excellence asset management pyramid, introduced by John D. Campbell in *Uptime*, provides a fundamental yet holistic approach to understanding where an organization is in its maintenance maturity and can act as the baseline for where it wants to be (Figure 1.2).
- The asset management life-cycle model, also introduced by Campbell, should be considered to ensure that each organization understands the full impact of an asset purchase or disposition and the role maintenance can play to promote the length and quality of the asset's life (Figure 1.3).

The challenge of maintenance excellence, and the goal of this book, is to provoke thought on strategic issues around maintenance and to develop tactics that will minimize breakdowns and will maximize the rewards of planned, preventive, and predictive work.

| Real Estate and Facilities | Plant and Production | Mobile Assets | Infrastructure | Information Technology |

FIGURE 1.1   Five distinct asset classes have developed. Courtesy of IBM Global Business Consulting.[6]

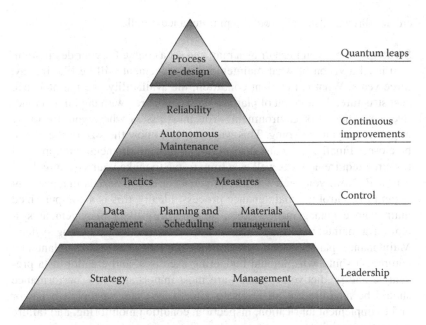

FIGURE 1.2   Maintenance Excellence Pyramid. Adapted from J. D. Campbell, Coopers & Lybrand Library, 1994.[7]

FIGURE 1.3   Total Lifecycle Asset Management, "Strategy through to disposition." Courtesy of J. D. Campbell, Coopers & Lybrand Library, 1994.[8]

There are three goals on the route to maintenance excellence:

- **Strategic:** First, you must draw a map and set a course for your destination. You need a vision of what maintenance management will be like in, say, three years. What is the plant condition, the availability, the maintenance cost structure, the amount of planned work compared with unplanned reactive work, the work environment? You must assess where you are today to get where you are going. This way, you will know the size of the gap to be closed. Finally, you must determine the human, financial, and physical resource requirements as well as a time frame to make your vision real.
- **Tactical:** Now, you need a work management and materials management system to control the maintenance process. Ideally, this is a computerized maintenance management system, an enterprise asset management system, or a maintenance module in an enterprise resource planning system. Maintenance planning and scheduling—for job work orders, plant and equipment shutdowns, annual budgeting exercises, and creation of a preventive and predictive program—are most important. Also, performance should be measured at all levels to effectively change people's behavior and to implement lubrication, inspection, condition monitoring, and failure prevention activities.
- **Continuous improvement:** Finally, if you engage the collective wisdom and experience of your entire workforce and adopt "best practices" from within and outside your organization, you will complete the journey to systematic maintenance management. But continuous improvement requires diligence and consistency. To make it work, you need a method, a champion, strong management, and hard work.

1.2   MAINTENANCE EXCELLENCE FRAMEWORK

As you read on, you will learn how to manage your equipment reliability and to optimize maintenance—the life cycle of your plant, fleet, facility, and equipment. The purpose of this book is to provide a framework. It is divided into four sections: (1) maintenance management fundamentals; (2) managing equipment reliability; (3) optimizing maintenance decisions; and (4) achieving maintenance excellence (Figure 1.4).

1.2.1   SECTION I: MAINTENANCE MANAGEMENT FUNDAMENTALS

Chapter 2 introduces an overview of the new basic understanding of how the asset classes have converged and how asset life-cycle dynamics have evolved. An overview of the basic strategies, processes, and approaches for managing equipment reliability through work management and leading maintenance management methods and tools is provided in Chapter 3.

Chapter 4 explores the following concept: if you can't measure it, you can't manage it. You'll learn how to monitor and control the maintenance process. We discuss the top-down approach, from setting strategic business plans to ensuring

FIGURE 1.4 Components of maintenance excellence.

that maintenance fully supports them. We also look at the measures needed on the shop floor to manage productivity, equipment performance, cost management, and stores materials management. When you pull it all together, it becomes a balanced scorecard.

Chapter 5 discusses the latest computerized maintenance management and enterprise asset management systems. We describe the basics of determining requirements and then of justifying, selecting, and implementing solutions that realize the benefits. Materials are the single biggest expense for most maintenance operations, so it's no wonder that poor maintainability is usually blamed on a shortcoming in parts, components, or supply.

Next, Chapter 6 describes how you can manage maintenance procurement and stores inventory to support effective and efficient work management and equipment reliability. We show you how to invest the limited parts inventory budget wisely to yield both top service levels and high turnover.

1.2.2 SECTION II: MANAGING EQUIPMENT RELIABILITY

The first chapter in this section (Chapter 7) addresses assessing risks and managing to international standards. We begin by looking at what equipment is critical based on several criteria and then by outlining methods for managing risk. To help you develop guidelines and easily assess the state of risk in an enterprise, we describe relevant international standards for managing risk.

A summary of the reliability centered maintenance methodology as the most powerful tool to manage equipment reliability is provided in Chapter 8. This often misunderstood yet incredibly powerful approach improves reliability while helping to control maintenance costs in a sustainable way.

Then, Chapter 9 examines reliability by the operator—what many leading companies are naming their total productive maintenance programs. We describe how equipment management and performance is everyone's job, especially the operator's.

1.2.3 SECTION III: OPTIMIZING MAINTENANCE DECISIONS

Chapter 10 provides the basics for using statistics and asset life cost in maintenance decision making. We hear about a company's Six Sigma quality management program, but the link between "sigma," or standard deviation, and maintenance management is often seen as subtle at best. In this chapter, we show you that collecting simple failure frequency can reveal the likelihood of when the next failure will happen. While exploring failure probabilities, we guide you through the concept of life-cycle costs and discounting for replacement equipment investments.

With the fundamentals in place and ongoing effective reliability management, Chapter 10 begins the journey to optimization. Here, we explore the use of mathematical models and simulation to maximize performance and minimize costs over equipment life. You'll find an overview of the theory behind component replacement, capital equipment replacement, inspection procedures, and resource requirements. Various algorithmic and expert system tools are reviewed along with their data requirements.

Chapter 11 focuses further on critical component and capital replacements. We include engineering and economic information in our discussion about preventive age-based and condition-based replacements. Then, Chapter 12 takes an in-depth look at how condition-based monitoring can be cost-effectively optimized. One of the biggest challenges facing the maintenance manager is figuring out what machine condition data to collect and how to use it. We describe a statistical technique to help make practical decisions for run to failure, repair, or replacement, including cost considerations.

Chapter 13 takes a look at a hybrid case study and how much of what is talked about in the preceding chapters can be analyzed and developed into a short- and long-term business transformation road map.

1.2.4 SECTION IV: ACHIEVING MAINTENANCE EXCELLENCE

This section introduces some of the challenges of real estate and information technology (IT) assets, summarizes concepts from this book, and prepares you for some of the evolving trends in the future of maintenance excellence.

Chapter 14 sets forth the challenges specific to driving to asset management and maintenance excellence in facilities and real estate assets. Chapter 15 introduces IT asset management (ITAM) and the IT information library (ITIL), after which Chapter 16 introduces ITSM and service delivery management concepts.

In Chapter 17, we show you how to apply the concepts and methods in this book to the shop floor. This involves a three-step process. The first is to determine your current state of affairs, the best practices available, and your vision. You need to know the size of the gap between where you are and where you want to be. The second step involves building a conceptual framework and planning the concepts and tools to execute it. Finally, we look at the implementation process itself and all that goes into managing change.

In the final chapter (18), we look at what we can expect to be many of the influences for asset management going forward. This is relevant for any organization striving for maintenance excellence.

1.3   THE SIZE OF THE PRIZE

We could debate at length about the social and business mandate of an organization with a large investment in physical assets. But this is certain: if these assets are unproductive, debating social or business mandates becomes irrelevant. Productivity is what you get out for what you put in. Maintenance excellence is about getting exemplary performance at a reasonable cost. What should we expect for investing in maintenance excellence? What is the size of the prize?

Let's look at capacity. One way to measure it is as follows:

$$\text{Capacity} = \text{Availability} \times \text{Use} \times \text{Process Rate} \times \text{Quality Rate}$$

If the equipment is available, being used, runs at the desired speed, and is precise enough to produce the desired quality and yield; we have the required maintenance output.

Now, look at cost. This is a bit more difficult, because cost can vary depending on many things, including the working environment, the resources and energy required to accomplish capacity, equipment age and use, operating and maintenance standards, and technology. One thing we know: a breakdown maintenance strategy is more costly than linking maintenance actions to the likely failure causes.

But by how much? Here's a helpful rule of thumb to roughly estimate cost-saving potential in an industrial environment:

$1 Predictive, Preventive, Planned = $1.5 Unplanned, Unscheduled = $3 Breakdown

Accomplishing "one unit" of maintenance effectiveness will cost $1 in a planned fashion, $1.50 in an unplanned way, and $3 if reacting to a breakdown.

In other words, you can pay now, or you can pay more later. Emergency and breakdown maintenance is more costly for a number of reasons:

- You must interrupt production or service.
- The site isn't prepared.
- Whoever is available with adequate skills is pulled from his or her current work.
- You must obtain contractors, equipment rental, and special tools quickly.
- You have to hunt down or air-freight in materials.
- The job is worked on until completed, often with overtime.
- There usually isn't a clear plan or drawings.

For example, if the total annual maintenance budget is $100 million and the work distribution is 50% planned, 30% unplanned and unscheduled, and 20% breakdown:

$(50\% \times 1) + (30\% \times 1.5) + (20\% \times 3) = 50 + 45 + 60 = 155$ "Equivalent Planned Units"

Planned Work Costs 50/155 × $100 million = $32 million, or $0.645 million
  per unit
Unplanned Work Costs 45/155 × $100 million = $29 million
Emergency Work Costs 60/155 × $100 million = $39 million

To compare the difference, imagine maintenance improvement yielding 60% planned, 25% unplanned, and 15% breakdown:

Planned Work Costs $0.645 × 1 × 60% = $39 million
Unplanned Work Costs $0.645 × 1.5 × 25% = $24 million
Emergency Work Costs $0.645 × 3.0 × 15% = $29 million
Total = $92 million
Savings Potential = $100 million – $92 million = $8 million

As you read through this edition, we hope that you will continue to let it stimulate thought on strategic issues around maintenance and help you to develop tactics that will minimize breakdowns and maximize the rewards of planned, preventive, and predictive work. As well, we hope that you will consider how the asset classes continue to evolve and how all of this now needs an element of "green" thinking across an assets life cycle. In other words, it is not enough that we are effective in achieving asset capacity; we also need to consider the resource and energy impacts the asset requires to reach capacity.

REFERENCES

1. Coetzee, J. L. *Maintenance*. South Africa: Maintenance Publishers, 1997.
2. Parkes, D. Maintenance: Can it be an exact science? In: A. K. S. Jardine, Ed. *Operational Research in Maintenance*. Manchester Univesity Press: Barnes and Noble, Inc., 1970.
3. Kelly, A. *Maintenance and Its Management*. Farnham, England: Conference Communication, 1989.
4. Moubray, J. M. *Reliability-Centered Maintenance*. 2nd ed. Oxford, England: Butterworth Heinemann, 1997.
5. Campbell, J. D. *Uptime: Strategies for Excellence in Maintenance Management*. Portland, Oregon: Productivity Press, 1995.
6. IBM Global Business Consulting, AMcoe, 2006.
7. Adapted from Campbell, J. D. Coopers and Lybrand Consultants Library, 1994.
8. Campbell, J. D. Coopers and Lybrand Consultants Library, 1994.

Section I

Maintenance Management Fundamentals

2 Asset Classes and the World of Life-Cycle Asset Management

Joel McGlynn and Frank "Chip" Knowlton

CONTENTS

2.1 INTRODUCTION

In most corporate organizations today, tangible assets in real estate, manufacturing, transportation fleets, physical infrastructures, and information technology dominate the balance sheet. They are frequently one of the two largest overhead costs (after personnel and benefits). In fact, according to the ARC Advisory Group:

> The monetary investment in capital assets on a global basis is staggering. Manufacturers in process industries, such as Chemical, Oil & Gas, Metals, and Pulp & Paper have billions of dollars invested in hundreds of plants worldwide. While the nature of the assets may differ, discrete manufacturers of items such as automobiles, semiconductors, aerospace equipment, and electronics can be equally asset-intensive. The cost for maintenance and replacement of capital assets can represent a major portion of total operating costs and limit the ability to compete.

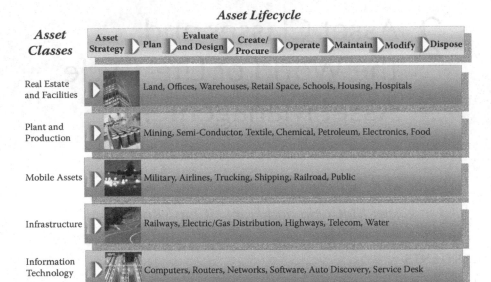

FIGURE 2.1

As an example, the direct costs of managing physical assets are huge:

Maintaining Plant as a Percentage of Total Operating Costs (Typical)

Steel mill = 29%	Underground metal mine = 36%
Nuclear utility = 30%	Nonferrous smelter = 32%
Bauxite mine = 52%	Pulp and paper mill = 26%
Petrochemical refinery = 28%	

Hidden Costs

Safety	Service interruptions
Environment	Shareholders
Legal compliance	

One of the first challenges for organizations in determining the impact of assets to the bottom line is to attempt to identify and categorize what really constitutes an asset. For the purposes of this book, we describe assets in a physical sense rather than from a financial portfolio perspective and, as shown in Figure 2.1, classify them into groupings such as real estate and facilities, plant and production, mobile assets, infrastructure, and information technology.

There are characteristics of each asset class that are unique to the assets typically found in that grouping (Figure 2.2). Conversely, some distinct similarities in the overarching processes must be addressed regardless of asset class. It is this fact that allows various asset management software providers to develop products that span all of the asset classes. Several examples include the following:

What is Unique in Some of the Asset Classes?

■ **RE and Facilities**

    - Asset hierarchies, value to stakeholders

    - Focus on location, construction, lease management

■ **Plant and Production**

    - RCM, TPM, Production ROA focus

    - Configuration management

■ **Infrastructure (Linear) Assets**

    - Asset hierarchies and data, by location

    - Depreciation and maintenance forecasting focus

■ **Mobile Assets**

    - Asset configurations, regulatory compliance

    - Tracking mobile asset locations, timing of planned maintenance

■ **IT Assets**

    Asset configuration version management change management

FIGURE 2.2 Comparing the asset classes.

- Asset management and configuration:
  - Track asset detail.
  - Establish asset location and hierarchy.
  - Monitor asset conditions.
- Work management:
  - Manage resources, plans, and schedules.
- Materials management:
  - Track inventory transactions.
  - Integrate work management with materials management.
- Procurement:
  - Vendor management, vendor performance analysis, and key performance indicators (KPIs).
  - Event-driven purchasing.
  - Enterprise-wide leverage in spending analysis.
- Contract management:
  - Manage vendor contracts.
  - Manage alerts and notifications to optimize vendor service-level agreements (SLAs).
- Service management:
  - Accept and manage new service requests.
  - Manage SLAs.

Companies increasingly face a competitive environment, requiring the development of more efficient and cost-effective operations than ever before. Many asset-heavy organizations are under intense pressures such as globalization, shifting markets, outsourcing, and external regulation. All of these factors drive organizations to increase productivity, reduce costs, and improve product quality. A 1% improvement in performance can be worth millions of dollars annually for a manufacturer. In addition, service rates are often regulated, making business survival dependent on efficient management of capital assets using best practices and standards. Organizations are now looking for ways to extend the capabilities of their existing systems. With ARC forecasting near double compound annual growth rate (CAGR) in certain asset-intensive manufacturing sectors, optimizing the maintenance excellence of all the assets in each asset class will be a continued high priority for years to come. ARC's latest reports collect data, compare market share, and forecast growth using equipment, facilities, fleet, information technology (IT), software, and other asset types. ARC estimates that the fastest-growing asset-type sectors are expected to be equipment assets and IT and software assets at 7.8% CAGR each. Enterprise asset management (EAM) and computerized maintenance management system (CMMS) software and services revenues for management of facilities, fleet, and other asset types are expected to grow at 7.4%, 7.0%, and 6.1% CAGR, respectively.

In the past, asset management was most often described in terms of maintenance management with an exclusive focus on the programs, procedures, and tasks necessary to optimize uptime of an organization's equipment. Today, it requires active life-cycle management of the major assets and components from design and inception to disposal to achieve an edge against competition. A more strategic view of asset management first requires new consideration of which assets are to be managed. In a traditional view, assets may include only items from a few categories, such as machines, factories, vehicles, or specific infrastructure. Alternatively, the responsibility for these items may have been lumped by their job function, financing scheme, or procurement categories. This old approach has a few weaknesses. By ignoring important categories, the company leaves its management to either chance or unstructured processes. By not taking a whole view of the portfolio, the company may have difficulty prioritizing investment or cost-savings decisions. In a traditional model, where different categories are managed separately, it can be near impossible to weigh decisions against one another. For example, a cost-reduction effort may be horrifically executed if the decision maker cannot balance equipment cost and equipment repair in the same analysis. It is easy to imagine internal turf wars and politics of asset category managers overriding decisions that should be based on business strategy. Imagine a forestry company mandated to cut costs. It decides to improve margins by purchasing less expensive (and less reliable) equipment but doesn't make adjustments to its equipment repair capabilities. At the same time, the logistics group under the same cost-reduction mandate reduces the number of vehicles available. In the short term, these decisions return the mandated cost reduction; however, soon an increased amount of equipment breakdowns results in logging stoppages, and the lack of trucking capacity handicaps the company's ability to bring in back-up equipment. The company then has to front money to rent emergency replacement assets, to pay repair teams huge premiums to work overtime, and to perform "damage control" for angry customers who are missing shipments. Production stalls frequently, and the company misses revenue opportunities while its field labor sits idle. All in all, the inability to analyze different asset attributes cripples the company's ability to drive cost reductions when they are needed the most.

With an expanded view of asset classes, the asset manager can have a wider and more complete influence over how the business spends and controls its key properties. This approach leaves fewer assets to be managed informally or by inconsistent procedures. By bringing more asset classes together (i.e., under a common purview and portfolio) the asset manager can make better decisions in support of the business, including investment decisions, performance decisions, or compromises across the entire portfolio.

2.2 TOTAL LIFE-CYCLE ASSET MANAGEMENT

An expanded view of classes brings new benefits to the completeness and rigor of asset management. Similarly, an expanded view of the asset life cycle provides a new level of rigor and understanding. The practice of total life-cycle asset management (TLAM) takes an expanded view of how assets are planned for, used, maintained, and ultimately disposed of. A traditional view often separates or ignores key phases within the asset life cycle. For example, in a conventional company a procurement officer may be in charge of buying new mobile assets, such as planes, trains, buses, or ships. He or she is motivated (and probably measured) on specific criteria for success, most likely negotiating cheap prices and meeting the needed number of vehicles. The maintenance of these vehicles is managed by someone else whose job is to keep repair costs down. The financing may be handled by another manager and the disposition and liquidation by yet another. While these job roles will always be needed, the company may have hurt itself by not taking a complete view of the entire cycle. When these roles are managed separately, we are inclined to ask the following: Were repair costs factored in at the time of purchase? Does the company know the total cost of ownership? Could smarter costing be possible if finance and procurement worked with the entire portfolio? Whether the company in our example suffered from a lack of knowledge is unclear, but the fact that it may not be able to find the answers at all demonstrates a primary shortcoming.

Figure 2.3 shows the TLAM framework that IBM has formulated. It breaks down the life cycle of assets into discrete phases of activity. In practice, companies should analyze their portfolio of assets (including the expanded view of asset classes) across the entire life cycle to make decisions and define asset strategy. The framework consists of eight life-cycle phases of use and planning, each of which has supporting financial management and technology attributes to consider.

2.2.1 Asset Strategy

Set an asset strategy that makes sense for the asset class and your company's business requirements. Activities may include assessing asset management practices, developing a comprehensive asset management strategy, and developing a measurement

FIGURE 2.3   Total life-cycle asset management.

program with key performance indicators (KPIs). Managers need to determine whether they own their primary assets or choose access them "on demand" or take a hybrid approach. For example, a chemical company might have a strategy where it owns and maintains all equipment that relates to core manufacturing but decides that all customized product development be manufactured with leased infrastructure.

2.2.2   PLAN

Clearly define asset targets, standards, policies, and procedures focusing on delivery of the asset management strategy. Companies may wish to develop policies and standards and conduct portfolio asset management planning. This enables them to plan across the entire portfolio of assets. For example, a petroleum company able to plan land acquisition and equipment construction and repair simultaneously may be more nimble in negotiations when purchasing equipment, better able to conduct discovery activities quickly, and better able to deal with emergencies.

2.2.3   EVALUATE AND DESIGN

Evaluate the assets if being purchased, or design the assets that need to be created. Activities in this phase include developing a capital program assessment model, which informs buying decisions. Computer-aided facilities planning can be used to reduce the complexity of managing buildings, storage, and plants. For example, consider a pharmaceutical company ramping up to manufacture a new drug and needing to build out completely new manufacturing facilities and processes around the product life cycle. The new product will require, for example, bioprocessing infrastructure (e.g., vats, bioreactors), manufacturing space, cold storage and shipping, new safety equipment, and process monitoring technology. By integrating the asset design plans with the product life cycle, a company will be able to better understand its infrastructure in regard to the overall product profitability as well as to ensure that the asset management activities support the time frames of the product launch.

2.2.4 CREATE AND PROCURE

This phase involves the act of creating, building, or procuring the planned assets. This phase may have one of the most visible impacts, since it is where the first significant money is spent in asset management. New practices in this area include capital project management, automated and computerized materials resource optimization (e-MRO), and new procurement and project delivery strategies. Imagine an asset procurement manager who is able to make purchasing decisions across all aspects of his or her production facilities globally with an integrated view. The manager is able to make purchases with few redundancies and fewer shortages. He or she would be able to negotiate with suppliers better and to manage installation, delivery, and deployment of assets in an integrated, coordinated fashion.

2.2.5 OPERATE

Operate the assets per the strategy, using the standards, policies, and procedures with feedback into the TLAM. The operation of assets is where performance is most affected (e.g., what value the assets deliver to the company). New practices in this area include formal IT asset management (ITAM), asset performance management strategies, and total asset visibility solutions. A mining company, for example, could track ore production to actual equipment ratios to understand which types of deployments are higher producing. This operational data could then be used when planning new asset acquisitions and deployments.

2.2.6 MAINTAIN

Maintain the assets in support of the strategy and targets using the standards, policies, and procedures in place with feedback into the TLAM. Maintenance costs and resources can wildly alter the total cost of ownership, from repair costs to downtime. New practices in this area include conducting process improvement workshops with multidisciplinary staff (e.g., users, technicians) and deploying EAM software systems. Predictive maintenance becomes a mainstay, based on understanding the past through failure databases and other tracking tools, which ultimately lower reactive maintenance allocations. EAM systems enable asset managers to track and manage assets across the enterprise, complete with centralized monitoring (even by mobile device). Radio-frequency identifications (RFIDs) and other "smart" technologies can be integrated into assets themselves. Imagine factory robots or pipelines that report their problems and remind owners of their maintenance schedules. Total productive maintenance (TPM) is a methodology deployed to manage maintenance and improve uptime and reliability of critical assets.

2.2.7 MODIFY

Modify assets when required. Ensure modifications are reflected in, for example, strategy, policies, and procedures. Some of the toughest modification decisions may come in IT-related assets, where changing requirements and options evolve rapidly. Many firms are deploying strategies that facilitate constant modification of systems, such as service-oriented architectures (SOA). Other practices include total life-cycle costing and performance improvement analysis. Modification can also be critical to the life extension of assets, as machines are retooled, facilities repurposed, and technology adapted to facilitate newer processes.

2.2.8 DISPOSE

This phase involves disposal, retirement, or liquidation of assets in accordance with the strategy, policies, and procedures. Disposal can have significant financial implications beyond replacement. For example, real estate calculations are in constant flux because of market variations. Some assets have environmental or regulatory costs to consider. Other disposal strategies are finding new pockets of income from online gray markets. Other programs, such as IBM's Global Asset Recovery Solutions (GARS) initiative, focus on refurbishing usable parts of disposed equipment to minimize the costs of their disposal. An emerging trend making headlines and driven by new regulation is the increased focus on "green" practices and operations. Practices such as sustainable facilities management, appropriate asset disposal, reduction of carbon footprint at manufacturing plants, and reduced carbon emissions for fleet are quickly becoming requirements the asset manager must consider. How assets are disposed of will be only the beginning of this trend, since green practices will need to move into every stage of the total life cycle for assets.

2.2.9 FINANCIAL MANAGEMENT

Each phase has financial management implications and planning requirements. These are often most pronounced during the create-and-procure and disposal phases, but of great importance are also the operate and maintain phases, where financial performance is also affected. Maintenance, for example, can be a massive contributor to total cost of ownership (TCO), and the performance of the operate phase can be a huge contributor to financial performance.

2.2.10 TECHNOLOGY

In this instance, we refer to technology as an asset management tool, not as the asset itself (although the asset management system is indeed an asset). Technology can transform how each of these phases is planned for and executed. In an EAM system, models for planning and management are resident within a common, centralized system. Active cataloging, monitoring, and measurement of assets is also tracked, often in real time, to aid repair actions, to enable quick procurement and replacement decisions, and to monitor performance. Technology is also used to integrate the EAM with other key systems, such as accounting, procurement, and business performance management (BPM) dashboards.

Operationally, this framework should be formalized and programmatic within the organization. This means applying a TLAM approach to asset management systems, integrating the approach into planning and strategy efforts, and using the framework to establish monitoring and metrics to gauge success and performance.

The TLAM approach is consistent with the elements of an asset-centric supply chain, in that decisions associated with strategic capital management (SCM) strategy, planning, product life-cycle management, logistics, procurement, and operations of an organization are impacted significantly by how that organization's assets are managed (Figure 2.4).

Pictured another way, the systems view of a basic asset life cycle and the components of the supporting infrastructure incorporate many stakeholders within the supply chain community. All of these elements contribute to and have a vested interest in an effective TLAM perspective (Figure 2.5).

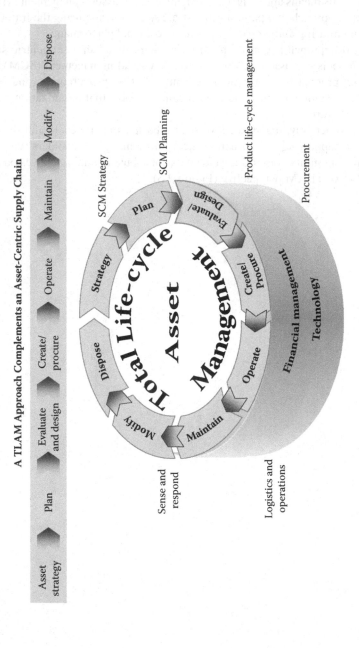

FIGURE 2.4 A TLAM approach complements an asset-centric supply chain.

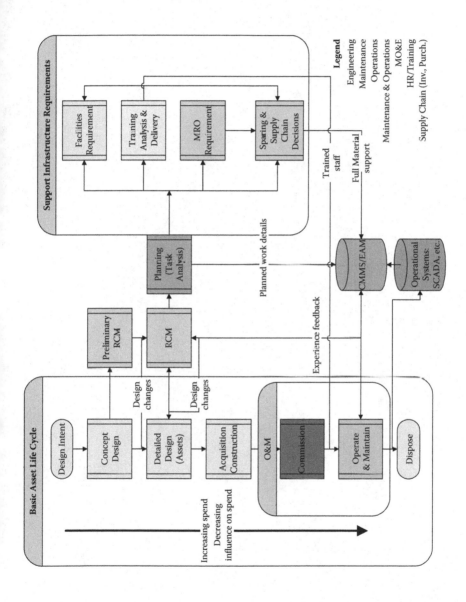

FIGURE 2.5

3 A Framework for Asset Management

Thomas Port, Joseph Ashun, and Thomas J. Callaghan

CONTENTS

3.1 INTRODUCTION

Today's maintenance and physical asset managers face great challenges to increase output, to reduce equipment downtime, to lower costs, and to do it all with less risk to safety and the environment. This chapter addresses the various ways to accomplish these objectives by managing maintenance effectively and efficiently within your organization's unique business environment.

This chapter presents an overview of the multiple aspects required for an effective and efficient maintenance management system. Of course, you must make trade-offs, such as cost versus reliability, to stay profitable in current markets. We show you how to balance the demands of quality, service, output, costs, time, and risk reduction. This chapter examines just how maintenance and reliability management can increase profits and add real value to the enterprise.

We discuss the levels of competence you must achieve on the road to excellence. There are clear evolutionary development stages. To get to the highest levels of expertise, you must ensure that the basics are in place. How can you tell if you are ready to advance? We provide a series of charts that will help you decide.

In the final sections of the chapter, we describe the methods used by companies that truly strive for continuous improvement and excellence. We will therefore briefly touch upon reliability-centered maintenance (RCM), root-cause failure analysis (RCFA), and decision-making optimization. This sets the stage for material presented later in the book.

3.2 ASSET MANAGEMENT: TODAY'S CHALLENGE

Smart organizations know they can no longer afford to see maintenance as just an expense. Used wisely, it provides essential support to sustain productivity and fuel growth while driving down unneeded and unforeseen overall expenses. Effective asset management aims to do the following:

- Maximize uptime (productive capacity)
- Maximize accuracy (the ability to produce to specified tolerances or quality levels)
- Minimize costs per unit produced (the lowest cost practical)
- Minimize the risk that productive capacity, quality, or economic production will be lost for unacceptable periods of time
- Prevent safety hazards to employees, and the public as much as possible
- Ensure the lowest possible risk of harming the environment
- Conform to national and international regulations on due diligence (e.g., Sarbanes–Oxley [SOX])

In today's competitive environment, all of these are strategic necessities to remain in business; the challenge is how to best meet them. In many companies, you have to start at the beginning—put the basics in place—before your attempts to achieve excellence and to optimize decisions will be successful. The ultimate aim is to attain a high degree of control over your maintenance decisions, and in this chapter we explore what it takes to get there.

3.3 OPTIMIZATION

Optimization is a process that seeks the best solution, given competing priorities. This entails setting priorities and making compromises for what's most important. Maximizing profits depends on keeping our assets in working order, yet maintenance sometimes requires downtime, taking away from production capacity. Minimizing downtime is essential to maximize the availability of our plant for production. Optimization will help you find the right balance.

Although increasing profit, revenues, availability, and reliability while decreasing downtime and cost are related, you can't always achieve them together. For example, maximizing revenues can mean producing higher-grade products that command higher prices. But that may require lower production volumes and, therefore, higher costs per unit produced.

Clearly, cost, speed, and quality objectives can compete with each other. An example is the improved repair quality from taking additional downtime to do a critical machine alignment correctly. The result will probably be longer run time before the next failure, but it does cost additional repair downtime in the short term.

The typical trade-off choices in maintenance arise from trying to provide the maximum value to our customers. We want to maximize

- Quality (e.g., repair quality, doing it right the first time, precision techniques)
- Service level (e.g., resolution and prevention of failures)
- Output (e.g., reliability and uptime)

At the same time, we want to minimize

- Time (e.g., response and resolution time and mean time to resolution [MTTR])
- Costs (e.g., cost per unit output)
- Risk (e.g., predictability of unavoidable failures)

Management methods seek to balance these factors to deliver the best possible value. Sometimes, however, you must educate the customer about the trade-off choices you face to ensure "buy in" to the solution. For example, a production shift supervisor may not see why you need additional downtime to finish a repair properly. You have to convince him or her of the benefit: extended time before the next failure and downtime.

Maintenance and reliability are focused on sustaining the manufacturing or processing assets' productive capacity. By sustaining we mean maximizing the ability to produce quality output at demonstrated levels. This may mean production levels that are beyond original design if they are indeed realistically sustainable.

3.4 WHERE DO MAINTENANCE AND RELIABILITY MANAGEMENT FIT IN TODAY'S BUSINESS?

The production assets are merely one part of an entire product supply chain that produces profit for the company. It is important to recognize that maintenance priorities may not be the priorities of the company as a whole.

In a very basic manufacturing supply chain, materials flow from source (suppliers) through primary, and sometimes secondary, processing or manufacturing and then outbound to customers through one or more distribution channels. The traditional business focus at this level is on purchasing, materials requirements planning, inventory management, and just-in-time supply concepts. The objective is to minimize work in process and inventory while manufacturing to ship for specific orders (i.e., the pull concept).

To optimize the supply chain, you optimize the flow of information backward, from customers to suppliers, to produce the most output with the least work in process. Supply chain optimizing strategies do the following:

- Improve profitability by reducing costs.
- Improve sales through superior service and tight integration with customer needs.
- Improve customer image through quality delivery and products.
- Improve competitive position by rapidly introducing and bringing to market new products.

Methods to achieve these include the following:

- Strategic material sourcing
- Just-in-time inbound logistics and raw materials management
- Just-in-time manufacturing management
- Just-in-time outbound logistics and distribution management
- Physical infrastructure choices
- Eliminating waste to increase productive capacity
- Using contractors or outsource partners
- Inventory management practices

Business processes that are involved include the following:

- Marketing
- Purchasing
- Logistics
- Manufacturing
- Maintenance
- Sales
- Distribution
- Invoicing and collecting

At the plant level, you can improve the manufacture part of the process by streamlining production processes through just-in-time materials flows. This way, you'll eliminate wasted efforts and reduce the production materials and labor needed.

In the past, maintenance received little recognition for its contribution to sustaining production capacity. It tended to be viewed only as a necessary and unavoidable cost. Even at the department level today, managers typically don't view the big picture—the entire plant; they focus only on their departmental issues. Unfortunately, maintenance is often viewed only within the context of keeping costs down.

In accounting, maintenance shows up as an operating expense and one that should be minimized. Maintenance is typically only a fraction of manufacturing costs (5% to 40%, depending on the industry). Those manufacturing costs are similarly a fraction of the products' total selling price.

Reducing maintenance expenses does indeed add to the bottom line directly, but, since it is a fraction of a fraction of the total costs, it is typically seen as less important, commanding less management attention. Most budget administrators don't seem to fully understand maintenance, judging by historical cost numbers. When you reduce a maintenance budget, service ultimately declines. Also, output is usually reduced and risk is increased when there isn't enough time or money to do the work right the first time.

Of course, the accounting view is one-dimensional because it looks only at costs. When you consider the value that maintenance delivers, it becomes much more important. By sustaining quality production capacity and increasing reliability, you generate more revenue and reduce disruptions. This requires the right application of maintenance and reliability. Certainly, doing maintenance properly means being proactive and accepting some amount of downtime. Effective maintenance methods are needed to make the best possible use of downtime and the information you collect to deliver the best value to your production customers.

3.5 WHAT MAINTENANCE PROVIDES TO THE BUSINESS

Maintenance enhances production capacity and reduces future capital outlay by

- Maximizing uptime
- Maximizing accuracy produces to specified tolerances or quality levels
- Minimizing costs per unit produced
- Sustaining the lowest practical and affordable risk to loss of production capacity and quality
- Reducing as much as possible the safety risk to employees and the public
- Ensuring the lowest possible risk of harming the environment

Notice the emphasis on risk reduction. This is why insurers and classification societies take a keen interest in their clients' maintenance efforts. Your maintenance reduces their exposure to risk and helps keep them profitable. Nearly every time a major accident involves a train, airplane, or ship, there is an in-depth investigation to determine whether improper maintenance was the cause of the disaster.

Maintenance can also provide strategic advantage. Increasingly, as companies automate production processes and manufactured goods are treated like commodities, the lowest cost producer will benefit. Automation has reduced the size of production crews while increasing the amount and complexity of work for maintenance crews. Maintenance costs will therefore increase relative to direct production costs. Even low-cost producers can expect maintenance costs to rise. That increase must be offset by increased production. You need less downtime and higher production rates as well as better quality at low unit cost—that means more effective and efficient maintenance.

Achieving all of this requires a concerted effort to *manage* and *control* maintenance rather than letting the assets and their random failures control costs. In today's highly competitive business environment, you cannot afford to tolerate that. Unfortunately, many companies do just that, allowing natural processes to dictate their actions. By operating in a "firefighting" mode, they merely respond to rolls of the dice, with random results. Without intervening proactively, these companies can react only after the fact—once failures occur. The consequences are low reliability, availability, and productivity: ingredients for low profitability.

3.6   READY FOR EXCELLENCE? THE PYRAMID

Optimizing your effectiveness cannot be accomplished in a chaotic and uncontrolled environment. Optimization entails making intelligent and informed decisions. That involves gathering accurate and relevant information to support decisions and to act in a timely manner. As the saying goes, "When up to the rear in alligators, it's difficult to remember to drain the swamp." You must have your maintenance system and process under control before you can optimize effectively. You need to tame the alligators with good maintenance management methods, followed in a logical sequence.

John Campbell, author of the book *Uptime: Strategies for Excellence in Maintenance Management,*[1] teaches that several elements are necessary to achieve maintenance excellence. These elements fall into four major areas, as illustrated in Figure 3.1 in the maintenance excellence pyramid:

- Exercise leadership at all times. Without it, change won't be successful.
- Achieve control over the day-to-day maintenance operation.
- Apply continuous improvement, once you have control, to remain at the leading edge of your industry.
- Continuous improvement activities will set the stage for quantum leaps in asset productivity.

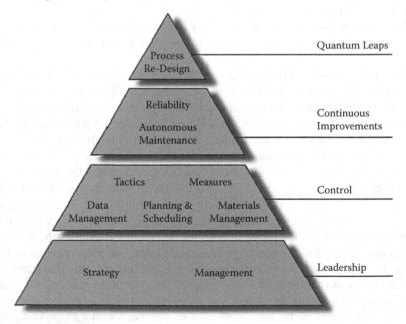

FIGURE 3.1   Maintenance excellence pyramid.

3.6.1 LEADERSHIP

We are frequently advised to have a personal physical exam by a qualified physician once a year. Most individuals would not disagree with this advice, but many fail to act upon it: I'm feeling just fine! Why bother going to the doctor? But we are more likely to visit the doctor when we are ill and expect to have whatever ails us remedied in quick order. Similarly, a physical asset manager is more likely to inherit a plant that is struggling to meet safety, production, and cost objectives than to inherit one where all systems are in control. This situation requires the manager not only to start his or her work from the bottom up but also to understand how each step contributes to the overall plan and moves the organization in the direction intended.

Asset management is a journey, not a destination. For every journey, leadership is crucial to success.

The essential elements of leadership include the following:

- Before starting off in a direction, a maintenance manager must know from what point he or she is starting: What resources are available? If we undertake this new work, what will not get done? How do we know where to start, the path forward, and where this journey should lead us?
- A physical asset manager starts with what he or she has in place and attempts to understand what is working and what is not working. There is no point in expending much time, effort, and resources redesigning a system that is already delivering the specified outcomes. If a system is in place, is it being implemented as designed? A baseline audit will identify which systems are working and, if they are not working, why. Is the problem with the system design or with the manner in which people in the organization are executing the design—or perhaps both? An audit will measure a range of criteria comparing the system design with the actual execution by various parts of the organization. It will measure overall and unit effectiveness. It will point in the direction the journey must begin.
- Leadership is required at all levels in the organization. Leaders must do the following:
  - Lead from the front
  - Remove barriers
  - Create a path for others to follow
  - Make room for others to contribute
- It is not necessary to be the chief executive officer or the senior manager to lead. A tradesperson, a foreman, a planner, or a middle manager can also lead. It is imperative that they do so.

To design and implement a physical asset management process appropriate for the business needs, the manager must understand the following:

- What is the plant meant to do? What are the key performance goals?
- Is the plant safe and reliable? Do employees, management, and the community have confidence that the presence of the plant is a benefit and not an unreasonable or unknown risk to their well-being?

- Does the organization achieve the goals with a common purpose, communication, and teamwork throughout all levels in the organization?
- Are people valued throughout the organization? Do they have the opportunity to reach their full potential regardless of whether they come from finance, maintenance, or operating?
- Does the plant have spare capacity, or must it run 24/7 to keep up to demand?
- Does the maintenance organization add value to the plant? Does it identify what needs to be done and then execute the work when it needs to be done without wasting resources such as manpower, materials, equipment, or process downtime?
- Does the organization optimize the use of its capital resources and infrastructure before replacement?
- What steps are necessary to establish some level of control? Can the maintenance manager reduce variation?
- If the organization was successful in implementing the intended systems, would we then achieve all of the desired results?

3.6.2 CONTROL

- Planning, scheduling, and execution practices to manage work and delivery of the service: Through careful planning you establish what will be done, using which resources, and provide support for every job performed by the maintainers. You also ensure that resources are available when needed. Through scheduling you can effectively time jobs to decrease downtime and improve use of resources. Execution delivers what was planned, when it was planned, and how it was planned.
- Materials management practices to support service delivery: Part of the job of a planner is to ensure that any needed parts and materials are available before work starts. Shutdown schedules cannot wait for the bureaucratic grind of materials delivery. Schedules are set and material procured as required by the schedule. To minimize operations disruptions, you need spare parts and maintenance materials at hand. Effective materials management ensures the right parts are available in the right quantities at the right time and distributed cost-effectively to the job sites.
- Maintenance tactics for all scenarios—to predict failures that can be predicted, to prevent failures that can be prevented and run-to-failure when safe and economical to do so, and to recognize the differences: This is where highly technical practices such as vibration analysis, thermographics, oil analysis, nondestructive testing, motor current signature analysis, and judicious use of overhauls and shutdowns are deployed. These tactics increase the amount of preventive maintenance that can be planned and scheduled and reduce the reactive work needed to clean up failures.
- Measurements of maintenance inputs, processes, and outputs to help determine what is and isn't working and where changes are needed: By measuring your inputs (e.g., labor, materials, services) and outputs (e.g., reliability or uptime and costs), you can see whether your

management is producing desired results. If you also measure the processes themselves, you can control them to align the execution with the system design. Statistical process control charts will tell the maintenance manager if a system is in control and at what point in time a process change took place. A change to a specific "input" such as labor or materials may not have the desired output. The annals of maintenance history are filled with evidence of throwing manpower at the problem with little to show for it.

- Systems that help manage the flow of control and feedback information through these processes: Accounting uses computers to keep the books. Purchasing uses computers to track orders and receipts and to control who gets paid for goods received. Likewise, maintenance needs effective systems to deploy the workforce on the many jobs that vary from day to day and to collect feedback to improve management and results.

3.6.3 CONTINUOUS IMPROVEMENTS

Continuous improvement involves a range of well-known methods and maintenance tactics. The best place to start is with the people closest to the work: the operators and the tradespeople. Through use of simple quality management techniques including on-site data collection, charting, fishbone diagrams, and Pareto charts, many problems can be solved. There is a time and place for the advanced techniques discussed in detail later in this book. But if the organization does not have the skills, knowledge, and discipline to execute the basic elements of the maintenance system design, it cannot be expected that more advanced techniques will have meaningful results. Improvement such as RCM and total productive maintenance (TPM) methods are described in detail in chapters ahead.

3.7 ADD VALUE THROUGH PEOPLE

One of the most important pieces of work for the physical asset manager is to add value through the development of people and assigning of tasks.

The manager adds value by

- Establishing and communicating the vision for the future
- Establishing the maintenance strategy
- Establishing the fundamental values by which the organization and people interact with each other
- Designing the physical asset management system
- Creating the organizational structure and roles descriptions that allow the maintenance system to be executed and people to reach their full potential
- Assigning tasks with consistent clarity and measurable outcomes
- Establishing the accepted standards for the physical plant
- Clarifying issues for middle management

The middle management adds value by

- Implementing the system as designed
- Acting in a manner consistent with the declared values
- Helping subordinates reach their full potential
- Assigning tasks with consistent clarity and measurable outcomes
- Achieving the established standards for the physical plant
- Removing barriers to successful execution
- Clarifying issues for the foreman and tradespersons

The foreman and tradespersons add value by

- Executing the plan and schedule
- Advising management of barriers to successful execution
- Participating in problem solving and continuous improvement
- Striving to reach their full potential

The work of the physical asset management leader is to effectively employ the available resources to optimize plant safety and reliability, leading to steady state capacity. This work is done best when people are valued and there is a systematic approach to designing the required commercial, technical, and social systems in the workplace.

3.8   LEVEL OF ASSET MANAGEMENT MATURITY

The degree to which a company achieves maintenance excellence indicates its level of maturity. A maturity profile is a matrix that describes the organization's characteristic performance in each of these elements. One example appears in Campbell's book (Figure 1.5).[1] Figure 3.2 presents another example of a profile that covers the spectrum of elements needed for maintenance excellence. It is presented in a series of profiles, with supporting details for a cause and effect diagram. The effect we want is maintenance excellence.

Every leg of the diagram comprises several elements, each of which can grow through various levels of excellence:

- Excellence
- Competence
- Understanding
- Awareness
- Innocence

Figure 3.3 through Figure 3.6 describe every level for each element from the cause and effect diagram shown in Figure 3.2.

In Figure 3.3, leadership and people are the most important elements, although they are not always treated as such. As you can see, the organization moves from reactive to proactive, depends more heavily on its employees, and shifts from a directed to a more autonomous and trusted workforce. Organizations that make

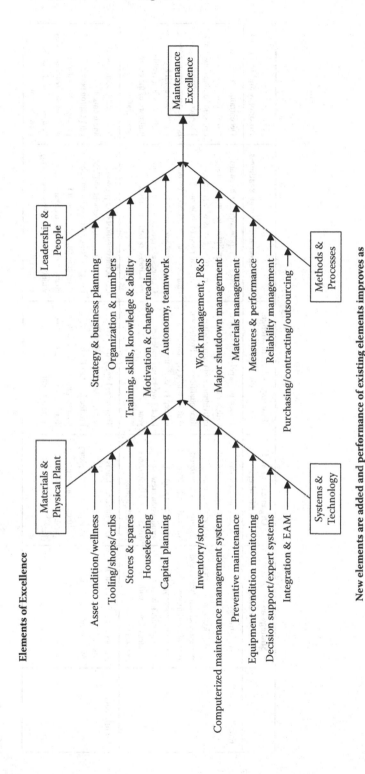

FIGURE 3.2 Elements of maintenance excellence.

	Strategy and Business Planning	Organization and Numbers	Training, Skills, Knowledge, and Ability	Motivation and Change Readiness	Autonomy and Teamwork
Excellence	Stated strategy with mission, long range vision, goals. Goals are specific, measurable, achievable, and realistic and timed (for two or more years). Actions match words. Strategy linked with corporate goals.	Decentralized teams operate independent of daily maintenance control and may report to production. Plenty of interaction with production crew members. Maintenance supports teams.	Trades are largely multiskilled with some multitrade qualified individuals and regularly use their qualifications. Production staff do minor equipment upkeep tasks. Training time at least 2 weeks per trade per year.	Trades' compensation has a reward component linked to business results. Competitive forces widely accepted as driving need for beneficial changes. Changes initiated by both management and workforce. Changes are usually successful and measurable benefits achieved.	Decentralized teams are self-directed and base decisions on business need. Excellent cooperation between maintenance and production at all levels. Teamwork is a visible hallmark of the entire organization.
Competence	Strategy (as above) but not linked to corporate goals. Actions close to the words.	Decentralized teams controlled by maintenance have plenty of interaction with production crew members.	Trades are largely multiskilled and regularly use their skills. Production staff do some minor equipment upkeep tasks. Training time 1 to 2 weeks per trade per year.	Cooperative atmosphere prevails, trust between management and labor is high. Change always initiated by management, and the need for changes explained in advance and widely accepted. Changes are usually successful.	Some self-directed workers and teams. Good cooperation between production and maintenance at all levels. Teamwork may be a feature of the entire organization.

FIGURE 3.3   Leadership and people.

Understanding	Some goal setting for long term, annual plans used.	Mix of decentralized teams reporting to maintenance and central shop structure.	Trades have some multiskilling and often use those skills. Production staff do minimal minor equipment upkeep tasks. Training time less than 1 week per trade per year. Training need analysis completed for all trades.	Some cooperation between management and labor exists and level of trust is moderate. Reason for change is usually explained in advance. Changes sometimes fail.	Directed workforce with some teamwork but little to no team training. Some cooperation between maintenance and production at the working level.
Awareness	PM program in place, benefits recognized.	Centralized structure based on trades breakdown. Control through maintenance supervisors/leads in response to production demands.	No multiskilling is used. Production staff do no equipment upkeep. Training time less than 1 week per trade per year. Some training need analysis performed.	Management motivation explained when questioned. Some distrust but desire to improve exists. Changes often fail.	Directed workforce with no attempt at teamwork outside of shop structure. Good cooperation between production and maintenance leadership.
Innocence	Breakdown maintenance, fire fighting, no stated goals.	Centralized structure based on trades breakdown. Action directed largely by operations supervisors.	No multiskilling is used. Production staff do no equipment upkeep. Training is driven by necessity only.	Highly resistive to change. Hourly workforce generally distrusts management motives. No visible desire to improve. Change initiatives usually fail.	Directed workforce with no attempt at teamwork outside of shop structure. Maintenance and production relationship is strained.

FIGURE 3.3 (continued).

	Work Management	Major Shutdown Management	Materials Management	Measures and Performance Management	Reliability Management
Excellence	Planning and scheduling are well established. All work except emergencies is planned at least 24 hours in advance. Priorities for work orders are used and respected. Work is usually scheduled at least 1 week in advance. Very few jobs are "held" because of material problems. "Wrench time" is high. Emergencies are few and far between. The atmosphere is orderly and controlled. Backlogs are managed at 2–4 weeks of work. Long-term planning for capital projects in place.	Shutdowns are planned over a 6 month period with lock down of work scope providing sufficient time for long lead item purchases. Formalized shutdown planning and management process in place. Heavy involvement by production, engineering, and maintenance in the process. Only emergency work arising is added during the shutdown. Shutdowns completed as or better than scheduled and achieve full work scope completion.	Automated inventory control and analysis+D12 are used in fully integrated system. Stock levels set based on sparing analysis with maintenance input. Automated management features used: stock reorders, grouping of purchases by vendor, pick list generation, bar code or other automated issue and return process. Stores personnel manages inventory fully. Maintenance not involved in obtaining materials beyond request/ specification stages.	Performance measures are a part of everyday life in the plant. All costs are captured and known by type of cost, area, equipment, work order. Company, plant, department, team performance is measured, known, and used to target improvements. Process measures are effective in driving behavior. External benchmarking is used to drive improvement targets.	Plant reliability is high. CMMS/ EAM are used as an effective tool to identify problem areas for resolution. Data are used from both CMMS/EAM and expert/ decision support systems to optimize decisions in maintenance. Use of these systems is by engineers, planners, and condition monitoring teams. Full analysis for condition monitoring and PM tasks completed in all plant areas. Task frequencies and tasks being refined through work order feedback. Root cause failure analysis used on all failures.

FIGURE 3.4   Methods and processes.

Competence	Materials and resources planned and usually staged before work is performed. Supervisors may do some planning but trades are now focused on the tools. PM and inspection results are finding problem areas before they become failures. PM and Inspection programs are being changed with more emphasis on inspection. Planners are involved in sourcing materials but not purchasing. Net capacity is used in scheduling. Work order priorities are generally respected. Daily planning meetings handle only a few adjustments to plan.	Production and stores involved in shutdown planning process. Production scheduling is under control, there is confidence in shutdown timing forecast. Shutdowns usually completed in scheduled downtime, but not all work is executed successfully. Minimal work added at last minute, but some added as arising during the shutdown. Finished crews usually used to get "extra" work done—risks extending the shutdown duration.	Stores computer records integrated with maintenance system. Using automated pick lists from planning of work orders. Materials are kitted for work orders/shutdowns/ projects for maintenance pick up. Service and inventory levels, stock turns known, monitored, and improved. All stock items including capital spares, bone yard spares, and locations are catalogued. Stock levels are set with maintenance input. Maintenance planning involved in sourcing of materials/parts. Performance measures evident on bulletin board for all maintainers to read. Measures include costs, results, and processes. Costs are broken down by areas in broad categories of labor, materials, and contracts. Some use is made of measures to drive improvement initiatives. Interest exists in comparisons with other plants.	Reliability is improving with some major gains known and possibly documented. Targeted improvement programs in place and generally regarded as successful in achieving their objectives. Some reliability improvement trends emerging. CMMS is being used to help identify problem areas. Formal analysis being used to target condition monitoring and PM tasks in critical plant areas. Some root cause failure analysis being applied successfully.	
Understanding	Planner or planning group in place. Technical support is available when needed but remains largely focused on projects. Scheduling is done weekly with daily adjustments. Weekly and daily "planning" meetings held—lots of adjustments daily. Work priorities changed frequently or disregarded. Little staging of resources before jobs are started. Resource planning is left to supervisors/lead hands and trades.	Shutdown work arising from PM and inspections is added to work list for shutdowns. Some PM work and inspection work is added to shutdowns. Planners support supervisors in resource planning and purchasing for shutdown work. Maintenance stages shutdown materials prior to shutdown. Planning may begin 1 to 2 months in advance if production can promise downtime window for the shutdown.	Parts records on computer. Stock levels set with no maintenance input—depend on vendor recommendations. Lead times and safety stock levels set but rarely changed. Analysis of inventory levels carried out. Some cataloging of spares in work areas. Some use of pre-expended/ready-use spares in shop areas managed by stores or shops themselves. Establishing stock items is Costs are tracked by labor, materials, and contract services but not analyzed. Downtime is measured overall and by area and by cause of the downtime. Bickering occurs between production and maintenance about "who's to blame" for downtime. Process	Reliability is low. Targeted improvement programs in place and being driven by data collected in failure databases by maintainers/engineers. Improvement programs viewed with skepticism because of short track record. Improvements largely credited to one or only a few "gifted" individuals who solved the problems.	

FIGURE 3.4 (continued).

	Work Management	Major Shutdown Management	Materials Management	Measures and Performance Management	Reliability Management
Understanding	Some feedback on PMs and inspections results in changes in PM program. Planners and supervisors may be doing a lot of purchasing.	Shutdowns don't complete work scope in alloted downtime or run overtime—often both. Some work added at last minute and during shutdown. Most work completed really requires shutdown.	onerous. Heavy involvement of maintenance supervision in managing parts and inventory and sourcing parts.	performance is not measured but it is judged in qualitative terms.	
Awareness	Daily scheduling is attempted but largely undermined by the high level of demand maintenance. Some technical help in troubleshooting exists. Inspection work and PM is scheduled.	Annual or other regular shutdown schedule to deal with equipment replacements, major overhauls, and capital project tie-ins. Planning is minimal and carried out by maintenance supervisors with some input from production. Shutdowns usually start late, finish late and don't get all the work done. Plenty of work arising added during the shutdown.	Parts records kept manually. Stock levels set but rarely changed—may be high or low. Zero stock on bin checks triggers stock orders. Very difficult to add items to inventory. Maintainers beginning to accept stores support.	Some downtime records kept on critical equipment. Costs are known but not under control. Budget overruns common.	Reliability is low. Little use of downtime records to target problem solving. Any record keeping is regarded as non-value added.

| Innocence | No serious attempt at scheduling or planning of daily work. | Annual shutdown with same work scope each year. Most work is overhauls. Planning is minimal and carried out by maintenance supervisors. Plenty of work arising added at last minute to clear backlogs (regardless of need for outage) and during the shutdown. | Sparing is minimal. Plenty of obsolete/unused spares. Some frequently used items stocked. Large disorganized bone yard with plenty of uncataloged material from old projects. Several or many unofficial caches of parts held by maintainers throughout the plant and shops. No or ineffective cataloguing. Maintainers looking out for themselves. | None. Budget overruns in maintenance are commonplace. | None. |

FIGURE 3.4 (continued).

	Inventory and Stores	CMMS/EAM (site)	Preventive Maintenance	Equipment Condition Monitoring	Decision Support and Expert Systems
Excellence	Highly integrated stores, maintenance, purchasing, finance, and other corporate systems. Full exploitation of automated features to remove manual effort.	Fully integrated maintenance, inventory, purchasing, and other systems in place. Information is managed and used as an asset by maintenance and engineering and production.	PM is used only where analysis dictates—it is targeted at preventable failures that are time or usage related.	All plant areas analyzed for condition monitoring and PM needs. Condition monitoring is used extensively—probably by a dedicated crew. Most corrective repair work arises from condition checks.	Analysis/decision support tools for capital planning. Condition monitoring data and CMMS data linked for optimum inspection/replacement decision making.
Competence	Stores systems linked with maintenance and purchasing. Some automated features and statistical analysis in use.	CMMS/EAM in place and linked with inventory and purchasing. Integration with other systems planned. Wide access to system by maintainers in shops. Training completed and users capable.	Some PM is dropped in favor of condition monitoring. Overhauls are infrequent.	Critical plant areas analyzed to determine condition monitoring needs. More condition monitoring than PM used. Inspections are revealing problem areas for correction.	Expert/analysis systems in place to help read/interpret condition monitoring data.
Understanding	Computerized stores inventory records. Inventory system may be linked with purchasing but not with maintenance.	CMMS/EAM in place or being implemented with outside professional help. Maintenance system not linked with inventory or purchasing, although this may be planned.	Experience is used to modify manufacturer recommendations for PM frequency.	Time and usage based inspections. Some condition monitoring (vibration, oil analysis, thermo graphic) based on manufacturer recommendations, monitoring equipment vendor recommendations or experience.	Contracted services interpret some condition monitoring data (e.g., vibration or oil analysis or thermographic). Manual interpretation of results used in house.
Awareness	Formalized but manual systems only.	Basic (daily) scheduling and work order tracking. No ability to check on parts inventory or other resource needs. CMMS or EAM may be planned.	Reliance on equipment manufacturer recommendations for PM.	Time based inspections outside of shutdowns.	Experience used to determine actions based on time-based inspections.
Innocence	Informal stock-keeping system.	No system at all. Most work is done on a demand basis. Backlog is managed in supervisor memories or on spreadsheets and in note books.	Overhauls used extensively. Annual shutdown for inspections and overhauls only.	None.	None.

FIGURE 3.5 Systems and technology.

	Asset Condition and Wellness	Tooling, Shops and Cribs	Stores and Spares	Housekeeping	Capital Planning
Excellence	Plant equipment is considered reliable, looks close to "new" even if old. Technology has been upgraded. Cleanliness of production equipment reveals sense of ownership.	Purpose built area centers. Tool crib attendant repairs tooling. Bays for fleet repair work by type of work with dedicated crews. Remote PLC programming. Separate lay down for work in process, materials receiving, and outbound completed work.	Access to stores is controlled but open to maintainers. Staging areas for prekitting of all parts ordered in and from stock for specific work orders/projects prior to delivery to shops/maintainers. Large capital spares located near point of use or in special storage.	Well kept and clean plant production, maintenance, and office areas. Area teams take pride in equipment upkeep. Cleaning is seen as an effective tool in keeping equipment in good condition.	Multiyear long range asset replacement strategy used for capital planning. Annual budget process refines long range plan.
Competence	Reliability is good but improvement is being sought. Plant appears to be in good operating condition. Asset replacement being done as needs arise with a one year look ahead at equipment condition as input to replacement decisions.	Area shops used. Shops adjacent to production areas. Central shops used to support area teams for jobs too large to handle in area. Fleet shops purpose designed. Area and central shops all have on-line access to maintenance management system. Tools in secure area adjacent shop/bay. Tool crib for special tools. Sign out system for tooling.	Access to stores is controlled but open to maintainers. Areas for prekitting of parts ordered in for specific work orders/projects. Little "dust" indicates few obsolete items. Bone yard is used only for oversized, weather proof items.	Production and maintenance areas receive daily clean up. Cleaning is still a "chore" and carried out to medium standards. Dedicated clean up crews may exist. Special cleaning routines used on a weekly or monthly basis to bring areas back to good condition.	One year look ahead for capital budget needs supported by equipment condition assessments. Capital expenditure history used to forecast replacement funding for equipment upgrades. Betterment projects treated as stand alone projects.

FIGURE 3.6 Materials and physical plant.

	Asset Condition and Wellness	Tooling, Shops and Cribs	Stores and Spares	Housekeeping	Capital Planning
Understanding	Reliability is improving quickly in the plant. Plant appears to be a bit run down but generally operable. Asset replacement considered only if need is evident and trigged by annual budget cycle. Repairs carried out to "as good as new" standard.	Formal tool replacement mechanism and tool crib manned on all shifts, controlled, and orderly. Tools tracked as stores items. Central shops designated for each trade or production area. Shop layout follows material flow. Shops well lit and ventilated. Area shops (if any) exist where space available. On-line access to maintenance management system in shops.	"Ready use" or "preexpended" high usage low value stock available in shop areas. Stores traffic is down and access is well controlled. Stores is orderly and parts can be found quickly. Shops. shipping and receiving areas handy to stores. Separate areas for quarantine items, warranty items, receiving and shipping and repairables awaiting work. Bone yard is catalogued and orderly.	Weekly clean up of production and maintenance areas used. Standards of cleanliness are good but enforced only in cyclical clean ups.	Asset replacements identified only when budget cycle calls for estimates.
Awareness	Reliability is low. Asset replacements occur only when uneconomic to repair any more. Repairs are carried out to "as good as before" standard not "as good as new." Plant is run down/tired. Some plant staff questions why the plant is kept open. Slow leaks allow contamination build up that could hide other equipment problems.	Tool crib exists with informal control by either maintenance supervision or stores and manned for one shift only with arrangement for off shift access. Informal mechanism in place to replace tooling. Trade tools stored in designated location. Calibration program for tooling/instruments in place and used. Central shop is close to stores, and layout mimics department organizational clusters not work flow. Shop may be untidy, lifting equipment and lay down areas marginal.	Stores appears orderly but parts can't be found without help from stores person. Location system is in place. Access is controlled but loosely. Lots of maintenance "traffic" in stores. Stores is close to central maintenance shops and may be close to shipping and receiving areas. Hazardous materials are segregated. Some parts packaging damaged. Large unkempt bone yard.	Upkeep of plant production and maintenance areas is generally poor and neglected. It's a mess! Any clean up is done only when repairs or other maintenance is performed.	No budgeting for capital for asset replacements. Each case is handled as it arises.

| Innocence | Reliability is low. Equipment is in disrepair. Numerous "band aid" type fixes. Technology is out of date. Safety guards may be out of place. Leaking equipment ignored or slow to be repaired. | Trades have own tools/share. No tool specs or standards. No tool crib or it is uncontrolled. Tools not stored in designated areas. Central shops generally in disarray. Unofficial production area shops may exist. Shop ventilation, lighting, lifting equipment, lay down areas inadequate. Shop far from stores. Work in process not identified by job/area or work order. | Stores is a mess—disorderly. Location system known only to stores keeper. Uncontrolled access—maintainers can find and take what they please. Hazardous materials not segregated from other stores. Parts not packaged properly or repackaged poorly. Lighting poor. Stores location is separate from receiving, shipping, and maintenance shop areas. Large unkempt bone yard. | Dirty, unkempt, uncleared; spills may be safety hazards. Minimal clean up done prior to or after repair work. | No budgeting for capital for asset replacements. Each case is handled as it arises. Replacements with cheapest available alternative are used. |

FIGURE 3.6 (continued).

these changes often use fewer people to get as much, or more, work done. They are typically very productive.

The next most important elements are usually the methods and processes (Figure 3.4). These are all about how you manage maintenance. They are the activities that people in the organization actually do. Methods and processes add structure to the work that gets done. As processes become more effective, people become more productive. Poor methods and processes produce much of the wasted effort typical of low-performing maintenance organizations.

In Figure 3.5, systems and technology represent the tools used by the people implementing the processes and methods you choose. These are the enablers, and they get most of the attention in maintenance management. Some organizations that focus tremendous energy on people and processes, with only basic tools and rudimentary technology, still achieve high performance levels. Other organizations, focused on the tools shown in Figure 3.5, haven't. Generally, emphasizing technology without excellence in managing methods, processes, and people will bring only limited success. It's like the joke about needing a computer to really mess things up. If inefficient or ineffective processes are automated and then run with ineffective and demotivated employees, the result will be disappointing

Ultimately, how well you manage processes, methods, and the people who use them comes down to the materials and physical plant you maintain (Figure 3.6). It is in the physical plant and materials area where you can best judge maintenance management effectiveness, which is why these descriptions are more detailed.

3.9   EFFECTIVE ASSET MANAGEMENT METHODS

Just as you must learn to walk before you run, you must be in firm control of maintenance before you can successfully begin continuous improvement. You do this by incrementally changing what maintenance is doing to strengthen choices that will optimize business objectives.

Methodologies are the systematic methods or procedures used to apply logic principles. The broad category of continuous improvement includes several maintenance methodologies, covered in-depth in this book:

- RCM focuses on overall equipment reliability
- RCFA
- TPM focuses on achieving high reliability from operators and maintainers
- Optimizing maintenance and materials management decisions

3.9.1   RCM

Reliability-centered maintenance aims to achieve maximum system reliability using maintenance tactics that can be effectively applied to specific system failures in the operating environment. The RCM process uses equipment and system knowledge to decide which maintenance interventions to use for each failure mode. That knowledge includes the following:

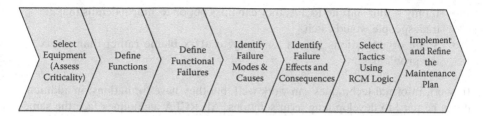

FIGURE 3.7 Moubray's work[2] will give you a deeper understanding of the concepts of reliability centered maintenance.

- System diagrams and drawings.
- Equipment manuals.
- Operational and maintenance experience with the system.
- Effects of individual failures on the system, the entire operation, and its operating environment.

The RCM basic steps are shown in Figure 3.7.

RCM results in a maintenance plan. The various decisions made for each failure are put into logical groupings for detailed planning, scheduling, and execution as part of the overall maintenance workload. Each task states what must be done to prevent, predict, or find specific failures. For each, there is a specified task frequency. You use optimization techniques to determine the best frequency for each task and decide on corrective actions.

3.9.2 ROOT-CAUSE FAILURE ANALYSIS

RCFA is one of the basic reliability enhancement methods. It is relatively easy to perform, and many companies already do it—some using rigorous problem-solving techniques and some informally. Later in the chapter, you'll learn about a formal method based on easy-to-follow cause and effect logic, but first we look at informal and other problem-solving techniques. In all methods, the objective is to completely eliminate recurring equipment or system problems, or at least to substantially diminish them.

Informal RCFA techniques are usually used by individuals or small groups to determine the best corrective action for a problem. Typically, this involves maintenance tradespeople, technicians, engineers, supervisors, superintendents, and managers. Drawing heavily on their own experience and information from sources like trade periodicals, maintainers from other plants, and contractors, they often have immediate success.

Plenty of pitfalls, though, can impair the informal approach:

- If only tradespeople do the RCFA, their solutions are often limited to repair techniques, parts, and materials selection and other design flaws.
- A restrictive engineering change control or spare parts (add to inventory) process can derail people who aren't skilled or accustomed to dealing with bureaucracy.

- If only senior staff do RCFA, they can miss out on technical details that the tradespeople would catch.
- Some organizations have a tendency to affix blame rather than to fix the problem.

In short, informal techniques can work well, but they have limitations; in addition, it can be hard to develop long-term solutions. All RCFA techniques face the same challenges, but they're greater if the process isn't formalized in some way.

More formal problem-solving techniques can be used very effectively. Consulting and educational organizations teach several techniques, two of which we examine here.

3.9.2.1   RCFA: What, Where, When Problem Solving

The first basic problem-solving process is relatively straightforward:

- Establish the problem, noting what has actually changed from "normal" to unacceptable.
- Describe the problem, asking what, where, and when questions to determine the extent of it. Quantify what went wrong and be specific so you solve the problem only where it exists. You need to understand both what is and is not happening now, as well as where and when.
- Identify possible causes.
- Identify the most likely cause. Test these possible causes against the is and is not criteria for the what, when, and where of the problem statement.
- Verify the cause. Test any assumptions you have made, looking for holes in the argument.
- Implement a solution that addresses the cause.

These formal problem-solving techniques, usually performed in a structured group with a facilitator, are very effective. To use this approach, set up a weekly or monthly meeting to identify and prioritize problems that need solutions. Although day-to-day maintenance and equipment issues will figure prominently, your mandate will likely extend beyond them.

Problem-solving groups should include a cross-section of interested stakeholders such as production, finance, human resources, training, and safety representatives as well as maintenance. Because it's so broad, the group is often most effective broken into smaller task teams. Each is assigned a specific problem to analyze and usually is made responsible for solving it completely. These teams report to the problem-solving group on progress and solutions.

Formal problem-solving groups usually produce credible results because of their broad representation and rigorous formal analytical processes. This thorough approach ensures solutions that work over the long term.

3.9.2.2   RCFA: Cause and Effect

The second RCFA technique we examine is based on cause and effect logic. Theoretically, all events are the result of potentially infinite combinations of preexisting conditions and triggering events. Events occur in sequence, with each event triggering other events, some being the failures that we are trying to eliminate.

Think of these sequences as "chains" of events, which, like chains, are only as strong as their weakest link. Break that link, and the chain fails—the subsequent events are changed. While you may eliminate the failure you're targeting, though, you could also trigger some other chain of events. Remember that the solution to one problem may well turn out to be the cause of another.

To perform a cause and effect RCFA, you need to do the following:

- Identify the unacceptable performance.
- Specify what is unacceptable (like the what, when, and where of the previous method).
- Ask, "What is happening?" "What conditions must exist for this event to happen?"
- Continue to ask this combination of what questions until you identify some event that can be controlled. If that event can be changed to prevent the failure from reoccurring, you have a "root cause" that can be addressed.
- Eliminate the root cause through an appropriate change in materials, processes, people, systems, or equipment.

By repeating the what questions, you usually get a solution within five to seven iterations. A variation of this process asks why instead of what—both questions work.

Because it is performed formally, with a cross-section of stakeholders exercising complete control over the solution, the success rate of cause and effect is also high.

3.10 OPTIMIZING MAINTENANCE DECISIONS—BEYOND RCM

Managing maintenance goes beyond repair and prevention to encompass the entire asset life cycle from selection to disposal. Key life-cycle decisions that must be made include the following:

- Component replacements
- Capital equipment replacements
- Inspection result decisions
- Resource requirements

To make the best choices, you need to consider not only technical aspects but also historical maintenance data, cost information, and sensitivity testing to ensure that your objectives are met in the long run. Jardine[3] describes several situations that can be dealt with effectively:

- Replacements when operating costs increase with use
- Replacements when operating costs increase with time
- Replacement of a machine when it is in standby mode
- Capital equipment replacements to maximize present value

- Capital equipment replacements to minimize total cost
- Capital equipment replacements considering technology improvements over time
- Optimizing replacement intervals for preventive component replacement
- Optimizing replacement intervals that minimize downtime
- Group replacements
- Optimizing inspection frequencies to maximize profit or to minimize downtime
- Optimizing inspection frequencies to maximize availability
- Optimizing inspection frequencies to minimize total costs
- Optimizing overhaul policies
- Replacement of monitored equipment based on inspection, cost, and history data

All of these methods require accurate maintenance history data that show what happened and when. Specifically, you need to make the distinction between repairs due to failures and ones that occurred while doing some other work. Note that it is the quality of the information that is important, not the quantity. Some of these decisions can be made with relatively little information, as long as it is accurate.

RCM produces a maintenance plan that defines what to do and when. It is based on specific failure modes that are either anticipated or known to occur. Frequency decisions are often made with relatively little data or a lot of uncertainty. You can make yours more certain and precise by using the methods described later in the present book and in Jardine's work.[3] Along with failure history and cost information, these decision-making methods show you how to get the most from condition monitoring inspections to make optimum replacement choices.

REFERENCES

1. Campbell, J.D. *Uptime: Strategies for Excellence in Maintenance Management.* New York: Productivity Press, 1995.
2. Moubray, T. *Reliability Centered Maintenance.* 1997.
3. Jardine, A.K.S. *Maintenance, Replacement and Reliability.* 1994.

4 Measurement in Maintenance Management

Edited by Don Barry
Original by J. Stevens

CONTENTS

Performance management is one of the basic requirements of an effective operation. Doing it well, though, isn't as straightforward as it may seem. In this chapter, we discuss effective tools to help you strengthen your maintenance performance measurement.

To start, we summarize measurement basics with an example that shows how numeric measures can be both useful and misleading. Measurement is important in maintenance continuous improvement and in identifying and resolving conflicting priorities. We explore this within the maintenance department and between maintenance and the rest of the organization. We look at both macro and micro approaches, using a variety of examples. This includes the balanced scorecard at the macro level and shaft alignment case history at the micro level.

Maintenance performance measurement is then subdivided into its five main components—productivity, organization, work efficiency, cost, and quality—together with some overall measurements of departmental results. You'll learn about consistency and reliability as they apply to measurement. A major section of the chapter is devoted to individual performance measures, with sample data attached. We summarize the data required to complete these measures so you can decide whether you can use them in your own workplace. We also cover the essential tie-in between performance measures and action.

The chapter concludes with a practical look at using the benefits/difficulty matrix as a tool for prioritizing actions. You will also find a useful step-by-step guide to implementing performance measures.

4.1 INTRODUCTION

4.1.1 MAINTENANCE ANALYSIS: THE WAY INTO THE FUTURE

Performance measurement has no mystique. The trick is how to use the results to achieve the needed actions. This requires several conditions: consistent and reliable data, high-quality analysis, clear and persuasive presentation of the information, and a receptive work environment. Since maintenance optimization is targeted at executive management and the boardroom, it is vital that the results reflect the basic business equation:

> Maintenance is a business process turning inputs into usable outputs.

Figure 4.1 shows the three major elements of this equation: the inputs, outputs, and conversion process. As shown, most of the inputs (e.g., labor costs, materials, equipment, contractors) are familiar to the maintenance department and readily measured. Some inputs are more difficult to measure accurately (e.g., experience, techniques, teamwork, work history), yet each can significantly impact results.

Likewise, some outputs are easily recognized and measured, and others are harder. As with inputs, some are intangible, like the team spirit that comes from completing a difficult task on schedule. Measuring attendance and absenteeism isn't exact and

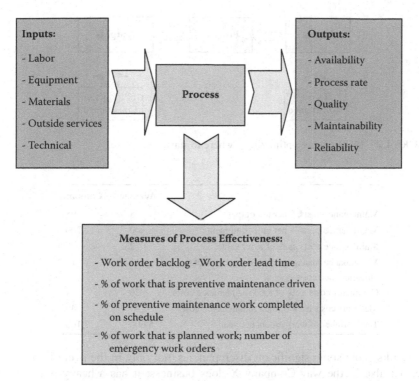

Inputs:

- Labor
- Equipment
- Materials
- Outside services
- Technical

Process

Outputs:

- Availability
- Process rate
- Quality
- Maintainability
- Reliability

Measures of Process Effectiveness:

- Work order backlog - Work order lead time
- % of work that is preventive maintenance driven
- % of preventive maintenance work completed on schedule
- % of work that is planned work; number of emergency work orders

FIGURE 4.1   Maintenance as a business process.

is no substitute for these intangibles but frequently can result in overall performance indicators being at much too high a level. Neither measurement is a substitute for intangible benefits. While intangibles contribute significantly to overall maintenance performance, though, the focus of this book is on the tangible measurements.

Converting the maintenance inputs into the required outputs is the core of the maintenance manager's job. Yet rarely is the absolute conversion rate of much interest in itself. Converting labor hours consumed into reliability, for example, makes little or no sense—until it can be used as a comparative measure, through time or with a similar division or company. Similarly, the average material consumption per work order isn't significant until you see that Press 1 consumes twice as much repair material as Press 2 for the same production throughput. A simple way to reduce materials per work order consumption is to split the jobs, which increases the number of work orders (but doesn't do anything for productivity improvement, of course).

The focus, then, must be on the comparative standing of your company or division, or on improving maintenance effectiveness from one year to the next. These comparisons highlight another outstanding value of maintenance measurement: it regularly compares progress toward specific goals and targets. This benchmarking process—through time, with other divisions, or companies—is increasingly being used by senior management as a key indicator of good maintenance management. Frequently, it discloses surprising discrepancies in performance. A recent benchmarking exercise turned up the following data from the pulp industry:

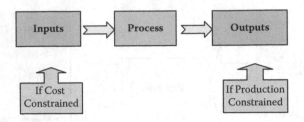

FIGURE 4.2   Maintenance optimizing—where to start.

	Average	Company X
Maintenance costs, $ per ton output	78	98
Maintenance costs, $ per unit equipment	8,900	12,700
Maintenance costs as % of asset value	2.2	2.5
Maintenance management costs as % of total maintenance costs	11.7	14.2
Contractor costs as % of total maintenance costs	20	4
Materials costs as % of total maintenance costs	45	49
Total number of work orders per year	6,600	7,100

The results show some significant discrepancies, not only in the overall cost structure, but also in the way Company X does business. It has a heavy management structure and hardly uses outside contractors, for example. You can clearly see from this high-level benchmarking that, to preserve Company X's competitiveness in the marketplace, something needs to be done. Exactly what needs doing, though, isn't obvious. This requires more detailed analysis.

As you'll see later, the number of potential performance measures far exceeds the maintenance manager's ability to collect, analyze, and act on the data. An important part, therefore, of any performance measurement implementation is to thoroughly understand the few, key performance drivers. Maximum leverage should always take top priority. First, identify the indicators that show results and progress in areas that most critically need improvement. As a place to start, consider Figure 4.2.

If the business could sell more products or services with a lower price, it is cost constrained. The maximum payoff is likely to come from concentrating on controlling inputs (e.g., labor, materials, contractor costs, overheads). If the business can profitably sell all it produces, it is production constrained. It's likely to achieve the greatest payoff from maximizing outputs through asset reliability, availability, and maintainability.

4.1.1.1   Keeping Maintenance in Context

As an essential part of your organization, maintenance must adhere to the company's overall objectives and direction. Maintenance cannot operate in isolation. The continuous improvement loop (Figure 4.3), key to enhancing maintenance, must be driven by and mesh with the corporation's planning, execution, and feedback cycle.

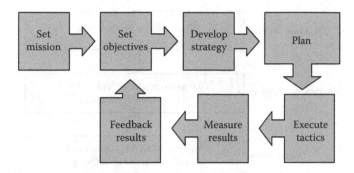

FIGURE 4.3 The maintenance continuous improvement loop.

Disconnects frequently occur when the corporate and department levels aren't in sync. For example, if the company places a moratorium on new capital expenditures, this must be fed into the equipment maintenance and replacement strategy. Likewise, if the corporate mission is to produce the highest possible quality product, this probably doesn't correlate with the maintenance department's cost minimization target. This type of disconnect frequently happens inside the maintenance department itself. If your mission is to be the best performing maintenance department in the business, your strategy must include condition-based maintenance (CBM) and reliability. Similarly, if the strategy statement calls for a 10% reliability increase, reliable and consistent data must be available to make the comparisons.

4.1.1.2 Conflicting Priorities for the Maintenance Manager

In modern industry, all maintenance departments face the same dilemma: Which of the many priorities should be put at the top of the list? Should the organization minimize maintenance costs or maximize production throughput? Should downtime be minimized, or should the concentration be on customer satisfaction? Should short-term money be spent on a reliability program to reduce long-term costs?

Corporate priorities are set by the senior executive and ratified by the board of directors. These priorities should then flow down to all parts of the organization. The maintenance manager must adopt those priorities; should convert them into corresponding maintenance priorities, strategies, and tactics to achieve the results; and then should track them and improve on them. Figure 4.4 is an example of how corporate priorities can flow down through the maintenance priorities and strategies to the tactics that control the everyday work of the maintenance department.

If the corporate priority is to maximize product sales, maintenance can focus on maximizing throughput and equipment reliability. In turn, the maintenance strategies will also reflect this and could include, for example, implementing a formal reliability enhancement program supported by condition-based monitoring. Out of these strategies, the daily, weekly, and monthly tactics flow. These, in turn, provide lists of individual tasks that then become the jobs that will appear on the work orders from the enterprise asset management (EAM) system or computerized maintenance

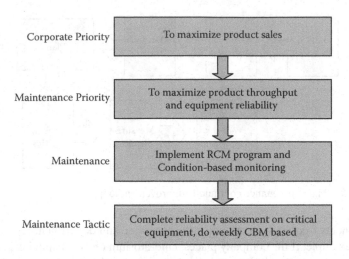

FIGURE 4.4 Interrelating corporate priorities with maintenance tactics.

management system (CMMS). Using the work order to ensure that the inspections get done is widespread. Where organizations frequently fail is in completing the follow-up analysis and reporting on a regular and timely basis. The most effective method is to set them up as weekly work order tasks, subject to the same performance tracking as preventive and repair work orders.

In trying to improve performance, you can be confronted with many, seemingly conflicting, alternatives. Numerous review techniques are available to establish how your organization compares with industry standards or best maintenance practice. The most effective techniques help you map priorities by indicating the payoff that improvement will make. The review techniques tend to be split into macro (covering the full maintenance department and its relation to the business) and micro (with the focus on a specific piece of equipment or a single aspect of the maintenance function).

The leading macro techniques are as follows:

- Maintenance effectiveness review: involves the overall effectiveness of maintenance and its relationship to the organization's business strategies. These can be conducted internally or externally, and typically cover areas such as the following:
  - Maintenance strategy and communication
  - Maintenance organization
  - Human resources and employee empowerment
  - Use of maintenance tactics
  - Use of reliability engineering and reliability-based approaches to equipment performance monitoring and improvement
  - Information technology and management systems
  - Use and effectiveness of planning and scheduling
  - Materials management in support of maintenance operations

- External benchmark: draws parallels with other organizations to establish how the organization compares with industry standards. Confidentiality is a key factor, and results typically show how the organization ranks within a range of performance indicators. Some of the areas covered in benchmarking overlap with the maintenance effectiveness review. Additional topics include the following:
  - Nature of business operations
  - Current maintenance strategies and practices
  - Planning and scheduling
  - Inventory and stores management practices
  - Budgeting and costing
  - Maintenance performance and measurement
  - Use of CMMS and other information system (IS) tools
  - Maintenance process reengineering
- Internal comparisons: measure a similar set of parameters as the external benchmark but draw from different departments or plants. They are generally less expensive and, if the data are consistent, illustrate differences in maintenance practices among similar plants. From this, you can decide which best practices to adopt.
- Best practices review: looks at maintenance's process and operating standards and compares them with the industry best. This is generally the starting point for a maintenance process upgrade program, focusing on areas such as the following:
  - Preventive maintenance
  - Inventory and purchasing
  - Maintenance workflow
  - Operations involvement
  - Predictive maintenance
  - Reliability-based maintenance
  - Total productive maintenance
  - Financial optimization (continuous improvement)
- Overall equipment effectiveness (OEE): measures a plant's overall operating effectiveness after deducting losses due to scheduled and unscheduled downtime, equipment performance, and quality. In each case, the subcomponents are meticulously defined, providing one of the few reasonably objective and widely used equipment performance indicators.

Following is a summary of one company's results. Remember that the individual category results are multiplied through the calculation to derive the final result. Although Company Y achieves 90% or higher in each category, it will have an OEE of only 74% (see Chapter 8 for further details of OEE). This means that by increasing the OEE to, say, 95%, Company Y can increase its production by (95 − 74)/74 = 28% with minimal capital expenditure. If you can accomplish this in three plants, you won't need to build a fourth.

	Target	Company Y
Availability	97%	90%
×		
Use rate	97%	92%
×		
Process efficiency	97%	95%
×		
Quality	99%	94%
=		
Overall equipment effectiveness	90%	74%

These, then, are some of the high-level indicators of the effectiveness and comparative standing of the maintenance department. They highlight the key issues at the executive level, but more detailed evaluation is needed to generate specific actions. They also typically require senior management support and corporate funding, which are not always a given.

Fortunately, many maintenance measures can be implemented that don't require external approval or corporate funding. These are important because they stimulate a climate of improvement and progress. Some of the many indicators at the micro level are as follows:

- Post (systems) implementation review to assess the results of buying and implementing a system (or equipment) against the planned results or initial cost justification
- Machine reliability analysis and failure rates, targeted at individual machine or production lines
- Labor effectiveness review, measuring staff allocation to jobs or categories of jobs compared with last year
- Analyses of, for example, material usage, equipment availability, use, productivity, losses, and cost

All of these indicators give useful information about the maintenance business and how well its tasks are being performed. You have to select those that most directly achieve the maintenance department's goals as well as those of the overall business.

Moving from macro or broad-scale measurement and optimization to a micro model can create problems for maintenance managers. You can resolve this by regarding the macro approach as a project or program and the micro indicators as individual tasks or series of tasks. Figure 4.5 shows how an external benchmark finding can be translated into a series of actions that can be readily implemented.

4.2 MEASURING MAINTENANCE—THE BROAD STROKES

To improve maintenance management, you need measurement capability for all of the major items under review. However, as previously mentioned, there usually aren't

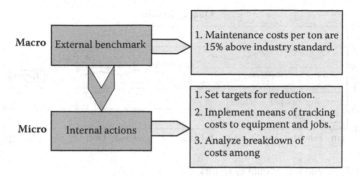

FIGURE 4.5   Relating macro measurements to micro tasks.

enough resources to go beyond a relatively small number of key indicators. The following are the major categories that should be considered:

- Maintenance productivity measures the effectiveness of resource use.
- Maintenance organization measures the effectiveness of the organization and planning activities.
- Efficiency of maintenance work is how well maintenance keeps up with workload.
- Maintenance costs is overall maintenance cost versus production cost.
- Maintenance quality is how well the work is performed.
- Overall maintenance results measures overall results.

Measurements are only as good as the actions they prompt; the results are as important as the numbers themselves. Attractive, well thought out graphics will help "sell" the results and stimulate action. Graphic layouts should be informative and easy to interpret, like the spider diagram shown in Figure 4.6. Here, the key

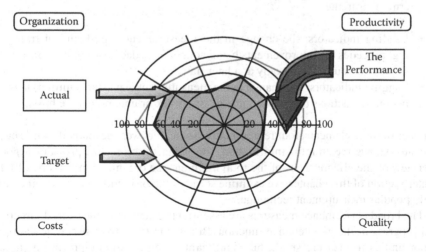

FIGURE 4.6   Spider diagram showing performance gaps.

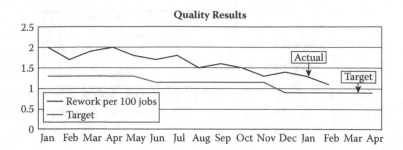

FIGURE 4.7   Trend line of quality results.

indicators are measured on the radial arms in percentage achievements of a given target. Each measurement's goals and targets are also shown, with the performance gap clearly identified. Where the gap is largest, the shortfall is greatest, and so is the need for immediate action.

The spider diagram in Figure 4.6 shows the situation at a point in time, but not its progress through time. For that, you need trend lines. These are best shown as a graph (or series of graphs) with the actual and the targets clearly identified, as in Figure 4.7. For a trend line to be effective, the results must be readily quantifiable and directly reflect the team's efforts.

4.2.1 BALANCED SCORECARD

Each of the previous examples shows only a single measure of performance. The balanced scorecard (BSC) concept broadens the measure beyond the single item. Each organization should develop its own balanced scorecard to reflect what motivates its business behavior.

Figure 4.8 shows an example that combines the elements of the input process–output equation referred to earlier with leading and lagging indicators and short- and long-term measurements:

- Leading indicators: the change in the measurement (e.g., hours of training) precedes the improvements being sought (e.g., decreases in error rates). Typically, you see these only at a later date.
- Lagging indicators: the change in the measurement (e.g., staff quitting) lags behind the actions that caused it (e.g., overwork or unappreciative boss).

For each of these elements, four representative indicators were developed. While each indicator shows meaningful information, the balanced scorecard provides a good overview of the effectiveness of the total maintenance organization (Figure 4.8). In a later section of this chapter, we examine some of the pros and cons of this increasingly popular measurement technique.

The broad performance measures are essential to understand the overall direction and progress of the maintenance function. But within this broad sweep lie multiple opportunities to measure small, but significant changes—in equipment operation,

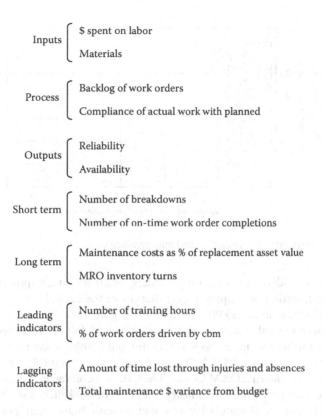

Inputs
- $ spent on labor
- Materials

Process
- Backlog of work orders
- Compliance of actual work with planned

Outputs
- Reliability
- Availability

Short term
- Number of breakdowns
- Number of on-time work order completions

Long term
- Maintenance costs as % of replacement asset value
- MRO inventory turns

Leading indicators
- Number of training hours
- % of work orders driven by cbm

Lagging indicators
- Amount of time lost through injuries and absences
- Total maintenance $ variance from budget

FIGURE 4.8 Example of a balanced scorecard.

labor productivity, contractors' performance, material use, and technology and management contribution. The next section examines some of these changes and provides examples that can be used in the workplace.

4.3 MEASURING MAINTENANCE—THE FINE STROKES

To understand individual elements of maintenance functions, you need analysis at a much more detailed level. To evaluate, predict, and improve the performance of a specific machine, you must have its operating condition and repair data, not only for the current period, but also historically. Also, it's useful to compare data from similar machines. Later in this chapter, we examine the sources of these data in more detail. For now, you should know that the best data sources are CMMS, EAM, CBM, supervisory control and data acquisition systems (SCADA), and process control systems currently in widespread use.

Among the many ways to track individual equipment performance are the measurements of reliability, availability, productivity, life-cycle costs, and production losses. Use these techniques to identify problems and their causes so that remedial action can be taken. An interesting case study examined a series of high-volume pumps to establish why the running costs (i.e., operating and regular maintenance

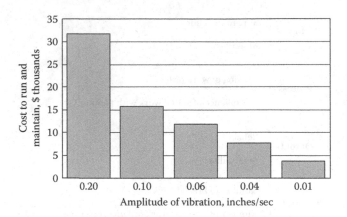

FIGURE 4.9   Internal pump operating and maintenance costs.

costs) varied so widely among similar models. Shaft alignment proved to be the major problem, setting off improvements that drove the annual cost per unit down from over $35,000 to under $5,000, as shown in Figure 4.9.

Despite this remarkable achievement, the company didn't get the expected overall benefit. After further special analysis, two additional problems were found. First, the operating and maintenance ("O&M") costs excluded contractor fees and so should have been labeled "Internal O&M Costs." Then, the subcontractor's incremental cost to reduce the vibration from the industry standard of 0.04 inches/sec to the target level of 0.01 inches/sec (Figure 4.10) was unexpectedly higher than predicted from earlier improvements. When the O&M costs were revised to include the subcontractor fee, the extra effort to move from 0.04 inches/sec to 0.01"/sec was considerably more expensive than the improvement in the operating costs. The optimum position for the company was to maintain at 0.04 inches/sec (Figure 4.11).

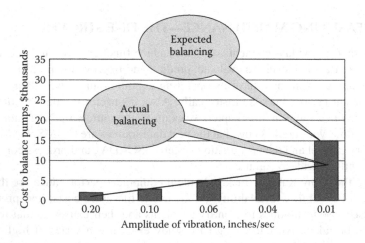

FIGURE 4.10   Subcontractor cost of pump balancing.

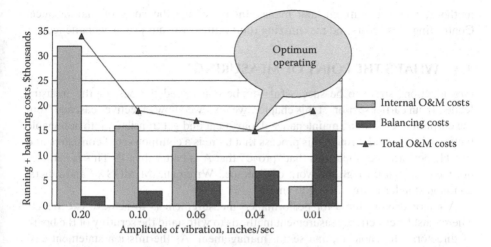

FIGURE 4.11 Total pump operating and maintenance (O&M) costs.

This type of quasiforensic analysis can produce dramatic savings. In the previously summarized case, the company had 50 of these pumps operating and was able to reduce costs from $1.6 million to $750,000, saving 53%. By taking the final step to refine the balancing from 0.04 to 0.01, the company spent an extra $200,000 for no additional tangible benefit.

You can use similar investigative techniques to evaluate and improve labor and materials consumption. Variable labor consumption among similar jobs at a number of plants can identify different maintenance methodologies and skill levels. By adopting the best practices and adding some targeted training, you can achieve significant improvements and avoid sizeable problems. In one example, a truck motor overheating was traced back through the CMMS work order to a badly seated filter. The maintenance technician responsible for the work had insufficient training for the task. Emergency recalls were issued for six other trucks on the highway that he had fitted the same way, and each one was fixed without damage. The technician received additional training and got to keep his job.

Variations in materials usage can be tracked from the work orders and the inventory records, leading to standardized methods. More recently, multiplant operations have been able to access data from multiple different databases for analytical and comparative purposes. Parts specification has become standardized, creating huge savings through bulk buying and centralized storage. There are several examples where a combination of reduced inventory, reduced supplier base, better negotiated prices, and removing many hidden purchasing costs through improved productivity has generated savings well beyond 15% in the in-bound supply chain.

In this chapter, we have introduced a wide variety of topics to set the scene for later and more detailed examination. The core issues are the same throughout this book: What should the maintenance manager optimize, and how should it be done? Another chapter in the book introduced the 7M model. We will return again and again to the elements of this model, which covers the entire maintenance domain, but from several different directions. Its components—machinery, manpower, materials, methods, measures, milieu, and management—create the costs of maintenance. Controlling these costs and maximizing their return are your primary challenges.

4.4   WHAT'S THE POINT OF MEASURING?

Organizations strive to be successful—to be considered the best or the industry leader. In an earlier section of this chapter, we reviewed how objectives cascade from the corporate level to the maintenance department and get translated into actionable tactics. It's assumed through this process that there is a common set of consistent and reliable performance measures that "prove" that A is better than B. This is actually not the case, but through the work of Coetzee,[1] Wireman,[2] MIMOSA,[3] and others, common standards are starting to emerge.

A major driving force for performance measurement is to achieve excellence. There must be effective measurement methods to withstand the scrutiny of the board of directors, shareholders, and senior management. As the mission statement cascades down to the maintenance department, so does the demand for accurate measurement. This is reason enough to measure maintenance performance.

There are many other reasons, though, to make improvements, including the following:

- Competitiveness: regardless of whether the goals are price, quality, or service driven, you must compare to establish how competitive you are.
- Right-sizing, down-sizing, and up-sizing: adjusting the size of the organization to deliver products and services while continuing to prosper becomes meaningless if you can't realistically measure performance.
- New processes and technologies are being introduced rapidly: not only in manufacturing but also in maintenance. To produce the expected improvements, you must keep careful track of the results.
- Performance measurement is integral when deciding to maintain or replace an item. In another chapter in the book we included an example of life-cycle costing. We use it again later in this chapter to determine whether to maintain or replace.
- The performance improvement loop (see Figure 4.12; reproduced here from earlier in this chapter, Figure 4.3) is the core process in identifying and implementing progress. Performance measurement and results feedback are essential elements in this loop.

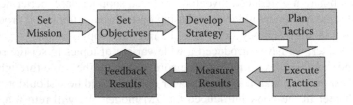

FIGURE 4.12   Measurements as a core part of the performance improvement loop.

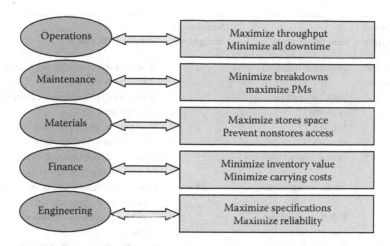

FIGURE 4.13 Conflicting departmental objectives.

4.5 WHAT SHOULD WE MEASURE?

In the next section, we look at a wide variety of performance measurements, with sources for further study. You must concentrate on the need for consistency and reliability. Comparisons over time and between specific pieces of equipment or departments must be consistent to be valid. You want the assessment of your operation to be reliably complete, without significant omissions.

To start, you must understand that maintenance management is a dynamic process, not static. It is inextricably linked to business strategy, not simply a service on demand. Finally, it is an essential part of the business process, not just a functional silo operating in isolation. Despite this, there are frequent conflicts in setting objectives for an organization. For example, the objectives in Figure 4.13, taken from a strategic review of Company C, will be confusing when translated into performance measurement and later action

From the discussion so far, you can see that there is no real consensus on the precise source data, what should be measured, and how it should be analyzed. Wireman[2] takes a hierarchical approach in his book *Developing Performance Indicators for Maintenance Management*. With comprehensive coverage, he develops performance measures based on a five-tier hierarchy: (1) corporate; (2) financial; (3) efficiency; (4) tactical effectiveness; and (5) functional effectiveness. Wireman covers the following areas.

Area Covered by Indicators	Functional Areas	Number of Indicators	Sample Indicator
Corporate	N/A	4	Return on net assets; total cost to produce
Financial	N/A	8	Maintenance cost per unit produced; replacement value of assets maintained

(continued on next page)

Area Covered by Indicators	Functional Areas	Number of Indicators	Sample Indicator
Efficiency and effectiveness	Preventive maintenance	6	% of total direct maintenance cost that is breakdown related
	Work order (WO) systems	3	% total work orders that are preventive maintenance
	Training	4	% of total maintenance work caused by rework due to lack of skills
	Operational involvement	2	Maintenance-related equipment downtime this year versus last year
	Predictive maintenance	1	Current maintenance costs versus those prior to predictive program
	Reliability-centered maintenance	7	Number of repetitive failures versus total failures
	Total productive maintenance	4	Overall equipment effectiveness combining availability, performance efficiency, and quality rate
Tactical	Preventive maintenance (PM)	2	% of total number of breakdowns that should have been prevented
	Inventory and procurements	4	Total of items filled on demand versus total requested
	Work order (WO) systems	4	Total planned WOs versus total WOs received
	CMMS	4	Total costs charged to equipment versus total costs from accounting
	Operational involvement	3	PM hours performed by operators as % of total maintenance hours
	Reliability-centered maintenance	2	Number of equipment breakdowns per hour operated
Functional	Preventive maintenance	3	% of total WOs generated from PM inspections
	Inventory and procurements	4	% of total stock items inactive
	Work order systems	7	% of total labor costs from WOs
	Planning and scheduling	2	% of total labor costs that are planned
	CMMS	6	% of total in plant equipment in CMMS
	Training	5	Training hours per employee
	Operational involvement	5	% of total hours worked by operators spent on equipment improvement
	Predictive maintenance (PdM)	2	PdM hours % of total maintenance hours
	Reliability-centered maintenance	3	% of failures where root-cause analysis is performed
	Total productive maintenance	2	% of critical equipment covered by design studies

(continued on next page)

Area Covered by Indicators	Functional Areas	Number of Indicators	Sample Indicator
	Statistical financial optimization	3	% of critical equipment where maintenance tasks are audited
	Continuous improvement	3	Savings from employee suggestions

With over 100 indicators listed, Wireman[2] clearly shows that you can measure just about anything. He also emphasizes the equal and opposite need to be selective about what is measured.

Coetzee's[1] approach in his book, *Maintenance*, centers around four pillars: (1) results; (2) productivity; (3) operational purposefulness; and (4) cost justification. The following approach expands his method by adding some extra measures and presenting them as calculations. Check the core data set in Section 3.7. We emphasize that new measurement formulae are being continually developed; those presented here are examples only, not to be considered as a complete or recommended set. What makes the best set of data will vary considerably with each situation.

4.6 MEASURING OVERALL MAINTENANCE PERFORMANCE

The measures here are macro level, showing progress toward achieving the maintenance department's overall goals. Later in this chapter, we cover some micro measurement that applies to individual equipment. Figure 4.14 summarizes the categories we use in the examples.

4.6.1 OVERALL MAINTENANCE

These indicators measure whether the maintenance department keeps the equipment productive and produces quality products. We look at these five measures:

1. *Availability* is the percentage of time that equipment is available for production, after all scheduled and unscheduled downtime. Note that idle time caused by lack of product demand isn't deducted from the total time available. The equipment is considered "available" even though no production is demanded.

FIGURE 4.14 Maintenance categories for macro analysis.

$$Availability = \frac{Total\ Time - Downtime}{Total\ Time}$$

Downtime includes all scheduled and unscheduled downtime, but not idle time through lack of demand.

$$Availability = \frac{8760 - 392}{8760} = 95.5\%$$

Total Time = 8,760 hours. Downtime = 392 hours.

2. *Mean Time to Failure (MTTF)* is a popular measure that will be revisited in some depth later. It represents how long a machine can be expected to run before it dies. It measures average uptime and is widely used in production scheduling to determine whether the next batch can likely be produced without interruption. In this example, the equipment is expected to run an average of 647 hours from the previous failure. If the runtime since the last failure has been 250 hours, expect the equipment to have 397 hours remaining operating life before the next failure. Note that this assumes the equipment has an equal failure probability throughout its 647 hours. This isn't really reasonable, since failure rates frequently approximate a normal distribution, which means failure probability increases as you approach the average.

$$MTTF = \frac{Total\ Time - Downtime - Nonused\ Time}{Number\ of\ breakdowns}$$

Total Time = 8,760 hours
Downtime = 392 hours
Nonused Time = 600 hours
Number of Breakdowns = 12

$$MTTF = \frac{8,760 - 392 - 600}{12} = 647\ hours$$

3. *Failure frequency or breakdown frequency* measures how often the equipment is expected to fail. It is typically used as a comparative measure, not an absolute, and therefore should be trended. It helps to regard it as the conditional probability of failure within the next time period. With this measure, it should be note that *failure* is generally an inexact term. For example, if a machine that is designed to run at 100 bits per minute is running at 85, is this deemed failure? Similarly, if it produces at 100 bpm, but 10 units are defective, is this failure? Adopt the reliability-centered maintenance approach—the run rate and the quality rates are given quantifiable

failure levels. Anything below is deemed to have failed, even though it may operationally still be struggling on.

$$Breakdown\ Frequency = \frac{Number\ of\ Breakdowns}{Total\ Time - Downtime - Nonused\ Time}$$

Total Time = 8,760 hours
Downtime = 392 hours
Nonused Time = 600 hours
Number of Breakdowns = 12/year
Production Rate = 10/hour

$$Breakdown\ Frequency = \frac{12}{8,760 - 392 - 600} = 0.0015\ (\text{Failures per hour})$$

Probability of failure within the next hour = 0.15%
Probability of failure within the next production run of 2,500 units
= 0.15% × 2500/10 = 37.5%

4. *Mean Time to Repair (MTTR)* is the total time it takes for the problem to be fixed and the equipment operating again. This includes notification time, travel time, diagnosis time, fix time, wait time (for parts or cool down), reassemble time, and test time. It reflects how well the organization can respond to a problem, and, from the list of the total time components, you can see that it covers areas outside the maintenance department's direct control. MTTR also measures how long operations will be out of production, broadly indicating maintenance effect on equipment production rate. Note that this can be used as a measure of the average of all MTTRs, as the MTTR for breakdowns, or as the MTTR for scheduled outages.

$$MTTR = \frac{Unscheduled\ Downtime}{Number\ of\ Breakdowns}$$

Unscheduled Downtime = 232 hours
Scheduled Outages = 160
Number of Breakdowns = 12
Number of Scheduled Outages = 6

$$MTTR\ for\ Unschedule\ Downtime = \frac{232}{12} = 19.3\ hours$$

MTTR for Scheduled Outages = 26.7 hours
MTTR for All Downtime = 21.8 hours

5. *Production rate index* is the maintenance impact on equipment effectiveness. As with the previous indicator, you must interpret results carefully, since operating speeds and conditions will have an impact. To minimize the effect of these variations, it is trended over time as an index and has no value as an absolute number.

$$Production\ Rate\ Index = \frac{Production\ Rate\ (Units/Hour)}{Total\ Time - Downtime - Nonused\ Time}$$

Production Rate = 10 units/hour
Total Time = 8,760 hours
Downtime = 392 hours
Nonused Time = 600 hours

$$Production\ Rate\ Index = \frac{10}{8,760 - 392 - 600} = 0.001287$$

4.6.2 MAINTENANCE PRODUCTIVITY

Maintenance productivity indices measure maintenance's use of resources, including labor, materials, contractors, tools, and equipment. These components also form the cost indicators that will be dealt with later.

- Manpower use is usually called *wrench time*, because it measures the time consumed by actual maintenance tasks as a percentage of total maintenance time. The calculation includes standby time, wait time, sick time, vacation time, and time set aside for things like meetings and training. It measures only time spent on the job. There are frequent problems measuring the results, because assigning time to jobs varies within organizations. For example, is travel time assigned to the job? For measurements within a single organization, the definitions need to be clearly defined, documented, and adhered to. In the following example, you'll see a wrench time figure of 69%: a fairly modest standard. High-performance, land-based factory operations will exceed 80%.

$$Manpower\ Use = \frac{Wrench\ Time}{Total\ Time}$$

32 Staff, *Total Time* = 32 × 2,088 = 66,816
Wrench Time = 46,100

$$Manpower\ Use = \frac{46,100}{66,916} = 69\%$$

- *Manpower efficiency* shows the extent to which the maintenance jobs completed matched the time allotted during the planning process. Although typically called an "efficiency" measure, it is also a measure of the planning accuracy itself. Many EAM systems can modify the job planning times for repeat jobs based on the average of the past number of times the job has been completed. The manpower efficiency measure becomes a comparison with this moving average. Many planners reject this measure because they don't want to plan a new job until the teardown and subsequent diagnosis has been completed. They'll apply it only to preventive maintenance or repeat jobs.

$$Manpower\ Efficiency = \frac{Time\ Taken}{Planned\ Time}$$

Time Taken = Wrench Time = 46,100
Planned/Allowed Time = 44,700

$$Manpower\ Efficiency = \frac{44,700}{46,100} = 97\%$$

- *Materials usage per work order* measures how effectively the materials are being acquired and used. Again, this is a composite indicator that you must further refine before taking direct action. The measure shows the average materials consumption per work order. Variations from job to job can occur as a result of changes in, for example, buying practices, pricing, or sourcing; in inventory costing or accounting practices; in the way the jobs are specified; and in parts replacement policy. As noted earlier, subdividing work orders greatly reduces the material cost per work order. Nevertheless, it is a simple trend to plot and, as long as you haven't made significant underlying changes, indicates whether material usage is improving.

$$Material\ Usage = \frac{Total\ Materials\ \$\ Charged\ to\ Work\ Orders}{Number\ of\ Work\ Orders}$$

Total Materials Consumed = $1,400,000
Total Work Orders = 32,000

$$Material\ Usage = \frac{1,400,000}{32,000} = \$44\ per\ work\ order$$

- *Total maintenance costs as a percentage of total production costs* indicates the overall effectiveness of resource use. It suffers from the same variations as the material usage measure previously shown, but if you maintain consistent underlying policies and practices, it will show overall performance improvements or deterioration. Although this indicator is shown only for total costs, similar indices can be readily created for labor, contractors, and special equipment—in fact, any significant cost element.

Total Maintenance Costs = $4.0 million
Total Production Costs = $45 million

$$\textit{Maintenance Cost Index} = \frac{4.0}{45} = 8.9\%$$

4.6.3   MAINTENANCE ORGANIZATION

Maintenance performance indicators measure the effectiveness of the organization and maintenance planning activities. This is frequently missed when considering the overall effectiveness of the department because it largely takes place at the operational end. Studies have shown, though, that effective planning can significantly impact maintenance's operational effectiveness. In fact, one of the early selling features of the CMMS products was that allocating 5% of the maintenance department work effort would increase the overall group's efficiency by about 20%. For example, dedicating one planner 100% from a maintenance team of 20 would increase the operating efficiency of the remaining 19 from 60% to 75%. Also, it would raise the overall weekly wrench turning hours from 20 × 0.60 × 40 = 480 to 19 × 0.75 × 40 = 570, for an increase of 18.75%.

1. *Time spent on planned and scheduled tasks as a percentage of total time* measures the effectiveness of the organization and maintenance planning activities. This places the focus on the work-planning phase, since planned work is typically up to 10 times as effective as breakdown response. The planning and scheduling index measures the time spent on planned and scheduled tasks as a percentage of total work time. Notice that emphasis is placed on both planning and scheduling. A job is planned when all the job components are worked out, for example, what is to be done, who is to do it, and what materials and equipment should be used. Scheduling places all of these into a time slot so that all are available when required. By themselves, planning and scheduling have a positive impact. This impact is greatly multiplied when they are combined.

$$\textit{Planning and Scheduling Index} = \frac{\textit{Time Planned and Scheduled}}{\textit{Total Time}}$$

Time Planned and Scheduled = 26,000 hours
Total Time = 32 employees × 2,088 hours each = 66,816 hours

$$Planning\ and\ Scheduling\ Index = \frac{26,000}{66,816}$$

2. *Breakdown time* measures the amount of time spent on breakdowns and, through this, indicates whether more time is needed to prevent them. Use this index in combination with other indices, because the numbers will improve as the organization becomes quicker at fixing breakdowns. This can lead to a different culture, one that prides itself on fast recovery rather than on initial prevention.

Breakdown Time = Time spent on Breakdowns as % of Total Time
Total Time = 32 employees × 2,088 hours each = 66,816 hours
Breakdown Time = 2,200 hours

$$Breakdown\ Time = \frac{2,200}{66,816} = 3.3\%$$

3. *Cost of lost production due to breakdowns* is measured because the acid test of the breakdown is how much production capacity was lost. The amount of time spent by maintenance alone shows only the time charged to the job through the work order, not necessarily how serious the breakdowns are. Wait time (e.g., for cooling off, restart, and materials) is not included but still prevents the equipment from operating. This measure, then, includes run-up time and the costs of lost production due to breakdown and fixing breakdowns.

Breakdown Production Loss = Cost of Breakdowns as % of Total Direct Cost
Cost of Production Lost Time = $5,140/hr
Maintenance Cost of Breakdowns = $135/hr
Total Direct Cost = $45 million
Breakdown Time = 232 hours

$$Breakdown\ Production\ Loss = \frac{232 \times (5,140 + 135)}{45m} = 2.72\%$$

4. *The number of emergency work orders* is an overall indication of how well the breakdown problem is being kept under control. To be effective, each "emergency" must be well defined and consistently applied. In one notable example, a realistic (but cynical) planner defined work priorities in terms of how high up the command chain the requestor was. It is best to note measures of work order numbers in relation to a previous period (e.g., last month or last year) and plot them on a graph.

Number of E-Work Orders This Year = 130
Number of E-Work Orders Last Year = 152

4.6.4 EFFICIENCY OF MAINTENANCE WORK

This set of measures tracks the ability of maintenance to keep up with its workload. It measures three major elements: (1) the number of completions versus new requests; (2) the size of the backlog; and (3) the average response times for a request. Once again, how the measurements are made will greatly affect the results: look for consistency in the data sources and how they are measured.

1. *Work order completions versus new requests* gives a turnover index. In the following example, an index of 107% shows that more work was completed than demanded. As a result, the backlog shrank so that overall service to customers rose. However, no account is taken of the size and complexity of the work requests or work orders.

   Work Order Turnover = Number of Tasks Completed as % of Work Requests
   Work Orders Completed Last Month = 3,200
   Work Requests Last Month = 3,000

   $$WO\ Turnover\ Index = \frac{3,200}{3,000} = 107\%$$

2. *Work order backlog* shows the relationship between overdue work orders and ones completed. As with most measures, you must clearly formulate and communicate definitions. Maintenance customers will be greatly frustrated if they don't understand how this measure is created. For example, is a work order "overdue" when it passes the requestor's due date or the planner's? If it is the requestor's, how realistic is it? The example shows 6.8 days; one to two weeks are generally targeted. If more than a week passes, users will complain that service is lacking. What's needed is better planning and more staff or better screening of work requests—any lower suggests overstaffing.

   Work Orders Overdue = 720
   Work Orders Completed This Month = 3,200

   $$Backlog = \frac{720}{3,200} = 22.5\ \text{of one month} = 6.8\ \text{days}$$

3. *Job timeliness* and *response times* are measured by the time it takes from when the request is received to when the maintenance technician arrives at the job site. Comparison with a standard is the best method here. The standard should depend on the level of service desired, distance from the

dispatch center to the job site, complexity of the jobs involved, availability of manpower, equipment and materials, and urgency of the request. There is little value in measuring response times for low-priority jobs, so the trends are usually limited to emergencies and high priorities, with the statistics kept separately. In many cases, an emergency situation response is prompted by a "work request" that comes over the phone, and the responder is dispatched by phone or pager. No actual work request will be prepared, and the work order will be completed only when the emergency is over.

High-Priority Response Time Standard = 4 hours
High-Priority Response Time Average Last Month = 3.3 hours

4.6.5 MAINTENANCE COSTS

More attention is probably paid to the maintenance cost indicators than any other set of measures. This is encouraging, since the link between maintenance and costs (profits) needs to be solid and well established. As you've seen, though, many factors affect the cost of delivering maintenance services—many of them almost completely outside the control of the maintenance manager. Driving down maintenance costs has in many companies become a mantra and, in some cases, rightly so. However, by reducing costs alone, you won't necessarily achieve your organization's objectives. Both the company's and maintenance's mission and objectives must be factored in. One way is to relate the maintenance costs to the overall cost of production or, where single or similar product lines are produced, to the number of units produced. For example, in interdivisional or inter-firm benchmarking, a maintenance costs per ton of output is a widely used figure. Within this category, many different measures are used. The examples here show four typical ones:

1. *Overall maintenance costs per unit output.* This measure keeps track of the overall maintenance cost relative to the cost of producing the product. In a competitive environment, this is very important, particularly if the product lines are more of a commodity than a specialized product. The process industries, for instance, use these measures extensively. You can also subdivide this into the major maintenance cost components, such as direct versus overheads, materials, manpower, equipment, and contractors.

$$Direct\ Maintenance\ Cost\ per\ Unit\ Output = \frac{Total\ Direct\ Maintenance\ Cost}{Total\ Production\ Units}$$

Direct Maintenance Costs Last Month = 285,000
Total Units Produced Last Month = 6,935

$$DMC = \frac{285,000}{6,935} = \$41/unit$$

2. *Stores turnover* measures how effectively you use the inventory to support maintenance. *Stores value* measures the amount of materials retained in stores to service the maintenance work that needs to be done. Company policies and practices will directly affect these numbers and limit how much the maintenance department can improve them. Similarly, if manufacturing is far from the materials source, this will also affect inventory turns. Some organizations argue that the cost of breakdown (in financial, environmental, or publicity terms) is so high that the actual stores value needed is irrelevant. Inventory turns are best measured against industry standards or last year's results. Inventory values are best measured through time.

$$Inventory\ Turnover = \frac{Cost\ of\ Issues}{Inventory\ Value}$$

$$Inventory\ Value\ Index = \frac{Inventory\ Value\ This\ Year}{Inventory\ Value\ Last\ Year}$$

Materials Issued Last Year = $1,400,000
Current Inventory Value = $1,800,000
Last Year's Inventory Value = $2,000,000

$$Inventory\ Turnover = \frac{1,400,000}{1,800,800} = 0.8\ turns$$

Inventory Value Index = $1,800,000 = 0.9 of base $2,000,000

Note that an inventory index of less than 1 indicates a reduced inventory value, or an improvement in performance, while an increase in the inventory turns shows better use of the organization's inventory investment.

3. *Maintenance cost versus the cost of the asset base.* This measures how effectively the maintenance department manages to repair and maintain the overall asset base. It uses the asset or replacement value to make the calculation, depending on available data. As with many of the cost measures, it can be cut several ways to select individual cost elements. The final measure shows the percentage of the asset's value devoted to repair and maintenance.

$$Direct\ Maintenance\ Cost\ Effectiveness = \frac{Total\ Direct\ Maintenance\ Cost}{Asset\ Value\ (or\ Replacement\ Cost)}$$

These can have quite significantly different values (e.g., book value compared with replacement cost).

Direct Maintenance Cost = $4,000,000
Asset Value = $40,000,000

$$DMCE = \frac{4,000,000}{40,000,000} = 10\%$$

4. *The overall maintenance effectiveness index.* This shows the rate at which maintenance's overall effectiveness is improving or deteriorating. It compares costs of maintenance plus lost production from period to period. The overall index should go down, but increased maintenance costs or production losses due to maintenance will force the index up. If a new maintenance program is introduced, this index will measure whether spending more on maintenance has paid off in fewer production losses.

Maintenance Improvement Index =

$$\frac{TotalDirMtcCost + ProdLossesPreviousMonth}{TotalDirMtcCost + ProdLossesLastMonth}$$

$$Maintenance\ Improvement\ Index = \frac{285,000 + 102,800}{270,000 + 128,500}$$

Note that results below 1.0 show an improvement in the index.

4.6.6 MAINTENANCE QUALITY

The pundits frequently ignore maintenance quality when looking at performance measures, yet most auto magazines feature this in their "Which Car to Buy?" columns. Use it to judge how often repeat problems occur and how often the dealer can fix the problem on the first visit.

The maintenance department can collect and measure these data in several ways. At least one CMMS has a special built-in feedback form sent automatically to the requestor when the work is completed:

1. *Repeat jobs and repeat breakdowns* generally indicate that problems haven't been correctly diagnosed or that training or materials aren't up to standard. Many maintenance departments argue that most repeats occur because they can't schedule the equipment down for adequate maintenance. The most effective way to get enough maintenance time is to measure and demonstrate the cost of breakdowns. Note that *repeat* refers only to corrective and breakdown work, not to preventive or predictive tasks.

$$Repeat\ Jobs\ Index = \frac{Number\ of\ Repeat\ Jobs\ This\ Year}{Number\ of\ Repeat\ Jobs\ Last\ Year}$$

Number of Repeat Jobs This Year = 67
Number of Repeat Jobs Last Year = 80

$$Repeat\ Jobs\ Index = \frac{67}{80} = 0.84$$

2. *Stock-outs* are one of the most contentious areas, reflecting tension between finance, which wants to minimize inventory, and operations, which to maintain output needs spares to support it. You can maintain the balance with good planning—predicting when materials will be needed, knowing the delivery times, and adding a safety margin based on historical predictions. Stock-outs are normally measured against the previous period but are best when tied to the higher-priority work. Zero stock-outs aren't necessarily good, indicating overstocking.

$$Stock\text{-}Out\ Index = \frac{Stock\text{-}Outs\ This\ Year}{Stock\text{-}Outs\ Last\ Year}$$

Stock-Outs This Year = 16
Stock-Outs Last Year = 20

$$Stock\text{-}Out\ Index = \frac{16}{20} = 0.8$$

3. *Work order accuracy* measures how closely the planning process from work request to job completion matches the reality. The core of the process is applying manpower and materials to jobs, and this is usually the focus of this measure. You can easily measure this through the comment section on the work order. Encourage the maintenance technician to provide feedback on job specification errors, skill requirements, and specified materials so corrections can be made for the next time around.

$$Workorder\ Accuracy = 1 - \frac{Number\ of\ Work\ Orders\ Completed}{Number\ of\ Work\ Order\ Errors\ Identified}$$

Number of Work Orders Completed = 1,300
Number of Work Order Errors Identified = 15

$$Work\ Order\ Accuracy = 1 - \frac{15}{1,300} = 98.85\%$$

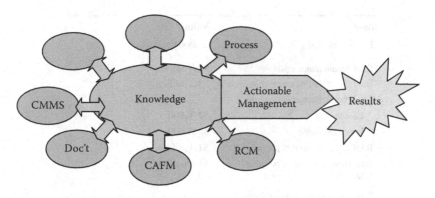

FIGURE 4.15 The knowledge base.

4.7 COLLECTING THE DATA

From the previous examples, you can select measures that will generate the right information to drive action. The data are drawn from condition-based monitoring systems, enterprise asset management systems, engineering systems, and process control systems. You don't need all of these systems, though, to start the performance evaluation process. Most of the data are also available from other sources, though computerized systems make data collection much easier. What follows is a core data set for all of the examples in Section 4.6.

Item	Value
Total time (full-time operation)	$7 \times 24 \times 365 = 8{,}760$ hours/year
Downtime	
Scheduled	160 hours/year
Unscheduled	232 hours/year
Nonused time	600 hours/year
Number of scheduled outages	6/year
Number of breakdowns	12/year
Production rate (units per hour)	10/hour
Annual production (units)	
Capacity	87,600 units
Actual	77,680 units
Units produced last month	6,935
Number of maintenance people	32
Working hours per person per year	2,088 hours
Maintenance hours – capacity	66,816 hours
Total wrench time	46,100 hours
Planned time	44,700 hours

(continued on next page)

Item	Value
Time scheduled	26,000 hours
Total maintenance costs	$4.0 million
Total direct maintenance costs/year	$3.2 million
Total direct maintenance costs	
Last month	$270,000
Previous month	$285,000
Total materials issued per year	$1.4 million
Total manpower costs per year	$1.6 million
Total work orders per year	32,000
Total work requests last month	3,000
Work orders completed last month	3,200
Breakdown hours worked	2,200
Work order errors	15
Repeat jobs	
This year	67
Last year	80
Overdue work orders	720
Emergency work orders	
Last year	152
This year	130
High-priority response time	
Standard	4 hours
Last month	3.3 hours
Maintenance cost of breakdowns	$135/hour
Stores value	
Last year	$1.8 million
Current	$2.0 million
Stock-outs	
This year	16
Last year	20
Total production costs	$45 million
Lost production time cost/hour	$5,140
Production losses	
Last month	20 hours
Previous month	25 hours
Asset value (replacement cost)	$40 million

The volume of data available is expanding rapidly. Historically, the maintenance manager's problem was not having enough data to make an informed decision. With today's various computerized systems, the problem is now too much data. One possible solution to this is the maintenance knowledge base, which is being developed by at least one EAM and one CBM company. This concept recognizes that not only must various data sources be identified, since more than a simple point to point linkage between these sources is needed; to achieve full value from the data, a knowledge base also must be constructed to selectively cull the data, analyze it, and use it as a decision support tool. That is where the real value lies—helping develop actionable management information that attains results. Without this, the data aren't particularly useful.

4.8 APPLYING PERFORMANCE MEASUREMENT TO INDIVIDUAL EQUIPMENT

So far, all of the performance indicators we have reviewed apply to the broader spectrum of maintenance costs and operations. Many indicators also effectively apply at the micro level to individual equipment or jobs. Particularly if the organization has numerous examples of the same piece of equipment running, comparative evaluations can show the varied operating results and costs caused by different running conditions and maintenance methodologies. Adopting an internal "best practices" approach in your organization can lead to significant improvements.

To do this type of forensic maintenance, the detailed data must be readily available, which almost requires that you use an EAM or CBM system. The data need to be accessed in large enough sample sizes to reduce the error probability to acceptable levels. For this, the data typically have to be aggregated from individual work orders, pick lists, condition reports, and process control data sheets. Manual collecting from these sources isn't really feasible. Two examples will be enough.

Looking at maintenance analysis at this level, you're typically seeking specific results, relating to optimum maintenance intervals, operating parameters, and cost savings. Figure 4.16 tracks the energy savings from compressors as various simple maintenance tasks were done. The improvement in energy costs (i.e., energy cost savings) totaled an annual $6,000, representing a savings of about 7.5% of the total annual "before" cost of $81,000.

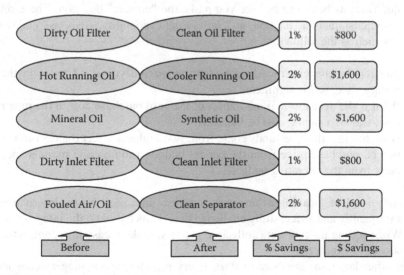

FIGURE 4.16 Energy cost savings generated from maintenance tasks.

Earlier in this chapter, we used an example of life-cycle costing to illustrate the pitfalls you need to be aware of. The following revisits this example but involves a replace versus repair decision. With the repair alternative, the slurry pump can continue to operate for five more years, but with a higher annual maintenance cost and likelihood of breakdown. To offset this, there will be no purchase or installation costs:

	Repair	Replace
Purchase cost	0	$20,000
Cost to install	0	$2,000
Annual running cost	$3,000	$4,000
Annual maintenance cost	$9,000	$8,000
Final disposal cost less scrap value	$1,000	$500
Pump life	5 years	12 years
Total life-cycle cost	$61,000	$167,000
Annual cost	**$12,200**	**$13,900**
Average throughput	100 gals/hr	175 gals/hr
Lifetime throughput	4.38 millions of gallons	18.40 millions of gallons
Cost	**1.39 cents/gal**	**0.91 cents/gal**
Average breakdown frequency	3 per year	2 per year
Average breakdown duration	1 day	1 day
Downtime cost	$1,000/hour	$1,000/hour
Downtime cost (total lifetime)	$360,000	$576,000
Downtime cost	8.22 cents/gal	3.13 cents/gal
Total operating costs	**9.61 cents/gal**	**4.04 cents/gal**

The case for a replacement appears clear-cut, but you need to ask the same qualifying questions as before to be sure you make the "correct" decision. These relate to capacity, customer satisfaction, and failure data reliability. Also, ask additional questions such as the following:

- The repair case covers a 5-year planning period versus 12 years in the replace case: is this significant?
- What is the decision-making impact of the zero purchase cost in the repair case: should a capital cost be included?
- Does the fact that the funds come from two budgets affect the decision— that is, purchase price from the capital budget, running and maintenance costs from the operating budget?

To do any kind of meaningful analysis, the base data must be readily available. Critics frequently and successfully challenge the results based on the integrity of the data. When setting up your data collection process, make it easy to record, reliable, and consistently analyzed.

The other key issue is where to start. Every maintenance manager needs more time and less work, but you know that making time comes only from doing things

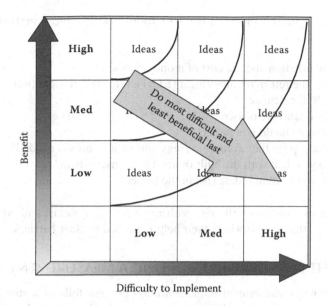

FIGURE 4.17 Benefits and implementation difficulty matrix.

more effectively. Figure 4.17 will help you sort out where to start. Grade the measures and actions high to low, based on how much benefit they create and how difficult they are to implement. Then start with those in the top left quadrant: highest payoff, least difficult to implement.

4.9 WHO'S LISTENING? TURNING MEASUREMENTS INTO INFORMATION

A frequent complaint from the maintenance department is that "they" won't listen. Who are they? Is it management deciding not to release funds for needed improvements? Is it engineering refusing to accept the maintenance specifications for the new equipment? Is it finance insisting on return on investment calculations at every turn? Is it purchasing continuing to buy the same old junk that we know will fail in three months? Or is it production not giving adequate downtime for maintenance?

In fact, it is all of them, and the core reason has usually been the same: maintenance hasn't been able to make its case convincingly. For that, you need facts, figures, and attractive graphs. The cynics claim that graphs were invented because management can't read. There is more than a grain of truth in this, although it stems from lack of priority, not time or ability. The trick is to make the results both attractive and compelling enough to make them a priority and involve the other departments in the process. Buy-in can work upward and sideways in your organization as well as downward.

For example, finance has become much more complex and sophisticated than it used to be. Many of the issues that make finance question the maintenance department's proposals are technical, requiring inputs from the accounting department.

Coopt a finance person onto the team to handle skill-testing questions such as the following:

- How are inflation and the cost of money handled?
- What are the return on investment (ROI) levels for threshold projects?
- How is ROI calculated?
- What discount rate is used?
- How far out should you project?
- What backup for the revenue, savings, and cost figures is needed?
- How do you deal with the built-in need for conservatism?
- How do you accommodate risk in the project?

A parallel set of questions will arise with engineering, systems, and other groups. Involve them in the process where their help is needed to clear barriers.

4.10   EIGHT STEPS TO IMPLEMENTING A MEASUREMENT SYSTEM

The performance measurement implementation process follows a straightforward and logical pattern. However, because many organizations don't do a full implementation, they get a methodology that can't meet the daily demands of the average maintenance department. The major steps are highlighted in Figure 4.18

1. Review and select the indices that make sense for your organization. As shown earlier in this chapter, literally hundreds of measurements and performance indicators can be used. The starting point is to understand the structure and intent of the more commonly used ones. Typically, you undertake a series of in-house round tables and workshops to initiate discussion among maintenance personnel.
2. Once the basics are reasonably well understood, choose a small cross-section of measurements. Others can be added later. Select the chosen few based on their fit with the department's objectives, measurability, and ability to impact the results. At the early stages, keep the amount of extra work to build the performance measures to a minimum. As the payoffs start to appear, the effort can be expanded. Also, make sure that the data for the early measurements are readily available and that the source data are

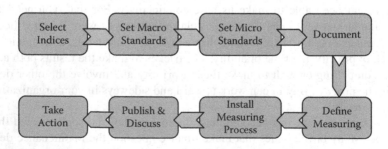

FIGURE 4.18   Steps to implement a measurement system.

accurate. For each of the chosen measurements, set standards, values, and targets. To ensure reasonably full coverage, choose at least one from each of the six macro categories:

a. Maintenance results
b. Maintenance productivity
c. Maintenance organization
d. Efficiency of maintenance work
e. Maintenance costs
f. Maintenance quality

You will have to do some research to come up with valid targets that relate to your organization's current numbers, plus the hoped-for future improvements. Beware of setting targets that are impossible to reach.

3. Supplement the macro measures with a small number of measures for critical equipment and bad actors at the micro level. Set standards and targets for each—again paying attention to the data availability and consistency.

4. Document the measures and targets, plus the interpretation of each number or trend. Absolute numbers may not mean anything, so whenever you use a trend or index, explain it (e.g., is up good or bad?).

5. Setting up the measuring process, make sure that the person responsible clearly understands the nature of the measurement, where the data come from, and what sort of analysis is required. Include how often the reading is to be taken and, if relevant, whether it should be at a specific time or event during the day (e.g., end of batch). Consistency needs to be stressed. Erratic results will not only sully the measurement, but may also induce the wrong action. With the widespread use of CMMS, EAM, and CBM systems, the amount of available data has grown dramatically. System capability to collect and maintain the right data, though, varies just as dramatically. Once you've established the data set, make sure the computer system makes it easy to accurately collect the data. The work order data fields, for example, should carry the same labels as the measurement process. They should be mandatory fields so the maintenance technician can understand why they are needed. Build in a value range so that any data entry outside of it will immediately be flagged.

6. Install the measuring procedure in the CMMS or EAM system as a regular weekly work order. This simple step is frequently missed and is a major reason measurement systems fail.

7. Publish the results so that all maintenance employees and visitors can see both the targets and achievements. This is best done on the departmental notice boards, although some maintenance departments have set up a special war room to display and discuss the results. Set aside time at the weekly meetings to review and discuss the results and especially to look for new ideas to achieve the results.

8. The point of measurements is to target trend implications and remedial actions to make improvements. A milestone flag on the trend chart is an effective way to show when a specific action was taken and the subsequent impact. Just like the measurement program implementation, the remedial

action should emphasize what has to be done, who does it and when, and what materials and tools are needed. In fact, what's needed is a regular work order to record the task for future evaluation.

Most often, measurement projects are seen as just that—measurement projects. But a measurement project can be much more than that. It can be a dynamic program for ongoing change. To make this happen, use the measurement results as the basis to reevaluate the macro and micro standards and targets. Establish a preventive maintenance type work order calling for an annual review of each measurement, and set new targets to introduce this feedback loop.

4.11   TURNING MEASUREMENTS INTO ACTION

Measurements are only as good as the actions they generate. If you fail to convert them into action, you miss the whole point of performance measurement. The flip side is the tendency for a good trend to lead to complacency. Some organizations have rejected the basic philosophy of benchmarking, for example, claiming that it forces the organization into a perpetual catch-up mentality. No company ever took the lead playing catch-up. The breakaway firm needs to think outside the box, and the measurement process helps to define the size and shape of the box.

The final word on this phase of the measurement issue goes to Terry Wireman:[2]

> Yesterday's excellence is today's standard and tomorrow's mediocrity.

4.12   ROLE OF KEY PERFORMANCE INDICATORS— PROS AND CONS

Management uses performance measurement primarily for monitoring purposes, and many performance indicators have been developed to support operational decisions. These indicators are, at best, descriptive signals that some action needs to be taken. To make them more useful, put in place decision rules, which are compatible with organizational objectives. This way, you can determine your preferred course of action based on the indicators' values.

To clarify trends when the activity level may vary over time, or when comparing organizations of different size, you can use indices to measure maintenance performance. Campbell[4] classifies these commonly used performance measures into three categories, based on their focus:

1. Measures of equipment performance (eg, availability, reliability, and overall equipment effectiveness)
2 Measures of cost performance (eg, O&M labor and material costs)
3 Measures of process performance (eg, ratio of planned and unplanned work, schedule compliance)

However, the underlying assumptions of these measures are often not considered when interpreting results, so their value can be questionable.

For example, traditional financial measures still tend to encourage managers to focus on short-term results, a definite drawback. This flawed thinking is driven by the investment community's fixation with share prices, driven largely by this quarter's earnings. As a result, very few managers choose to make (or will receive board approval for) capital investments and long-term strategic objectives that jeopardize quarterly earnings targets.

Income-based financial figures are lag indicators. They are better at measuring the consequences of yesterday's decisions than indicating tomorrow's performance. Many managers are forced to play this short-term earnings game. For instance, maintenance investment can be cut back to boost the quarterly earnings. The detrimental effect of the cutback will show up only as increased operating cost in the future. By then, the manager making the cutback decision may have already been promoted because of the excellent earnings performance. To make up for these deficiencies, customer-oriented measures such as response time, service commitments, and satisfaction have become important lead indicators of business success.

To assure future rewards, your organization must be both financially sound and customer oriented. This is possible only with distinctive core competencies that will enable you to achieve your business objectives. Furthermore, you must improve and create value continuously, through developing your most precious assets: your employees. An organization that excels in only some of these dimensions will be, at best, a mediocre performer. Operational improvements such as faster response, better quality of service, and reduced waste won't lead to better financial performance unless the spare capacity they create is used or the operation is downsized. Also, maintenance organizations that deliver high-quality services won't remain viable for long if they are slow in developing expertise to meet the emerging needs of the user departments. For example, electromechanical systems are being phased out by electronic and software systems in many automatic facilities. In the face of new demand, the maintenance service provider has to transform its expertise from primarily electrical and mechanical trades to electronics and information technology.

Obviously, you won't fulfill all these requirements by relying on a few measures that represent a narrow perspective. You need a balanced results presentation to measure maintenance performance. The balanced scorecard proposed by Kaplan and Norton[5] offers the template for the balanced presentation. Many organizations have done balanced scorecards. It translates a business unit's mission and strategy into objectives and quantifiable measures built around four perspectives:

1. *Financial* (the investor's views)
2. *Customer* (the performance attributes valued by customers)
3. *Internal processes* (the long- and short-term means to achieve the financial and customer objectives)
4. *Learning and growth* (capability to improve and create value)

The balanced scorecard (BSC) focuses managers on a handful of critical measures for the organization's continued success

The BSC had been implemented in numerous major engineering, construction, microelectronics, and computer companies. Their experience indicates that the scorecard's greatest impact on business performance is to drive change process. The balanced scorecard promotes a strategic management system that links long-term strategic objectives to short-term actions.[5]

A strategic management system built around a BSC is characterized by three keywords: focus, balance, and integration. Ashton[6] explains these three attributes:

> *Focus* has both strategic and operational dimensions in defining direction, capability and what the business or its activities are all about, while *balance* seeks an equilibrium for making sense of the business and to strengthen focus. *Integration* is critical, ensuring that organizational effort knits into some form of sustainable response to strategic priority and change.

The BSC is a complete framework for establishing performance management systems at the corporate or business-unit level. When the approach is applied to managing maintenance performance, follow this process:[7]

1. Formulate maintenance operation strategy: Consider strategic options such as developing in-house capability, outsourcing maintenance, empowering frontline operators to practice autonomous maintenance, developing a multiskilled maintenance workforce, and implementing condition-based maintenance. Get others involved in making decisions.
2. Operationalize the strategy: Translate the maintenance strategy into long-term objectives. Identify the relevant *key performance indicators* (KPIs) to be included in the BSC, and establish performance targets. Suppose an electric utility company has chosen to outsource its maintenance and repair of generic and common equipment and vehicle fleets so that it can manage its core transmission and distribution system. The KPIs and performance targets that relate to this strategic objective are "outsource 20% of maintenance work" and "reduce maintenance costs by 30%" in two years. The former indicator belongs to the internal processes perspective and the latter the financial perspective. To achieve vertical alignment, these objectives, KPIs, and targets are aligned into team and individual goals.
3. Develop action plans: These are means to the ends stipulated in the targets established in step 2. To reach the outsourcing targets in the previous example, your company may develop capabilities in the following outsourcing areas: contract negotiation, contract management, and capitalizing on emerging technology and competitive opportunities in maintenance. These action plans should also include any necessary changes in your organization's support infrastructure, such as maintenance work structuring, management information systems, reward and recognition, and resource allocation mechanisms.
4. Periodically review performance and strategy: Track progress in meeting strategic objectives, and validate the causal relationships between measures at defined intervals. After the review, you may need to draft new strategic objectives, to modify action plans, and to revise the scorecard.

Some of the scorecard KPIs that measure the maintenance performance of an electricity transmission and distribution company may include the following:[8]

Perspective	Strategic Objectives	Key Performance Indicators (KPIs)
Financial	Reduce O&M costs	O&M costs per customer
Customer	Increase customer satisfaction	Customer-minute loss
		Customer satisfaction rating
Internal processes	Enhance system integrity	% of time voltage exceeds limits
		Number of contingency plans reviewed
Learning and growth	Develop a multiskilled and empowered workforce	% of cross-trained staff
		Hours of training per employee

Since these measures reflect an organization's strategic objectives, the BSC is specific to the organization for which it is developed.

By directing managers to consider all the important measures together, the balanced scorecard guards against suboptimization. Unlike conventional measures, which are control oriented, the BSC puts strategy and vision at the center and emphasizes achieving performance targets. The measures are designed to pull people toward the overall vision. To identify them and to establish their stretch targets, you need to consult with internal and external stakeholders—senior management, key personnel in maintenance operations, and maintenance users. This way, the performance measures for the maintenance operation are linked to the business success of the whole organization.

The theoretical underpinning of the balanced scorecard approach to measuring performance is built on two assertions:

1. Strategic planning has a strong and positive effect on a firm's performance.
2. Group goals influence group performance.

The link between strategic planning and company performance has been the subject of numerous research studies. Miller and Cardinal[9] applied the meta-analytic technique to analyze empirical data from planning performance studies of the last two decades, establishing a strong and positive connection between strategic planning and growth. They also show that, when a company is operating under turbulent conditions, there is a similar link between planning and profitability. A similar study on previously published research findings[10] confirms group goal effect.

Although industry commonly agrees that strategic planning is essential for future success, performance measures and actual company improvement programs are often inconsistent with the declared strategy. This discrepancy between strategic intent and operational objectives and measures is reported in a recent Belgian manufacturing survey by Gelders.[11] You can ensure you don't make the same mistake by introducing the balanced scorecard.

REFERENCES

1. Coetzee, J. L. *Maintenance*. Maintenance Publishers, 1997.
2. Wireman, T. *Developing Performance Indicators for Managing Maintenance*. Industrial Press Inc, 1998.
3. MIMOSA: Machinery Information Management Open Systems Alliance. http://www.mimosa.org
4. Campbell, J. D. *Uptime: Strategies for Excellence in Maintenance Management.*
5. Kaplan, R. S. and D. P. Norton. *The Balanced Scorecard*. Boston, MA: Harvard Business School Press, 1996.
6. Ashton, C. *Strategic Performance Measurement*. London: Business Intelligence Ltd., 1997.
7. Tsang, A. H. C. A strategic approach to managing maintenance performance. *Journal of Quality in Maintenance Engineering* 4(2):87–94, 1998.
8. Tsang, A. H. C. and W. L. Brown. Managing the maintenance performance of an electric utility through the use of balanced scorecards. Proceedings of 3rd International Conference of Maintenance Societies, Adelaide, 1998, pp. 1–10.
9. Miller, C. C. and L. B. Cardinal. Strategic planning and firm performance: a synthesis of more than two decades of research. *Academy of Management Journal* 37(6):1649–1665, 1994.
10. O'Leary-Kelly, A. M., J. J. Martocchio, and D.D. Frink. A review of the influence of group goals on group performance. *Academy of Management Journal* 37(5):1285–1301, 1994.
11. Gelders, L., P. Mannaerts, and J. Maes. Manufacturing strategy, performance indicators and improvement programmes. *International Journal of Production Research* 32(4):797–805, 1994.

5 Information Management and Related Technology

Don Barry, Brian Helstrom, and Joe Potter
Original by B. Stevens

CONTENTS

The goals in maintenance often focus on four keys business outcomes:

- To improve the quality output of an asset
- To ensure maximum equipment availability or uptime, or, conversely, minimal downtime
- To improve the optimal life of the asset
- To ensure a safe operating environment for both the operator/maintenance worker and the environment

Many tools have been developed to assist in or accelerate the attainment of these goals within the maintenance process. Fortunately, we live in an information age, and today, more than ever, decisions are driven by hard information, derived from data. In asset management, particularly maintenance, you need readily available data for the kind of thorough analysis that produces optimal solutions. The world is growing increasingly intolerant of asset failure, particularly where environment and safety is concerned, so the importance of having good maintenance processes and effectively selecting, planning, executing, and recording maintenance activity is also growing. In this chapter, we discuss the key aspects of modern computer-based maintenance management systems:

- Maintenance management systems: computerized maintenance management system (CMMS) and enterprise asset management (EAM)
- Evolution and direction of maintenance management systems
- Technology's impact on maintenance systems
- Emerging technical enhancements to maintenance management processes

First, we explore CMMS and EAM—how they work and how they can help you attain maximum results. Often, the terms CMMS and EAM are used interchangeably. But you will see that CMMS, the packaged software application, is really the enabler for EAM, just as a word processor enables you to manage text on a computer.

Next, we examine how maintenance management systems have evolved to their current state, both functionally and technically. As these systems continue to evolve, so, too, do their potential benefits to you increase.

Once you decide to acquire a maintenance management system, you will want to ensure it will produce tangible business improvements. For this we look at the two crucial areas that you must get right: selecting and then implementing the system to achieve the results you need.

Finally, since many current and evolving CMMS and EAM solutions are still somewhat limited in what is included in their packaged solution, we will explore some of the many technical enhancements that can be applied to a maintenance management system or process.

5.1   INTRODUCTION: DEFINING MAINTENANCE MANAGEMENT SYSTEMS

First of all, what is a maintenance management system? Nearly everybody is familiar with an accounting system, and there is not a chief financial officer (CFO) or controller who does not use accounting data daily. Investment decisions are made with a keen eye on the balance sheet, income statement, and, of course, shareholder perception. The accounting system, in effect, monitors the health of the business, and there are dozens of standard measures to work with.

Likewise, at an individual level, banks have moved toward "personal" financial systems that ideally record and categorize every expenditure, investment, and income stream. They provide software and Internet transactions so that you can now better plan expenditures, budget for vacations, and track savings. The growth of computer-based personal financial systems has resulted in, and encouraged, greater numeracy.

Similarly, a maintenance management system enables you to monitor the asset base and the activity planned and performed on the asset base, the value of which, especially for resource development and manufacturing enterprises, can add up to billions of dollars. The maintenance management system allows for the monitoring of the health of the assets and provides the standard measures to work with to optimize the life of these assets.

Maintenance management systems are often referred to as computerized maintenance management systems and enterprise asset management systems. The difference is essentially one of scale, and there is no clear dividing line. The choice usually comes down to an organization's asset management philosophy and strategies. To help you decide, look at maintenance management systems as an enabler for total enterprise asset management. If your enterprise spans many plants and several jurisdictions and nationalities, you need sophisticated, high-end CMMS applications. At this level, CMMS can seamlessly integrate with other business applications (e.g., financial, human resources, procurement, security, material planning, capital project management) to produce a total business information solution.

Figure 5.1 clarifies the similarities and differences between CMMS and EAM. When it comes to maintenance management, some systems, though properly called enterprise resource planning (ERP), do not include this function, so the component is shown partially enclosed.

5.2　EVOLUTION OF MAINTENANCE MANAGEMENT SYSTEMS

Maintenance management systems have developed, over a long time, out of focused business needs. Table 5.1 shows how key functions have evolved. The earliest maintenance management applications were custom-built to unique business specifications. From these pioneer systems, often well built but costly, most current packaged applications descend. There was a huge difference between these initial systems in their functions and implementation, just like the first personal computer (PC) applications. At first, there were clear leaders, but, over time, the best features of each product became standard. Vendors consolidated and the selection process became more complex. Today, you need to do a detailed comparison to detect the differences.

The earlier systems were built to suit each customer's specific functional and technical needs. As a result, they could not easily be deployed elsewhere. This changed with the advent of the many de facto standards (e.g., MS DOS, Windows, relational databases, Structured Query Language [SQL]). Today, it is mandatory for all CMMS

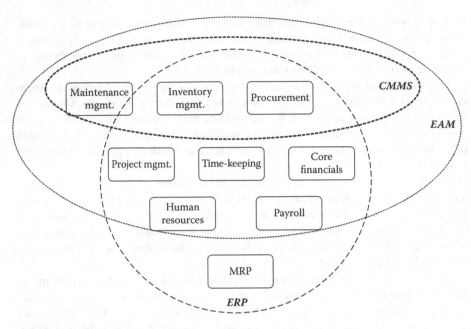

FIGURE 5.1 Comparisons among CMMS, EAM, and ERP.

TABLE 5.1
Functional Developments and Trends in CMMS

From	To
Custom built	Package solution
Clear leaders	Many common features
Difficult to interface with other business applications	Integration with ERP systems
Single "site"	Enterprisewide
Narrow focus (work management)	Total asset management functionality
Difficult to modify	Easily customized and configured
Added functionality	Deployed with embedded/integrated predictive maintenance (PM), condition-based monitoring (CBM), and reliability-centered maintenance (RCM) functions; mobile workforce; planning solutions; document management solutions

applications to share their data with other business applications, except in the smallest of CMMS implementations.

As organizations grow and evolve over time, so should their core business systems. Consolidation, particularly, has driven the need for business systems to be true enterprise applications, dealing with multiple physical plants, sites, currencies, time zones, and even languages. World-class ERP systems offer this kind of location transparency, and the leading maintenance management applications are now comparably complex.

The scope of leading maintenance management applications is very wide, incorporating modules for the following:

- Maintenance (work order management, scheduling, estimating, workflow, preventive maintenance, equipment hierarchies, equipment tracking, capital project management)
- Inventory management (parts lists, repairable items management, catalogs, warehouse management, inventory replenishment algorithms, parts kitting, parts reservations)
- Procurement (purchase order processes, vendor agreements, contractor management, and administration)
- Human resources (health and safety, time control, skills management, payroll, benefits, recruitment, training)
- Financials (general ledger accounts, payable accounts, receivable, fixed assets, activity-based costing, budgeting)

Also, the leading maintenance management vendors offer sophisticated performance measurement and reporting capabilities. This wide range contrasts with earlier maintenance management applications, which focused solely on tactical work management, usually consisting of work order initiation, resources required, processing, and closing.

Finally, modern asset management is turning to optimization methods to improve maintenance effectiveness. Techniques such as condition-based monitoring (where maintenance is condition driven rather than time interval driven), reliability-centered maintenance (described in Chapter 8), and optimal repair and replace decisions draw on rich stores of historical information in the maintenance management system database. Using these techniques requires additional software modules, now being embedded in the leading maintenance management products.

Technically, there have been several significant transitions, as shown in Table 5.2. From the previous list, you can see a gradual standardization process developing. Certainly, adopting the relational database "standard" has made intersystem communication much easier. As well as a new architecture for distributed systems, the Internet provided Transmission Connection Protocol/Internet Protocol (TCP/IP), the

TABLE 5.2

Technical Developments in CMMS Applications

From	To
Mainframe	Micro/mini
Data files	Relational database
Terminal/file server	Client/server
Proprietary	Open standards
Dedicated infrastructure	TCP/IP/Internet enabled
Paper help	Context sensitive/online help
Classroom training	Computer-based training

current communication standard for networks. This, in turn, has allowed systems to interconnect as never before. It used to be that racks of paper information for large business systems would sit in the system administrator's office, virtually inaccessible to users. This has entirely disappeared, and current applications now have extensive online help systems. Nevertheless, as application complexity increases, training continues to be a problem.

You cannot consider current business systems without including the Internet and electronic (e)-enabled initiatives (in this chapter, the word *Internet* is used for all related network terms such as intranet and extranet). Today, it would be difficult to find a maintenance management system request for proposal (RFP) that did not specify the need for Web-enabled functionality. But one of the greatest difficulties facing maintenance management system specifiers is to understand what that really means and the associated tradeoffs. There are many opportunities for Web enabling a maintenance management system, notably the following:

- Complete application delivery, including application leasing
- Internet-enabled workflow (e-mail driven)
- Management reporting
- Supplier management and procurement
- Standardized human-to-computer interaction
- Facilities management

The following sections give a brief description of each initiative.

5.2.1 APPLICATION DELIVERY

As long as your hardware infrastructure is sufficiently robust (e.g., bandwidth, reliability, performance), you can deploy some CMMS products completely through a standard Web browser. The advantages of this include version control, potentially smaller (less powerful and therefore cheaper) desktop machines, and ease of expansion. A recent twist on this is to lease, not buy, the CMMS application. This is akin to mainframe "time-sharing," which is still used for functions like payroll processing.

5.2.2 INTERNET-ENABLED WORKFLOW

One of the most popular maintenance management system features is comprehensive workflow support so that business rules can be written into the system and changed at will. By including Internet technology, you can expand the scope of your business rules to cover the globe and, if you can imagine, perhaps even further. There is ongoing research into expanding the Internet to include spacecraft and extraterrestrial sites!

5.2.3 MANAGEMENT REPORTING

Closer to home, the Internet is ideal for disseminating management reports, and, in fact, it is a growing part of true enterprise asset management. You need current data to make better business decisions. In some cases, data even a week old are out of date. Being able to access an up-to-the-minute corporate database, from any Internet connection, is becoming a competitive necessity.

5.2.4 SUPPLIER MANAGEMENT AND E-PROCUREMENT

This has received a lot of attention, primarily because it is so Internet driven. Whether you're buying capital items or office supplies, the basic procurement cycle is essentially the same: e-procurement automates time- and labor-intensive processes as well as enforces a single enterprisewide policy through a single buying interface.

The process of e-procurement includes online supplier catalogs (several new companies have sprung up to provide them), requisitions sent over the Internet, purchase orders, receipt and billing confirmation (essentially electronic data interchange, or EDI, with a new twist), and guaranteed security for financial transactions. We are only skimming the surface here; entire books have been written on these topics.

5.2.5 STANDARDIZED HUMAN-TO-COMPUTER INTERACTIONS

This has always been a fundamental issue, since the people who manage the system and its data need to interact with some desktop or workstation to gather the information. Browser-based interaction is somewhat intuitive in its application and supports ubiquitous ease of use and implementation. The features of all and any application can be managed easily through this comfortable interaction, with which the vast majority of individuals are familiar. This reduces the training time of users and implementation strategies for technology changes. Further, with the advent of newer handheld devices, this browser-style interaction has become a necessity to support these smaller ubiquitous devices.

5.2.6 FACILITIES MANAGEMENT

This has been around for some time and was always considered in a different arena; however, as we are now seeing, facility management is not that much different from managing other assets. Within facilities, inventory and equipment require work management to maintain them. Now it is recognized that this is not any different from what is managed within CMMS systems. Locations are the key aspect of CMMS systems, but within facilities these become fundamental items for proper management. As such, the consolidation of facilities management software products are now aligning with the CMMS solutions to offer a complete integrated solution to facilities and maintenance of the facilities, the associated equipment, and inventory.

5.3 TECHNOLOGY-ASSISTED MAINTENANCE

Using technology in the 21st century to assist with maintenance is not new; however, it definitely is in its infancy when you consider what can be applied and how few organizations have actually applied it. Most organizations that manage assets do have an implemented CMMS solution in place. Most do not fully use the software solution they have implemented, and most also are working with dated or old versions of the processes and applications.

One industry that has demonstrated it is ahead of the maturity curve when it comes to applying technology to asset and maintenance management is, not surprisingly, the computer industry. The concept of collecting data to help the asset manager/ maintainer understand mean time to resolution (MTTR) and mean time between failures (MTBF) data has been around since the latter half of the 20th century in the computer industry. Having a system or even the asset itself collect and analyze its failure data or its "potential to fail" data was well in place in the late 1960s and 1970s. Converting the data so they would flag or even initiate a request for service through an external telephone system was well exercised in the late 1970s. A major computer services provider implemented a maintenance management and response system to support the 1984 Olympic installation in Los Angeles. It took advantage of its existing analysis process capabilities at that time and introduced a radio-frequency (RF) communications process to ensure that the maintenance craftsmen who serviced the 1984 Olympics could minimize MTTR. This RF terminal technology introduced and effectively automated the asset "call in process" to a central dispatch system and facilitated two-way communications between the technical craftsman and the supported work management processes.

This work management process worked in an automated fashion as follows:

1. The machine asset determines it has a problem and signals the predefined fault code to the supporting service organization.
2. The service organization receives the call and registers it as a specific supported asset with a specific fault code, determines which available craftsman with the appropriate skills can support this fault code, and sends the appropriate craftsman a notification through the RF system.
3. The assigned service craftsman receives the notification through the personal portable RF terminal, acknowledges the call to the dispatch system, and interprets the fault code and the received recommended action. The craftsman then decides if an order should be placed for the recommended parts that are associated with the fault code.
4. The assigned technician orders parts via the same RF portable terminal and informs his or her dispatch system of the planned estimated time of arrival (ETA) and then proceeds to the machine location to support the service call.
5. The parts organization receives the parts order, fills the order, and ships it so it will arrive at the machine location within one hour from the time it was sent.

Although many tactics and strategies ensure the availability of a producing asset, the Los Angeles Olympics example proves that technology has been significantly leveraged even more than 20 years ago to minimize response time to the process components of a work management process (a maintenance call) (Figure 5.2) and also used effectively to accomplish the following:

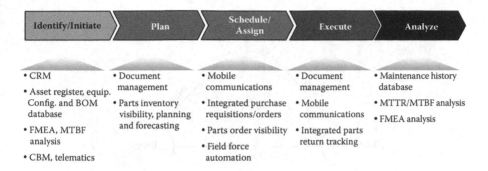

FIGURE 5.2 Technology solutions known to support the 1984 Los Angeles Olympics.

1. Track potential failure conditions and initiate a call or work request to the service organization based on a predetermined criteria established in the assets' expert system
2. Accept the call in a work management dispatch process that has predetermined who the best resource is to accept the maintenance call based on skill and availability
3. Provide two-way communications via an RF technology for the call management process as well as the parts ordering and second-level maintenance support
4. Provide a predetermined course of action, including a recommended list of repair parts that could solve the issue, based on the asset fault code

This set of call–response and work flow accelerators was subsequently established across all locations this major service provider supported in North America and in many other countries. Over time, the tools used were upgraded to reflect more effective and updated technologies. This set of initiatives created significant service enhancements for the provider's clients while ensuring it could deliver this service with an optimal set of delivery and support resources. In other words, the users of the assets got a proactive asset management service and world-class emergency call support, and the delivery costs for the service provider were significantly reduced through the use of this integrated set of technology solutions.

5.3.1 CMMS Implementation Success Factors

The implementation of a CMMS and EAM solution has many considerations that must be managed in a concurrent fashion if the success of any technological implementation solution is to be ensured. The typical business transformation triangle of any implementation strategy encompasses the concerns of process, technology, and people; however, the important issues of sound leadership, training, and security must be included to support them. Once all these are in place and managed, a final concern remains, which can have a major impact on success: the culture of the corporation and its people in adopting the new processes and technology (Figure 5.3).

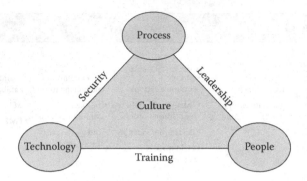

FIGURE 5.3   Points of involvement.

Multiple work streams of effort are typically required to ensure an effective implementation process. These work streams will typically be applied across four stages of a software implementation (see Figure 5.4). This includes working through project initiation and stakeholder engagement analysis (as Stage 1), solution design (Stage 2), the actual configuration, test, and rollout (shown as Stage 3 or "implementation"), and, finally in the fourth stage the assessment for managed sustainability (which includes assessing the effectiveness of the implantation by looking at the expected rewards of delivery). So what work streams of the implementation should be managed and included to ensure project delivery success and CMMS sustainability? To discover this, we must revisit the business goals and issues by highlighting the same set of concerns and issues. As we begin any project we can recognize the need for project management, but in addition we must also look at the work streams for managing the technology and its integration into the business. With the understanding that CMMS cannot stand alone as a technological implementation if it is to be successful, we must consider the stream of the business process that delivers to the people their "software managed tasks" such that they align with the enabling technology and its ability to execute the desired tasks. This further implies that we are educating and training people how to use the technology and the business processes to change the ways they will be executing these tasks (doing business). However, such change cannot happen without understanding the culture and impact these planned changes will have on the culture. Consideration needs to be taken to ensure that the folks affected by the new solution will adapt or that the solution will be adopted into the culture. This requires change management. Change management is a critical work stream as it works to affect the people's side of the transformation. It is not unusual to observe that a full successful software implementation of a CMMS solution is 60% people oriented, 25% business process oriented, and 15% technology oriented. With all these works streams in place, the last stream to be managed to ensure that the efforts of implementation do not end at go-live is the "support stream." The Support Stream along with Change Management work to ensure that the technology, process and people are supported and recognize the benefits we are expected to achieve by this planned initiative. It is recommended that you engage a support model early to align with and build to a sound support solution (see Figure 5.4).

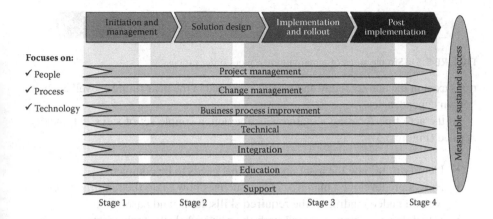

FIGURE 5.4 CMMS stages for success.

5.3.2 MEASURING SUCCESS

So how do we know we are successful and have obtained the targeted realization of value for all the effort undertaken to implement a CMMS or EAM solution? The important thing is to know where you are starting from and be able to assess the key performance metrics that can be extracted from the daily performance that will show the return on investment to the business. That is, we must select our measure earlier and perform the measurements in the old system to establish a baseline from which we started. Hence, in the early stages of the project, we need to select what indicators will best measure success—engaging the business along with the technology stream to assure not only that the metrics can identify the success of the new maintenance efforts but also that they can in fact actually be measured and trended. This can sometimes be the greatest challenge of implementing the measures for success, so careful consideration is required in selecting these metrics. This of course is followed up by prudent reporting and review.

5.4 EMERGING TECHNICAL ENHANCEMENTS TO THE MAINTENANCE MANAGEMENT PROCESSES

As stated earlier, many of the leading CMMS/EAM solutions today provide the basic requirements to support the basic maintenance management processes and some form of integration to parts, procurement, human resources, and financials. However, the need to meet the goals listed at the beginning of the chapter and to optimize MTBF and MTTR in pursuit of these goals has driven many organizations to search for and implement enhancements to their packaged CMMS/EAM solution.

To capture some of the technology approaches that have been successfully applied across many industries, we will discuss some of the key applications, features, and benefits in which technology has been applied to contribute to the goals listed at the beginning of the chapter. In addition, we will briefly explore how some of these technologies have evolved. We will describe what variations of the technical solu-

FIGURE 5.5   Simple steps of a maintenance work order process.

tions exist in the market today and what is expected in the near future as well as their known dependencies.

In the act of performing a maintenance action, a simple set of steps is typically applied (Figure 5.5):

- Identify or initiate the asset and work that needs to be completed.
- Have a predeveloped plan that clearly defines how to do the required work with an understanding of the required skills, tools, and parts.
- Schedule and assign the work with the required skills, tools, parts, operational alignment, and management approvals.
- Execute the work per the provided plan, document what was done, and return used parts and tools.
- Analyze the maintenance activity history to look for unusual activity per each asset, skill, part, or resource.

CMMS software often provides fundamental solutions for the work management process. Many CMMS software vendors have improved their products to differentiate them from their competitors with features that promote the following:

- Effective listing of supported assets
- Efficient work management processes
- Control through enhanced asset, resource, parts, and tools management
- Superior analysis of asset, resource, parts, and tool activity so that improvement can be made to further improve asset availability

Because of this, new and exciting technologies are being deployed to better assist the operations and maintenance solutions by aligning with existing CMMS software solutions and enhancing capabilities to end users and maintenance staff.

5.4.1   MOBILE TECHNOLOGIES

5.4.1.1   History

Field operations tend to be paper based with little or no data analysis being performed on field data (refer to box in Figure 5.6). Information is collected for different assets through readings that are gathered and logged using log sheets and notations about required maintenance items obtained through observation of personnel operational deficiencies. These may be addressed through issuance of work orders or scheduled routine checks. Following this effort, the information is assessed, which results in generated maintenance work requests that will likely be entered into some form

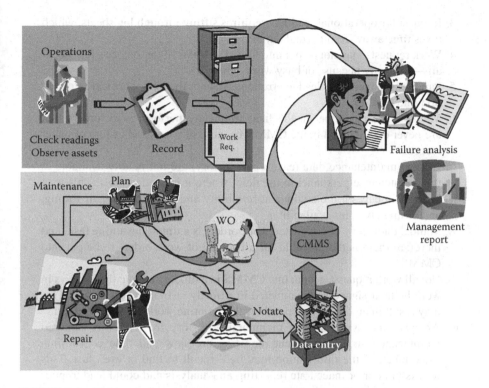

FIGURE 5.6 Field operations/maintenance of today.

of record-keeping or CMMS system. The planner then schedules and assigns these work requests to maintenance personnel or maintenance activities for issuance.

Alternatively, if manual data entry is used, field maintenance workers would receive a paper copy of the work order (whether from some form of CMMS or not) and then would assign these tasks out for action. The work orders may correspond to a work request created by operations personnel or another maintenance person. Once the work has been assigned and completed by the maintenance personnel, it is documented as work performed, likely on the same sheet of paper. The completed work orders are then manually updated or filed (preferable to a CMMS system) for future analysis and management reporting (Figure 5.6).

5.4.1.2 Limitations and Issues

1. Critical operational data are not immediately available to the field personnel where they are needed most, thus impacting equipment reliability and efficiency and ultimately equipment or plant availability.
2. Field operations are entirely paper based and full of handwritten readings that are filed away in a cabinet with little or no analysis for preventive maintenance (PM). Most analysis is performed after the fact or as needed for investigations.

3. Request for operational readings requires sifting through log sheets, which takes time away from normal duties.
4. Work requests may not be put into CMMS for days—or, on occasion, not at all—primarily because of busy work schedules.
5. Paper-based processes lead to erroneous data or the loss of critical information into CMMS.
6. Maintenance personnel in the field frequently perform work orders that are never documented in the CMMS because of work load or loss of paper records.
7. Critical maintenance data (e.g., repair history) are not readily available to the maintenance personnel in the field, where it is needed most. This may have an impact on the equipment reliability and efficiency, repair timing, and ultimately plant availability.
8. Manual data entry of completed work orders is a time-consuming task, and therefore there is the potential for a backlog of work orders to be put into CMMS.
9. Not all work requests get put into CMMS in a timely fashion; thus, delays in work being assigned and completed in a timely fashion are inevitable. This may result in emergency work rather than routine work being performed.
10. Analysis and assessment of maintenance history for reliability-centered maintenance and management reporting are slow and time consuming, since much of the information may be difficult to find or even lost. This results in poor or inaccurate reporting and analysis and could lead to poor decision making regarding an asset.

5.4.1.3   What Is Expected in the Near Future

If we look at the foregoing, it is probably not that much different from what has been experienced in your plant. However, now with current technological advancements, some of these problems can be addressed through implementing handheld or mobile devices to support in-the-field personnel. Mobile system architecture with handheld computing technology to complement the data currently collected via remote-site monitoring methods and support makes having critical information in the field possible. Combining data collection and availability of key maintenance information will provide a more accurate picture for enabling asset preservation, preventive maintenance, and other field-related planning tasks.

Handheld computing devices enable field operations and maintenance processes to be streamlined by gathering and putting information into the hands of the field person. This simplifies the field operations process (Figure 5.5) and provides assurance that the information is in the right place at the right time. Using the handheld device, field personnel can collect their current operational readings as well as create and observe equipment and enter necessary work requests. They can then transfer their operational readings and routes for the day directly into and from the handheld device, either through some form of docking or, if feasible, through a wireless connectivity network. This information would then be transferred back into the CMMS system for action without additional handling by data entry or other personnel. This

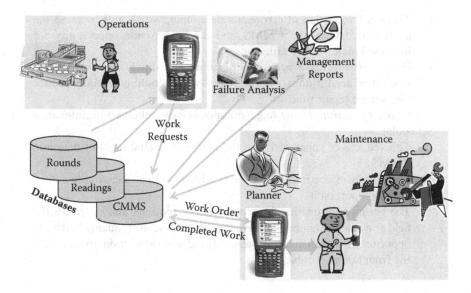

FIGURE 5.7 Field operations/maintenance of tomorrow.

reduces information loss caused by human error and supports a better integration of information into a CMMS solution. This will provide improved overall management of maintained assets (Figure 5.7).

This process applies equally to maintenance personnel, who can perform a similar set of steps to obtain work orders and information about the asset and then use the handheld device to further document and complete as well and create new work orders. This information can either be transmitted directly into the CMMS system through wireless if available or be held and transferred directly into CMMS at the end of the day through some docking station. This then provides a more current view of an asset's state and provides further readiness for the maintenance personnel to perform new tasks.

Subsequent reporting and analysis of critical operational and maintenance data will be able to be managed and fully collected by using the information within CMMS. Report analysis and performance characteristics are readily stored and available for better analysis and generation of failure reporting and management reports.

5.4.1.4 Business Benefits

In general, applying handheld computing to field operations and maintenance can offer numerous advantages that are aligned with key business drivers, such as the following:

1. *Improving operational availability* (e.g., equipment reliability and efficiency, planned maintenance outage, information availability):
   a. The CMMS has accurate and timely data on all work orders and equipment status, thereby enabling more accurate and timely reporting and the organization of maintenance schedules.

   b. There is a reduction of erroneous data input, which will increase the quality of data in the CMMS and thereby provide opportunity for increased maintenance efficiency.

   c. Field personnel have access to accurate and timely data where and when they need it most—in the field—which could potentially reduce the average repair time.

  2. *Managing operational cost* (e.g., manpower cost, planned maintenance schedule):

   a. This reduces of paperwork and the need for manual input of data into CMMS.

   b. This improves accuracy and time needed to collect data and to subsequently report on it.

   c. The ability to immediately generate work requests and orders in the field as needed enhances performance of the asset by reducing "missed" opportunities for repair and potentially reducing lost productivity resulting from failures (Table 5.3).

TABLE 5.3

Sample Mobility Solution Business Benefits Review

Business Drivers	Today—Issues	Tomorrow—Benefits
Plant Availability: Availability of information	Limited operational and/or maintenance available in the field	Maintenance and operational data made available on handheld devices
Plant availability: Analysis/reporting of critical operaitonal data	Minimal analysis performed on critical operational data	Reports can provide up-to-date operaitonal data quickly, and subsequent analysis can be performed by relevant parties (e.g., engineers, site management)
Plant availability: Accurate and timely input of critical data	Paper-baed processes lead to: 1. Erroneous data entry in CMMS 2. Loss or delay of capture of field information (work requests/orders) relevant database or CMMS 3. Backlog of work orders to be input	Work requests/orders can be 1. Captured and completed in the field iteration 2. Uploaded via device with no need for manual entry
Plant cost: Time savings	Operator foreman spends time: 1. Analyzing paper log sheets for weekly replenishment ordering 2. Creating work requests for operators and/or maintenance 3. Gathering operational readings for engineering	1. Operator foreman can use CMMS report to determine weekly replenishment orders 2. Engineers can view operation readings reports on the web 3. Field operators can upload field-generated work requests 4. Maintenance personnel can now complete most work orders

3. Other features of benefit from introducing mobile technology may include the following:

    a. Mobile technology allows replacement of obsolete technology and infrastructure (e.g., hardware).

    b. Mobile technology affords an opportunity to integrate warehouse material management functions onto a common handheld device.

    c. Mobile technology can extend the use of bar code infrastructure to other functional areas (e.g., bar coding equipment).

5.4.2 Data Management

Within any system or process, underlying maintenance management are the data. These are the heart of any CMMS, since they are the foundation by which we can evaluate the impact and measure the successes of achieving better maintenance processes and improving the life value of the assets that are maintained. Good data allow for the ability to extract information to make good decisions. Good maintenance data come from data that follow the business processes of sound maintenance solutions. Therefore, good data management will greatly benefit managing through an effective maintenance solution.

So let us spend some time looking at what constitutes good data management within any CMMS. It is important first to ensure that there is a clear definition for each piece of data that is stored, used, or entered into the system. Data must be representative of assets or attributes of the asset or their associated transactional information against these assets. Therefore, it is very important that the data in a CMMS system be of value to the management of the assets.

So how do we know with some assurance that we are managing the data within the enterprise asset management solution? There are tools that can help us to evaluate and ascertain that the data we have are of importance to us. However, they are only tools, and the true analysis has to come from the people who use the data, who ensure that the information entered is correct and useful. Data used on a regular basis will tend to be good data since they are representative of the assets in use.

So how do we make these assurances? The best method is usually the simplest: Limit permission on who can update, enter, or modify data items only to people who actually need to modify these data elements. Many of the CMMS tools have such internal systems to control data change, and some time should be spent to assure that these security features align with the business processes.

In addition to security to guarantee data, we can institute limitations on the contents that can be entered into data fields. Making the number of choices that can be entered to a data field not only keeps the data useful, but also enables analysis based on these features.

Open text is very hard to manage; therefore, restricted choices and restricted access can actually improve the data values you enter. However, there are times when open text is all we can use to handle the needs for recording information. In these cases you might wish to establish standards of how this open text can be entered and then have periodic samplings of the data to assure the standards are being practiced.

Obviously, managing data within the CMMS solution is a very involved process. What we have described here is only the aspect of managing the value of the data we have in the system. Although this is one of the most important factors, some other issues fall more into the technology realm for data management, including storage management, archiving, backup, and data recoverability. So as computer technology changes, so, too, do these technology issues.

Historically, maintenance data were managed on paper and, over time, were moved onto large computer systems or mainframes. Then, data solutions were migrated from these monolithic data files onto more versatile storage solutions and were implemented into relational databases on servers located throughout the organization. Now as we move to a more Internet-based solution, the methods of data access are changing ever still.

Today, enterprise asset management solutions are run on a number of platforms, many of which include a relational database engines in the background. Some solutions run on a single PC, and some run on a client-server configuration. Still others now run in a distributed Web-based architecture using browser technology. All of these offer differing opportunities for managing the CMMS solution data.

So what can you expect in the future? The wireless capabilities of the future may offer us ways to connect to the data stores more instantly and globally. As bandwidth continues to improve and the performance of handheld devices and workstations continues to expand, we can expect to find that the data availability will be closer to the source of use. Furthermore, as assets become more evolved and computerized, they may actually become data entry points in themselves and be able to update the data directly into the enterprise asset software.

5.4.3 PLANNING AND SCHEDULING TOOLS

The concepts of planning and scheduling related to managing in an efficient and timely manner the completion of any undertaking are not new, especially because of the many intricacies of aligning and coordinating the delivery of material, labor, tools, and services. In the past, however, being a good planner and scheduler was an art form. Many hours of organizing and coordinating were required, with a full understanding of what to do and how to accomplish it. But as more sophisticated methods and tools are evolving, the art of planning and scheduling is changing into more of a science. The level of skill and experience needed to build and manage a schedule can now be supported with tools that take its complexities and simplify it into a process of routines.

We are all familiar with Gantt charts and organizing our tasks in visible fashion for easy analysis. There are a number of technology tools that provide assistance in building charts for simplified planning analysis. But which tools and how we use them are always questions that must be considered. Planning and scheduling take time even with the correct tool. So how do we identify which tool to choose?

The planning tools available on the market have been developed over many years to support the planning and scheduling effort for projects and work that have a clear start and finish. The effort involved in such planning and scheduling can be anything from simple to complex, depending on the nature and issues of the work to be done. It is important to remember that it is not the tool that perfects the planning but the people using the tool, since the tool can be only as good as the information it is given. Given this, tools like Microsoft® Project or Primavera have been around for many years and can aid in both developing and managing these planned undertakings. These tools currently not only support the planning of tasks through sophisticated algorithms used for balancing and managing task scheduling within the undertaking, but now also provide a way to manage time, resources, skills, and costs. This gives the planner and scheduler information that can be applied and reviewed to fine-tune and improve upon repeating work.

Synchronizing your team performance through such tools goes far beyond planning and scheduling: these tools can provide a common comprehensive platform for maintenance and project management alike. Using these advanced collaboration tools, with role-based features and action alerts to support centralized information that aids in keeping efforts on track with cross-functional activities in sync, is done through integrating other critical systems where data have to be entered only once and then reused, thereby providing trusted forecasts and reports for maintenance turnaround solutions. This is what the focus of a planning and scheduling tool should be (Figure 5.8).

Planning systems are evolving to support the integration with both the financial systems and the asset managements systems (CMMS; Figure 5.6) to produce a completed solution for the delivery of managed and maintained assets to the organization. Since the tools for planning and scheduling are much more entrenched and have been developed around such common practice principles, it makes sense for CMMS tools to use these capable process systems in developing the schedule for the

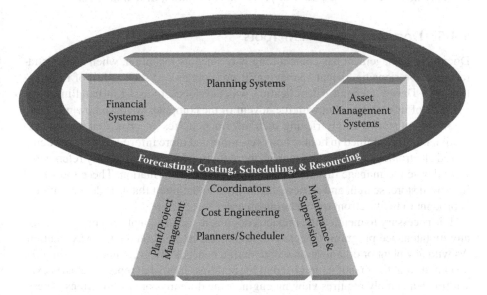

FIGURE 5.8 Asset management, finance, and planning integration.

maintenance activities too. Forecasting, costing, resource management, and scheduling encapsulate all of the needs in support of finance, planning, and asset management. Therefore, it is important that we integrate through these processes in a bidirectional focus to take the work requests from the CMMS solution into the planning and scheduling tool only to return the successful solution back into the CMMS system for tracking and gate approval management. The effective solution for such planning and scheduling involves a number of groups within the organization. The planning and scheduling tools assist us with communication among management, coordinators, planners, technicians, and field operation foremen in assuring that the right resources are aligned to complete the undertaking in the best cost-controlled manner. Why would we not want to use these in an integrated fashion within any maintenance strategy?

5.4.4   EXPERT SYSTEMS AND ASSET MONITORING TOOLS

Expert systems can be augmented to the asset and maintenance management arena by effectively using statistical process controls or some other forms of programmed automation. Meters are installed on the assets to continuously measure such things as vibration, wear, viscosity, or simply time. The sensors provide feedback to a central processor that makes calculations to assess the condition of the asset. Then this will, depending on the state it encounters, automatically request service in the form of a part swap or maintenance of some form. The example provided earlier is a rudimentary type of expert system. With today's technology we can hope to get more sophisticated assessments and to track history such that the expert system can diagnose the problem, can attempt to correct it by itself, and, failing that, can issue a request for service. Expert systems continue to evolve as the capabilities of understanding the asset are increased and the power of processor calculation time improves.

5.4.5   DOCUMENT MANAGEMENT TOOLS

Document management has come a long way since its inception, when administrators and librarians first began managing filing cabinets and shelves full of information. The Dewey Decimal Classification was developed to assist in handling large volumes of information, but with the advent of the computer we have now moved to a new way of storing and retrieving such information. Yes, it is true that much information is still contained in books; however, more and more information is now being stored electronically. Document management tools, which are frequently referenced, were devised to manage this electronic storehouse of information. These tools have improved store, search, and retrieve capabilities to the point that this documentation is brought to the forefront of managing an asset.

It is necessary to mention the impact a document management system can have on any maintenance program, especially one that has implemented a CMMS solution. As with any plant or organization where maintenance is involved, understanding the asset and what the manufacturer used to build the asset or what repairable parts exist within an assembly requires viewing engineering drawings or specifications. These, along with the reliability specifications, are now typically stored in electronic pieces that can now be managed through some tool that understands both the document and the need to access it by content.

5.4.6 INTEGRATION

Many books and papers have been written about integration, and we will not attempt to cover all the facets of such a topic. Instead, we will view some of the directions integration is taking in CMMS application with other application areas.

The design of integration has been in a state of constant change. As technology improves, more efficient means are being developed to support the heavy demand for integrated business. Within CMMS there is integration demand for financials, documentation, planning, scheduling, mobility tools, and more, which pose not only technological but also process-related problems. Integration is about technology and also must include the issues associated with the business processes that are being integrated.

Originally, integration meant large amounts of data exchange: extracting data from one application and then importing them into another. Since many applications were using a proprietary structure, this was indeed a sophisticated task and was best handled only by very experienced programmers. But along came relational databases, which paved the way to better application integration. Then, SQL queries could be generated to extra data, and, since the data were already in a standard database format, they could be easily parsed and imported into another database. This enabled backend processes to be created that moved information from one system to another. Although the process of manually triggering these activities was eliminated, it still left the duplication of data across multiple systems, even though it may not have appeared as such.

The newest of architectures now leverages the capabilities of the relational database. This new architecture focuses the content of data to be related to the object of its service. As developers build their applications to suit this new architecture, growth of services within the application and the true marriage of information across the enterprise can be seen. The service-oriented architecture supports a services approach to the design and building of an application. This "Web Services" approach results in the integration layer becoming more complex but with more advantageous features and control. The clear separation of its parts has granted greater capability in the integration space. Integration is now the assembly of many parts and components to deliver an end-to-end solution (Figure 5.9).

Integration of the CMMS solution is just one component of managing the organization's assets through technology.

5.5 EAM AND ITIL

Information technology (IT) management has been included in an international standard under ISA20000, of which the Information Technology Infrastructure Library (ITIL) is a major component in defining processes and a framework in which it can operate. Of course, within IT are assets that must be maintained in the same fashion as all other enterprise assets.

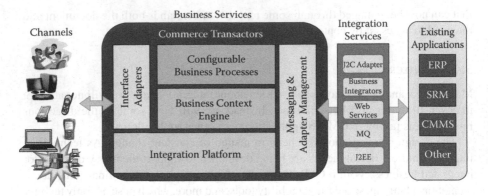

FIGURE 5.9   Integration landscape.

ITIL is a collection of books that identify a framework of processes established for managing and maintaining the IT assets and services of any IT-based organization. These books include many concepts and processes that can likewise be adapted for any asset maintenance. Once you get past semantics, you will discover that ITIL and EAM have a lot in common from the process and goal purposes. EAM aims at driving to better service and up-time of assets to be more productive, and so does ITIL. Because of this, some tools that manage IT assets have crossed over with the tools that manage other maintainable assets. Thus, it is quite likely that you will begin to see more and more that tools being used for EAM or for IT are being cross-purposed to perform much of the same functions in managing the assets of the enterprise.

Here also, as the tool developers continue to refine their tools and processes for EAM and the tools for ITIL, it is likely that the tools will become the same. So get ready for the infusion of IT assets into your maintenance space as we move to a more integrated world within the enterprise. This truly will become an enterprise asset management solution.

5.6   CONCLUSION

By now, it must be clear that a CMMS is an indispensable tool for today's asset manager. In fact, organizations are looking at asset management as a core competency, and, in many cases, the asset manager is on par with the CFO.

The asset management systems solutions now available cover all organizations—from small, single-plant operations to multiplant, multinational companies. These modern systems use the latest technological advances, such as the Internet. Functionally, they are very rich and provide features for both the strategic aspects of asset management (long-range budgeting, capital project management) as well as the tactical and operational aspects of maintenance. The latest CMMS advances include built-in "intelligence" to customize maintenance for a particular area of the plant (predictive maintenance). The potential cost savings are huge, and availability is usually improved as well.

More universally, it is not just maintenance management but also asset management. Further, as the worlds of differing assets in the organization start to converge into a single solution of both processes and framework, it becomes even clearer that it will become a single enterprise tool. So selecting the right tool for the job becomes even more important, and so too are the additional features that you add. The best advice we can give is to spend the time and do it right.

During the selection process, it's most important to know both what you are asking for and why. In other words, you need to understand the requirements and have a sound business case for buying a CMMS. It is not unusual for organizations to spend huge sums on asset management systems and not see any improvement in equipment availability or maintenance costs!

Not everything you are looking for in a CMMS solution may be found in a single package. Many emerging leading processes have been around in limited use for a number of years but are just coming available in a CMMS/EAM package or as a "bolt-on." Care should be taken to consider whether implementing a "bolt-on" enhancement to your selected asset management solution is worth the additional process and systems management time. With large ERP solutions growing in their asset management capability and "Best of Breed" asset management solutions continuing to lead in functionality, there continues to be a gap between integrated CMMS and functionality or the "user friendliness" of a "Best of Breed" package.

The implementation process is where "the rubber meets the road." Even the best CMMS can be crippled by poor implementation decisions, training, and support. Like any large system, operating a CMMS is a process-driven exercise. It should come as no surprise that if users don't follow the process, the results will be unsatisfactory.

Business processes will always change as a result of industry conditions, personnel issues, and so on. Recognize that the CMMS is, at heart, just a computer system. Like our automobiles, it needs to be regularly tuned up, with process changes, to operate at maximum efficiency.

REFERENCES

1. Brodie, M. and Stonebraker, M. *Migrating Legacy System: Gateways, Interfaces, and the Incremental Approach.* San Francisco: Morgan Kaufman Publishers, 1995.
2. Linthicum, D. S. *Enterprise Application Integration.* Reading, MA: Addison-Wesley, 1999.
3. Mylott, T. R. III. *Computer Outsourcing: Managing the Transfer of Information Systems.* Englewood Cliffs, NJ: Prentice Hall Direct, 1995.
4. Wireman, T. *Computerized Maintenance Management Systems.* New York: Industrial Pr, 1994.
5. Primavera Systems Inc., Power, Energy and Process, http://www.pllogic.com. 2005.
6. Maximo Systems Inc., Workforce Mobilization with Maximo, http://www.maximo.com. 2005.

APPENDIX A: ASSET MANAGEMENT SYSTEM SELECTION AND IMPLEMENTATION

SYSTEM SELECTION

If your organization decides to acquire a new asset management system and makes you responsible for selecting and deploying it, this could be the career opportunity of a lifetime. What you need, however, is to ask some key questions before you proceed further. Once you receive satisfactory answers, the next requirement is a robust system selection methodology roadmap.

Preliminary Considerations

For the most part, you use computer-based business solutions to increase your organization's effectiveness. Put simply, the solution has to enhance profitability. Normally, you make a business case to describe, justify, and financially estimate the expected benefits. As shown in Figure 5.10, acquiring a business system like a CMMS is usually part of an improvement cycle. In this cycle, the business case is part of the remediation plan, along with other future success measures.

For example, management may expect the CMMS to improve productivity and overall equipment effectiveness (OEE) by a certain percentage, to increase inventory turns, and so on. The team responsible for selecting the CMMS needs to understand not only these specific goals but also the overall strategic context for acquiring the CMMS.

System Selection Process

Most packaged systems are selected in a similar manner. Unlike custom systems, when you procure a packaged system, you decide what you need, and then evaluate various vendor offerings to see which fits best. The general approach is shown in Figure 5.11.

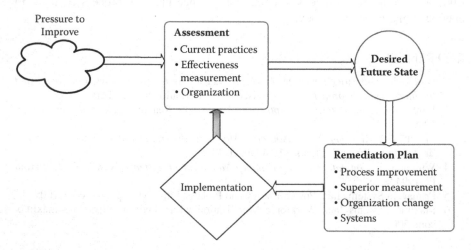

FIGURE 5.10 Business improvement cycle.

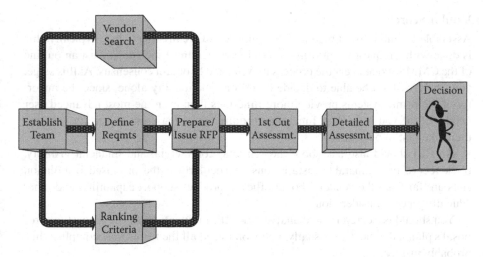

FIGURE 5.11   System selection roadmap.

In the sections that follow, we describe each step in the process. In practice, most organizations use an outside consultant to either help them execute the process or take it over completely. The advantage is that the consultant can fast-track many tasks, minimizing the cost of disrupting the organization.

Establish Teams

Your goal at this point is to establish working teams to set requirements and validate vendor offerings against them. Teams usually include users, maintenance managers, and others who have a large stake in the success of the CMMS.

If the scope of the planned CMMS is truly enterprisewide, you may need several teams, representing different plants, sites, or locations. If the CMMS will be complex, consider forming teams with particular domain expertise. There could be teams from maintenance, inventory management, procurement, and so on. Clearly, as the number of teams grows, so does the task of coordinating their outputs into cohesive requirements. You need, then, to form each team carefully, with a clear definition of what it is expected to deliver.

The key deliverables are as follows:

- Project "charter": This defines what teams are expected to deliver and in how much detail. The charter should also define each team's specific responsibilities.
- Task schedule: This defines the time line for each team. Do not expect results after a couple of meetings, since capturing requirements can be onerous.
- The overall manager of this task should recognize that team-building skills will be needed and that he or she will frequently have to adjudicate in situations where responsibilities aren't clear.

Vendor Search

Assemble an initial list of prequalified vendors you are inviting to bid. Typically, this is done with a request for information (RFI), which must include at least an outline of the CMMS scope to get the process moving and to obtain consensus. At this stage, you likely will not be able to decide based on functionality alone, since the majority of the prime systems provide more functions than even the most advanced user needs. But if you have clear functional requirements that point to certain vendors, then take that into account.

An RFI should also ask about the vendors' commercial and financial viability, track record of comparable installations (particularly if the proposed installation sites are far from the vendors' home offices), product support capability, and other "due diligence" considerations.

You should issue, return, and analyze the RFI in time to meet the request for proposal's planned issue date. Usually, you won't need all the teams to accomplish this, probably just team leaders.

To recap, typical deliverables from this task are as follows:

- Request for information
- Initial vendor list

Define Requirements

The quality of vendor proposals largely will reflect how complete and clear your requirements document is. Also known as the system specification, it is the core document against which the CMMS application is acquired, implemented, and tested. Obviously, it needs to be assembled with care.

At the highest level, group requirements into major categories. Then break them down into subcategories and again, if necessary, into very specific requirement criteria. Here is an example of what a requirements hierarchy could look like:

Category	Subcategory	Subsubcategory/Issue
Operations	Equipment	Asset hierarchies
Data analysis	Drill down, graphical, history	
	Work management	Blanket WOs, approvals, resources, scheduling, safety, crew certification, contractors, condition reporting
		Can labor hours charged to a work order be broken down into regular and overtime hours?
	Preventive maintenance	*Can the system trigger an alarm when equipment's inspection measurements are trending outside a user-defined criterion?*
	Inventory	Reordering, vendor catalogs, multiple warehouses, repairable spares, multiple part numbers, ABC support, service-level costs

(continued on next page)

Category	Subcategory	Subsubcategory/Issue
		Can the system support multiple warehouses? Is a warehouse hierarchy supported?
	Procurement	
	Resources	
Financial	Electronic data collection	*How would bar-coding support issues and receipts?*
	Reporting	*How does the system use/generate a cycle count report?*
	Accounting methods	
Technical	Concurrent users, No. of licenses, "power" users vs. casual users, architecture, scalability, performance, security/audit logs, databases supported, integration with other systems, data import/ export, workflow solution, application architecture, database management, client configurations, development tools, interfaces, capacity performance	*Does the system use constraints (also cascading)?*
Human	Documentation	
	Training	
	User interface	*Can a user have multiple screen access? If so, how?*
	Services	

Note that where detailed requirements are stated as questions, they should be presented in an RFP in the form "the system shall...." Otherwise, if you are uncertain about specific needs then pose the requirement as a question, inviting vendor comment.

Requirements are best gathered using business process maps as the context. You should have generated these as part of the Assessment and Desired Future State work shown in Figure 5.10. For example, consider work order processes, raising, approving, executing, closing, and reporting. With these maps, the teams responsible for requirements gathering should set up workshop interviews with user groups. A workshop format, bringing together different perspectives, stimulates maximum input.

Define Ranking Criteria

At the end of the exercise, the requirements hierarchy will be very large, with potentially hundreds of detailed system requirements across all of the major categories. You will need a quantitative approach to compare all the vendor responses.

Two numeric scores are relevant to each specification item: degree of need and degree of compliance. Degree of need represents how badly you must have the specification item, from mandatory to "would be nice." Clearly, a mandatory requirement should be scored higher than one that is optional. In the previous sample, the requirement for the system to support multiple warehouses would probably be mandatory and scored, say, 5 in degree of need. In contrast, if supporting a warehouse hierarchy is optional, it could be scored 3, representing a "highly desirable" requirement. An unimportant specification item would be scored at only 1. To make the rating even simpler, you can use a binary score: 5 for mandatory, 0 for nonmandatory. Choose your approach by how detailed your evaluation needs to be.

The other score to set up is the degree of compliance. In the previous example:

	Relative Importance (Need)	Vendor Compliance
Multiple warehouses	5	
Warehouse hierarchy	3	

Rate vendor compliance scores on the following criteria:

5 fully compliant with the current system, no customization required
3 compliant with the current system, customization included by vendor
1 not compliant without third-party customization
0 the requirement cannot be met

Here, it is better to use a range of scores rather than the binary approach. Why? Vendors can often supply the needed requirement with minor customization. In this case, they would score 3, and you need to be able to distinguish among minor, vendor-provided customized, and significant third-party add-ons. When you issue the RFP, give vendors the previously provided list as well as the degree of need scores so that completed bid sheets have two scores for each requirement. When evaluating the bids, multiply both scores together to get a "raw" score for the requirement.

What about nonfunctional issues, such as vendor track record, support, financial stability, nonfixed price arrangements, or implementation partners? These are as important, sometimes even more, than functional requirements. You need a scoring scheme to compare vendor offerings in these areas. Check with your organization's procurement department. It should have guidelines to follow as well as standard scoring rules.

Prepare and Issue the Request for Proposal

The RFP is a system specification of the CMMS's functional requirements. However, depending on your organization's procurement practices, your department will have to assemble, check, and issue standard terms and conditions, forms of tender, bid bonds, guarantees, and so on. For public organizations, the RFP issue and management process has to conform to procurement rules in that jurisdiction. For example, if one vendor raises a query during the bidding period, you may have to formally issue clarifications to them all. Bid opening can be public, with formal processes to manage appeals.

Once you've issued the RFP, the selection teams should develop demonstration scripts for vendors selected for detailed assessment. Demonstration scripts provide a common basis to judge how the candidate systems and vendors operate and perform. Without any constraining requirements placed on them, vendors will naturally showcase the best features of their system, and you will inevitably wind up with an "apples to oranges" comparison. The requirements should focus on important functional issues that are mandatory and reference back to business process maps, if available. Typically, the scripts should be comprehensive enough to cover two to three days of detailed product demonstrations, a reasonable time for modern CMMS applications.

To summarize, task deliverables are the following:

- Request for proposal, reviewed and approved by all involved teams
- Clarifications issued during the proposal preparation period
- Communications from vendors
- Detailed demonstration scripts used during assessment

Initial (First-Cut) Assessment

The initial assessment is reasonably mechanical. As we mentioned, the scores for each functional requirement are multiplied together, and the result is used as that requirement's raw score from each vendor. Calculate the ideal scores (degree of need score times the fully compliant score) to calibrate all bids. If the bid results are significantly lower than the ideal, it does not necessarily mean a poor response. Perhaps the requirements list was extremely detailed in areas outside the CMMS market. It is for this reason that it is a good idea to use an expert consultant to build the system specification. He or she should be extremely familiar with each vendor's product and know whether certain requirements can be easily met.

Tabulate the nonfunctional responses (e.g., commercial, financial), and, if a scoring scheme has been set up, apply initial scores. Often, organizations visually inspect the results, which can lead to interpretation problems. For example, which of the following responses to the track record question is better?

- We have 10 installations in your industry sector, three of which match your user count.
- We have six completed installations in your industry sector, each of which matches your user count.

Although trivial, this example illustrates the potential for making decisions based on qualitative assessments.

At this time, distribute the initial results to the selection teams, and seek opinions. This is not always easy. Ideally, the rules for joint decision making should be defined up front as part of the project charter. Does a majority decision carry? Is a majority defined as 50% plus 1, or should it be a significant majority? This kind of critical question should be addressed early on, before the decision needs to be made.

The output of this task includes the following:

- A documented initial assessment, reviewed and signed off on by each team leader
- A short list of vendors (we suggest a maximum of four) who will be evaluated in detail, with supporting documentation for their inclusion

Detailed Assessment

We suggest a two-stage approach. First, each of the short-listed vendors should give a presentation that concentrates on product overview, corporate background, financial stability, and ability to deliver high-quality services. Follow this with a detailed scripted demonstration and presentation by the two best vendors, emphasizing both product software and services. Of course, you do not have to limit the demonstrations to two vendors. It isn't unusual for three vendors to be involved. The process is time-consuming, though, and expensive. Weigh the benefits of having more than two vendors involved at this stage against the cost of the extra effort. As we continue describing the detailed demonstration steps, for clarity, assume that only two vendors are involved.

Invite the short-listed vendors (again, we suggest no more than four) to display their credentials in a three-hour presentation. This is to ensure that the vendor's philosophy and way of doing business is consistent with yours, in key areas such as services, support, and company background. Firmly steer the vendor away from detailed software demonstrations at this stage and instead encourage him or her to concentrate on, for example, his or her approach to implementation, experience in the industry sector (e.g., manufacturing, resource development, utilities), training methods. A typical agenda would include the following:

- General introduction (15 min.)
- Company overview (45 min.)
- Questions from selection team (60 min.)
- Software demonstration (30 min.)
- Wind-up and remaining questions (30 min.)

To help you decide on the two best finalists, use a scoring scheme for each of these topics. In particular, team questions and a response rating system should be decided in advance. This is complicated because the answers will be delivered interactively. Also, ensure that each selection team has the same set of expectations from the brief software demonstration so that they're looking for the same thing. Clearly, this is an inexact process that will require a lot of discussion to work out.

Based on the presentation and evaluation criteria, select the two best vendors and prepare written justifications. Also, notify the losing bidders, clearly spelling out why they were eliminated. In fact, everyone involved in the process should be advised about who made it to the final selection and why.

From here on, you begin detailed assessment in earnest. Specific steps include the following:

- Invite each of the two vendors to a site visit. You want them to better understand your operating needs, to collect data to use in the final detailed demonstration, and to reflect your processes in the final software review.
- Undertake initial reference checks, simultaneously with the site visit, if you wish. This can include conference calls or visits to each reference. You want to ensure that the vendors' information is consistent with the reference user's experience. References must be chosen carefully, since their

operating environments must be relevant to yours. You should advise the reference in advance about the nature and length of your call, so that he or she can adequately prepare. Naturally, the vendor isn't included in the actual reference call or visit.

- Invite each finalist for a detailed presentation and software demonstration, following scripts prepared and supplied in advance. There are two major objectives: (1) to ensure that the software is truly suitable; and (2) to ensure that the vendor can provide high-quality and effective implementation. Vendors, naturally, will demonstrate software attributes that show their system in the best light. So that they address your needs, predetermine that the demonstrations must reflect how the system will be used in your application. Similarly, you want to know how the system implementation would be delivered, either by the vendor or a business partner.

Again, this kind of interactive process is best served by preparing in advance. For selection, team members need a common understanding of what they are looking for and some agreed pass/fail criteria for the scripted demonstrations.

A critical part of the evaluation is to analyze the implementation and postimplementation services required, together with the vendor's (or implementation partner's) ability to supply them. Include items such as customer support; system upgrades; training quality; postimplementation training; user group meetings, conferences, and Web sites; location; and support quality. The selected vendor should provide a sample implementation plan as part of the final presentation and demonstration, followed by a detailed plan, which is to be approved by the selection teams before awarding a final contract.

At this point, the winning vendor will, most likely, be apparent. Carefully document your justification for this, and present it to senior management for ratification. Notify the successful vendor as well as the second-place candidate, who should also be advised that he or she may be invited to continue the evaluation process if a final agreement cannot be reached with the preferred vendor.

What if there is not a clear winner? You could do a detailed functionality test of the two finalists to filter out the best solution. You develop a test instruction set, based on each of the criteria. Because of the functionality depth in most leading CMMS applications, this is a major task, taking several weeks of detailed analysis. To keep it within reasonable bounds, we recommend that only mandatory functions be included. Consider this optional analysis only if after the detailed demonstrations, reference checks, site visits, and selection team discussions you're still deadlocked over the final choice.

Contract Award

Before proceeding to contract, hold final discussions with the successful vendor to clarify all aspects of the proposed scope, pricing, resources, and schedule. While it is unlikely that anything major will be uncovered at this stage, remember that, up to now, the primary focus has been on functionality. This is your opportunity to deal with other important aspects of the vendor's proposal, which also demand your full attention. Once completed, the next step is usually to issue a purchase order. The selection team should also prepare a detailed selection and justification record.

SYSTEM IMPLEMENTATION

To effectively implement packaged computer systems, three elements must work together:

- People
  - Willingness to change
  - Role changes (eg, planners, schedulers)
  - Organization change ⇒ reporting line change
  - Training effectiveness
- Processes
  - How business is done now
  - How business should be done
- Technology
  - Hardware and operating systems
  - Application software
  - Connectivity (network)
  - Interfaces
  - Data

A well-designed implementation project addresses each element so that the system will be effective and accepted by users. You can apply the outline steps that follow to most packaged business systems. However, the detail applies specifically to CMMS implementations.

Readiness Assessment

Before the implementation teams arrive on site with software and hardware, some preparatory steps should be taken. The first should be to conduct what we term a readiness assessment, covering the following:

- Organization and culture issues: This review asks questions such as the following: Is this company ready for system and process change? Is there a consistent sense of excitement, or is there tangible resistance? Is senior management supportive of the initiative and prepared to act as change agents throughout the implementation?
- Business processes: Are they documented, practiced, and understood? Is process change necessary?
- Technology: Is there a need for remedial work to be done before the system is deployed (e.g., network, communications, staffing)?
- Business case: Is the conclusion understood and appropriate key performance indicators agreed upon? How can we be sure the CMMS is delivering the expected benefits?

- Project team: Has it been formed? Do the members understand their roles and responsibilities? If they are drawn from operational staff, do they have the commitment to deal with project, not operational, issues?

Clearly, some of these topics will have been (or should have been) addressed before the RFP was issued. However, it is good practice for the implementation project manager to review them again.

Implementation Project Organization

Several different user groups are needed to successfully implement an enterprise-level CMMS. If you think about where the CMMS functions in your organization, this should not be a surprise. Maintenance certainly is front and center, but other skills and staff also need to be included from warehouse and inventory, procurement and purchasing, accounting, engineering, and project support, vas well as IT support to configure and manage the system. And that is just for a "routine" CMMS!

Figure 5.12 shows the relationship between the project teams.

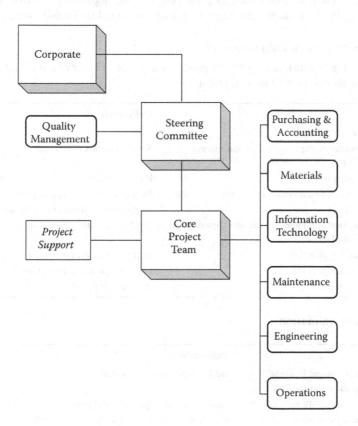

FIGURE 5.12 Typical project organization.

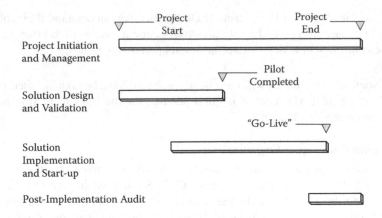

FIGURE 5.13 Typical CMMS project plan.

Implementation Plan

A CMMS implementation generally proceeds with the high-level timeline shown in Figure 5.13. Each of the project stages is briefly explained in the following sections.

Project Initiation and Management

This is an ongoing task lasting for the project's duration. The key activities and deliverables are shown in the following table.

Activity	Deliverable
Prepare services contract	
Confirm objectives, expectations, critical success factors (CSFs)	Services contract document
Finalize project budget	Project cost report
Prepare project schedule and develop for initial work	Project schedule—expanded to detail level for system configuration and validation
Define and document project procedures (reporting, change control, etc.)	Project procedures document
Detailed activity planning	Updated schedules (rolling wave)
Budget and change control	Change reports and budget/actual reports
Project kickoff meeting	Working project plan

Design and Validation

Activity	Deliverable
Define and document business processes and user procedures	Business processes document
Configure system with sample data	Conference room pilot (CRP) system ready for validation
Provide infrastructure support	Database sizing sheets, network recommendations

Activity	Deliverable
Train implementation team members	Training services/materials
Conduct the CRP	Modified system configuration and associated documentation
Complete the CRP and sign off	CRP completion document signed off by the core team members

You may find that developed and documented business processes from previous work can be excellent in helping you select the new system. However, if the implementation team is not familiar with them, conduct a review to ensure they can be configured into the system. Where there are fit problems, you will have to make some process changes (hopefully minor—underlining the advantage of having reasonable process maps for the selection teams). Of course, these will be developed in conjunction with user groups.

The conference room pilot, also referred to as a proof of concept configuration, is where all business processes and user procedures defined earlier are tested and validated. It is here that you will implement most configuration changes. The CRP environment is ideal because the data volume is low, users are knowledgeable, and the impact of configuration changes is minimal.

CRP sign-off, which documents that the system adequately meets the defined functional requirements, typically follows. It is also important to document shortcomings that can be addressed in subsequent phases to ensure that client-raised issues are tracked and managed throughout the implementation.

Implementation and Start-up

Activity	Deliverable
Detailed activity/task planning	Updated project schedule
Defined data conversion requirements	Data mapping documents
Define interface requirements	Technical design documents, test plans
Develop user training materials	Training package
Develop and implement system testing	System test plan, test results
Deliver user training	Trained users
Final conversion of production data	Converted data on target application
Verify interfaces	Interface sign-offs
Final readiness checks/define resources, etc., disaster planning	Go-live checklists, resource list, contingency plan
Go live	Production system

It is during this stage that you assemble the production system, having previously validated the base configuration, and undertake support activities, not directly shown in the previous table, including infrastructure changes (usually network, hardware, and database related).

The importance of thorough system testing cannot be overstated. It is often inadequate, causing frustration among the user community after go-live. One reason for this is that implementation team members do not have the time, inclination, or training to develop and conduct detailed test cases. Depending on staff availability, there is a good argument for bringing in fresh minds to focus on testing. While this can be expensive, in the long run it's often the cheapest alternative, particularly where there are several integration paths between the CMMS and other systems.

Postimplementation Audit

Activity	Deliverable
Monitor system	Performance reports, database tuning changes, etc.
Obtain user feedback	
Measure achievements against critical success factors (CSFs) and key performance indicators (KPIs)	Analysis of results, where available (may be time dependent)
Implement required changes, where possible	Configuration changes

After initial production operation (go-live), put in place a rigorous monitoring process to ensure that the system is technically stable (performance, availability), being used correctly, and, after a period of operation, producing business benefits. Although this is the reason that the CMMS was procured in the first place, it is often given scant attention. However, if you set up business measures at the outset of the project, they can easily be measured after an appropriate time.

APPENDIX B: INTRODUCTION

Radio-frequency identification (RFID) is an enabling technology that can radically change and improve the way that organizations track and manage assets. While RFID is still in the early stages of adoption, it is clear that it stands at the brink, poised for widespread implementation across multiple industries. As the price of RFID tags and readers continues to drop and the technology continues to advance at the rate of Moore's Law (doubling in functionality every two years), there is little doubt that RFID will soon become an industry standard. This subchapter briefly explains what RFID is and how it is changing the way some industries manage their assets.

Every so often, an enabling technology will disrupt an industry or multiple industries. RFID is on the verge of doing this, but, before we explain its impact, first we will break down the technology to have a better understanding of its components. Later we will discuss its asset management applications.

AUTOMATIC IDENTIFICATION

RFID is an example of automatic identification (Auto-ID) technology by which a physical object can be identified automatically. Other examples of Auto-ID include bar code, biometrics (e.g., fingerprint and retinal scan), voice identification, and optical character recognition (OCR) systems.

Bar codes are the most familiar, so let's use them as a point of comparison as we look at RFID.

Comparison of Bar Code and RFID Advantages

Bar Code[1]	RFID
Lower cost	Support for nonstatic data
Comparable accuracy rate	No need for line of sight
Unaffected by material type	Longer read range
Absence of international restrictions	Larger data capacity
No social issues	Multiple simultaneous reads
Mature technology with large installed base	Sustainability/durability
	Intelligent behavior

Bar Code

On June 26, 1974, Marsh's Supermarket introduced the first bar code. The first product to contain the new technology was a pack of Wrigley's Juicy Fruit gum. Some thought that the new technology would never pay off. In fact, at the time, a Midwest grocery chain executive said, "I think the industry has sold itself on a program that offers so little return that it simply won't be worth the trouble and expense."[2]

Before the bar code, inventory management consisted of manually counting and recording items on a ledger. This system was very inaccurate and time-consuming. Bar codes were invented to make the system of tracking inventory faster and more accurate.[3] This technology was very expensive to implement, but over time it has provided tremendous savings. Bar coding automated the recording process, lowered recording errors, made inventory management economical, and reduced the number of workers needed to handle inventory.[4] Bar coding was so effective that entire supply chains were adapted to embrace the technology.[5]

Scanning bar codes generally takes between 4 and 10 seconds per pallet because line of sight is required to obtain valid reads.[6] In addition, a bar code reader can read only a single bar code at a time. When you think of high-volume warehouses, which may "scan bar codes up to 25 times between inbound receiving and outbound shipping, you start to understand the magnitude that traditional bar codes often impose."[7]

Another drawback of the bar code is its inability to store large amounts of data. A bar code is able to store only enough data to allow it to be identified. This identification at a stock-keeping unit (SKU) level is not unique, however, which is an important benefit of RFID. A bar code can tell you general characteristics about a product but not specifics, such as when it expires.

RFID

"Radio frequency identification (RFID) technology uses radio waves to automatically identify physical objects (either living beings or inanimate items)."[8] A radio device called a *tag* is attached to the physical object that needs to be identified. Unique identification data about this object are stored on the tag. When such a tagged object comes within range of a RFID reader, the tag transmits these data to the reader, which captures the data and forwards them over suitable communication channels, such as a network or a serial connection, to a software application running on a computer.

RFID Tag

It is important to distinguish among the types of RFID tags: passive, active, and semiactive (also known as semipassive).

- A **passive** tag does not have an onboard power source and uses the RF signal emitted from the reader to energize itself and transmit its stored data to the reader.
- An **active** tag has an onboard power source (a battery or a source of power, such as solar energy) and electronics for performing specialized tasks. An active tag uses its onboard power supply to transmit its data—either autonomously or under system control—to a reader, typically over larger distances than passive tags. It does not use the reader's emitted RF power for data transmission. The onboard power supply enables (1) several readers to determine a tag's specific spatial location; and (2) functionality of microprocessors, sensors, and input/output ports.
- A **semiactive** tag has an onboard power source and electronics for performing specialized tasks. The onboard power supply provides energy to the tag for its operation. However, for transmitting its data, a semiactive tag uses the reader's emitted RF signal.

FUTURE AND BENEFITS

Imagine walking into a supermarket, filling up your cart, and walking directly out of the store without stopping to check out. Imagine receiving reminders from medicine bottles to take prescribed medications. Imagine a business owner forecasting items' expiration in real time. Imagine your house automatically setting a room's mood based on an individual's preferences. With the way businesses are innovating with RFID, each of these situations could be become reality.

When a customer wants something that a retailer does not have, the retailer loses a sale. RFID's greatest realized benefit in the retail market today is decreased stock outs based on longer lead time.[9] In this case, RFID enables decision makers with increased supply chain visibility. According to Paul Fox, director of global external relations for Gillette, "in the United States alone, billions of dollars are lost each year as a result of supply chain inefficiencies, with product being lost, misplaced or ordered inaccurately."[10]

RFID will enable businesses to lower operating expenses and maximize profitability by doing the following:

- Reducing inventory and shrinkag
- Lowering store and warehouse labor expenses
- Ensuring fewer out-of-stock items

According to Vic Verma of Savi Technology,[11] "the benefits of RFID in the end-to-end supply chain solution reduce costs in the following areas: inventory holding, labor, maintenance, insurance, reverse logistics, damage, and pilferage." RFID will also increase accounting accuracy. Accountants will be able to determine ending inventory levels more accurately. Also, cost of goods sold will be easier to calculate.[12]

Additionally, RFID reduces the cost of counting inventory items, making it possible to operate with fewer employees.[13] Auditors will be able to catch obsolete inventory items faster, which will allow them to value the inventory more easily. RFID also eliminates the chance of counting an item twice.[14]

Another effect on profitability when assets are enabled by RFID is service documentation and record keeping. According to a study by AMR Research, businesses that show leadership in automating their service operations are 25% more profitable than average companies in the same industry.[15] RFID enables automatic identification to build service records that can help improve service efficiency, lower costs, and position the company to gain additional service revenue.

The benefits are limitless. Now, let's take a closer look at how RFID affects asset management.

ASSET MANAGEMENT[16]

What characterizes a RFID-enabled *asset management* application?

- The asset needs to be managed through its inventory characteristics (manufacturer, model, serial number, description, configuration level, if appropriate, and storage or installation location).
- A means of identification that contains a unique identifier associated with a background database of information containing specific asset information should be attached to the asset; this tag may be a bar code, RFID tag, or combination tag.
- The location and other properties and states of this asset should be detected in real time by attempting to read the tag data on a periodic as well as an on-demand basis.

Asset management applications can tie the unique identity of an asset to its location. This can be accomplished using RFID tags; passive RFID tags will provide information about location based on when the tag was last read (zoned location or proximity), and active RFID tags can determine more refined location in real time because tags can be pinpointed using triangulation techniques (also called real-time location systems [RTLSs]).

As an example, the American National Standards Institute (ANSI) INCITS 371 standard, developed by the International Committee for Information Technology Standards, enables users to locate, manage, and optimize mobile assets throughout the supply chain. Stationary active RFID readers read the asset tags as they pass through zoned locations in a facility or yard. These data and the reader's location information are transferred then into an asset management system. Both local and global/wide-area monitoring is possible. Global asset monitoring is using satellite communication networks to link RFID systems at remote sites.

Fleet Management

Used as a fleet management tool, RFID tags are mounted on transportation items such as power units, trailers, containers, dollies, and vehicles. These tags contain pertinent data about the item by which it can be identified and managed. Readers, both stationary and mobile, are placed at locations through which these tagged items move (e.g., access controlled gates, fuel pumps, dock doors, maintenance areas). These readers automatically read the data from the tags and transmit them to distributed or centralized data centers as well as an asset management system. This system can then allow or deny a vehicle access to a gate, fuel, maintenance facilities, and so on. Thus, using the data from the tagged items and vehicles, an asset management system can locate, control, and manage resources to optimize use on a continuous, real-time basis. The data captured from the tagged items can be timely and accurate, resulting in elimination of manual entry methods, which in turn reduces wait times in lanes and dwell times for drivers and equipment.

An extension of this would be the collection vehicle diagnostic data in combination with the vehicle's unique ID to improve fleet life-cycle management. This would be an example of fleet monitoring and management.

Benefits of Asset Management

- **Better use of assets**. The ability to locate, control, and use an asset when needed allows fleet asset optimization.
- **Improved operations**. Accurate and automatic data capture coupled with intelligent control leads to better security of controlled areas, provides proactive vehicle maintenance, and enhances fleet life.
- **Improved communication**. Real-time, accurate data provide better communication to customers, management, and operation personnel.

Caveats

Initial investment may be required for hardware and infrastructure. Cost increases with the fleet size, the number of data capture points, and the amount of custom implementation services required. In addition, for geographically dispersed operations, wide-area wireless communications such as satellite communication may be needed, thus increasing the infrastructure cost.

Implementation Notes

Semiactive, read-only, and read-write tags with specialized onboard electronics (e.g., to indicate the status of a data transaction) are generally used. Most importantly, such a tag can be integrated with a vehicle's onboard sensors to relay critical vehicle information such as fuel level, oil pressure, and temperature to a reader. The fleet management system uses these data to determine proactive maintenance on vehicles, resulting in a longer fleet life.

APPLICATIONS

The applications of RFID are widespread; this section addresses those that pertain to asset management. The entities most active in implementing RFID technology are Wal-Mart and the U.S. Department of Defense (DoD). Other recognizable companies beginning to implement RFID technology are Target, Albertsons, Best Buy, Tesco, K2, DOW Chemical, Metro, and UPS.[17]

Retail

Wal-Mart has begun a rollout of passive RFID at the pallet and case level in many of its distribution centers and retail locations. Testing began in January 2005 at three of its Texas warehouses using 137 suppliers and 150 Walmart stores. The company plans to reengineer its supply-chain management process with RFID technology giving the company, which is already considered by many to have the strongest supply chain in the world, even more of an advantage over its competition.

Department of Defense

Among the first to implement RFID in the supply chain, the DoD issued an RFID policy affecting many of its 43,000 suppliers. The policy includes passive as well as active tags. In fact, "every container shipped to Afghanistan and Iraq includes an RFID tag that helps improve the military's ability to track supplies and their condition."[18] Having spent $100 million to date, the DoD expects to fix RFID tags on each individual item in the future.[19]

U.S. Food and Drug Administration

In addition to the DoD, the U.S. Food and Drug Administration (FDA) is encouraging use of RFID by U.S. drug suppliers in 2007 (1) to prevent counterfeit drugs from infiltrating the pharmaceutical supply chain[20]; and (2) to control the diversion and theft of drug shipments. The industry loses between $10 billion and $30 billion annually to counterfeit drugs.[21]

Drugs can be authenticated using RFID's ability to provide unique serialization and identity. Some companies, such as Pfizer, Purdue Pharma, and GlaxoSmithKline, are already using RFID with their high-risk drugs.[22] The technology shows a lot of promise for tracking the pedigree of drugs and may be an extremely useful tool for fighting counterfeiting and ensuring drug safety.[23]

Automotive Industry

The automotive industry uses RFID to track materials as they move through the supply chain. The result of improved visibility enables improved just-in-time inventory capabilities and therefore lower levels of spare parts inventory.

Forecasting is a particularly challenging issue for most auto companies, stemming from the large number of automobile and feature configurations available. By using RFID tags, auto companies may be able to reduce the amount of time it takes to produce and deliver a car of a particular configuration to an end customer.

The manufacturing process can be improved by RFID also. Faster than bar code technology, RFID tagged parts enable manufacturers to locate needed parts and to know when a part's quantity is low. Using RFID technology for an auto manufacturer is likely to enable faster time to market and lower inventory costs.

Personnel Emergency Location

Imagine a fire breaks out in a building of 600 employees. How would rescuers know if everyone had escaped? Even if RFID is able to determine that three people are still inside, how would the people be located?

Active RFID tags combined with location awareness and safety software can provide location identification to a high degree of accuracy (using specialized technologies like ultra wideband). Imagine the power of knowing where each employee is located and being able to grant or deny access to authorized users. In the event of an emergency, RFID may even save lives.

Container Shipping

RFID is revolutionizing the global supply chain by enabling status updates on cargo containers as well as near real-time, wireless global access to their content's status and location.

In general, the RFID-enabled container—offered by IBM, Lockheed Martin, and General Electric—has the potential to turn every container into a moving virtual warehouse. Imagine having a container contact you with the status of its contents as soon as it arrives at its destination.

As a result, supply-chain stakeholders can receive and respond to product data. The benefits spread across stakeholders from consignees to operation managers to customs' authorities: lowered costs to market, improved control of just-in-time delivery of components for assembly, and reduced warehousing overheads.

RFID enables the required conditions from source to market to be monitored, maintained, and verified. The sharing of this diverse information on containers and their contents is truly changing the global transportation industry.

DATA COLLECTION STANDARDS

There is a lot of activity in the RFID industry around data collection standards. Different standards bodies and organizations are focused on resolving and standardizing different aspects of the technology. A list of the major standards organizations involved in data collection is listed as follows for reference:

- American National Standards Institute (ANSI)
- Automotive Industry Action Group (AIAG)
- European Article Numbering Association International, Uniform Code Council (EAN.UCC)
- EPCglobal leads the development of industry-driven standards for the electronic product code (EPC) to support the use of RFID in today's fast-moving, information-rich trading networks.[24] Their approach is recognized as smart and proactive, since the number of companies using RFID is expected to explode, including collaboration with business trading partners. EPCglobal exists to prevent a massive reconstruction effort to get down the standards to where trading partners can collaborate and share information (see http://www.epcglobalinc.org)

- International Organization for Standardization (ISO)
- Comité Européen Normalisation (CEN; European Committee for Standardization)
- European Telecommunications Standards Institute (ETSI)
- European Radio Office (ERO)
- Universal Postal Union (UPU)
- American Society for Testing and Materials (ASTM)

CHALLENGES

As with any emerging technology, there are challenges. RFID's largest challengers are information privacy and accuracy of reads. Despite these challenges to RFID's immediate, widespread adoption, industry experts say there is no reason to stall RFID implementation projects. "Taking appropriate measures, however, will mitigate risk and preemptively address any fears that may be raised by employees or consumers."[25] Rather than reflect radio waves, liquids and metals absorb them, making it generally more difficult to read tags close to liquids or metals. Innovative manufacturers are countering this limitation by extensive research to design an easy-to-read tag in these special environments.

Privacy for the consumer and the corporation is one of the most discussed topics, and developers are working on ways to secure sensitive information. In the effort to mitigate RFID concerns among retail consumers, it is recommended that companies who use RFID tags do the following:

1. Inform customers of the presence of RFID tags in purchased items.
2. Obtain customer permission to use the tag data.
3. Destroy (optionally) the tags before customers leave the retail environment.

To enhance consumer privacy, the "Clipped Tag" was developed and could possibly allow RFID tags to be used to tag individual consumer items. "The 'Clipped Tag' has been suggested for individual items in order to enhance consumer privacy. It allows the consumer to tear off a portion of the tag in order to transform a tag that may be read at a range of 10 meters to one than can only be read at a few cm. The use of these tags puts privacy protection in the hands of the consumer, provides a visual indication that the tag has been modified, but makes it possible for the tag to be used later for returns, recalls, or recycling."[26]

It could take years before every risk is mitigated completely, but the technology in its current form is still widely successful.

CONCLUSION

After reading this subsection on RFID, you should be able to understand and evaluate the merits of the technology in asset management. RFID stands, as bar codes did in the early 1970s, on the brink of mass implementation. It will impact and transform business by providing greater asset visibility and improving asset management. As improvements are made and the technology drops in price, the market will innovate ways to integrate RFID into many, cross-industry applications. Marketplace leaders are already embracing RFID—the technology that stands on the cutting edge of making significant improvements in asset management. We stand at the beginning of the "Internet of Things."

ENDNOTES

1. Lahiri, S., *RFID Sourcebook,* Upper Saddle River, NJ, IBM Press, 2006, p. 128.
2. Kinsella, B., Delivering the Goods: RFID Conveys a Sharp Signal to the Supply Chain World to Come, *Industrial Engineer,* 37(3). 2005. 24+.
3. Ibid.
4. Ibid.
5. Ibid.
6. Ibid.
7. Ibid.
8. Lahiri, *RFID Sourcebook,* p. 1.
9. RFID News: Now Wal-Mart CIO Rollin Ford Rebuts *Wall Street Journal* Article. Editorial. *SCDigest,* March 8, 2007, 1 par. April 4, 2007. http://www.scdigest.com/assets/newsViews/07-03-08-5.cfm
10. Atkinson, W., Tagged: The Risks and Rewards of RFID Technology, *Risk Management,* 51(7). 2004. 12+.
11. Verma, V., Buzz on RFID Gets Louder, *Insights,* Vol. 3(3), Lockheed Martin.
12. Michael, S.L., and H.E. Davis, Radio Frequency Identification: The Wave of the Future; as Bar Codes Give Way to RFID Tags, Accounting for Inventory Will Go High-Tech, *Journal of Accountancy,* 198(5). 2004. 43+.
13. Ibid.
14. Ibid.
15. Increasing Profits and Productivity: Accurate Asset Tracking and Management with Bar Coding and RFID, 2007, Zebra Technologies Corporation, April 4, 2007. http://www.zebra.com
16. Lahiri. *RFID Sourcebook,* pp. 74–77, Direct excerpt.
17. Kinsella. Delivering the Goods.
18. Ibid.
19. Ibid.
20. Koroneos, G. FDA Opens Door to RFID Packaging, *Pharmaceutical Technology,* 28(12). 2004. 17+.
21. Kinsella. Delivering the Goods.
22. Koroneos. FDA Opens Door.
23. Ibid.
24. Overview, 2007, EPCglobal, April 4, 2007. http://www.epcglobalinc.org/about/.
25. Parker, B., Analyst Viewpoint: Take Appropriate RFID Precautions, March 28, 2007, Manufacturing Insights, April 12, 2007. http://www.rfidupdate.com/articles/index.php?id=1327.
26. Moskowitz, P., S. Morris, and A. Lauris, A Privacy-Enhancing Radio Frequency Identification Tag: Implementation of the Clipped Tag, Proceedings of Fifth Annual IEEE International Conference on Pervasive Computing and Communications Workshops, PerCom Workshops, PerTec 2007, pp. 348–351, IEEE Computer Society, 2007.

6 Materials Management Optimization

Don Barry and Eric Olson
Original by Monique Petit

CONTENTS

6.1 INTRODUCTION

For many organizations, maintenance, repair, and operations (MRO) materials are not always available when and where they are required, despite significant expenditures to stock them locally and despite heroic efforts by inventory managers and procurement to meet unpredictable demands.

When looking at materials management that supports a maintenance organization, the complexities can range from the very simple to the very complex. As an example, are we supporting the functionality for assets on one campus or across

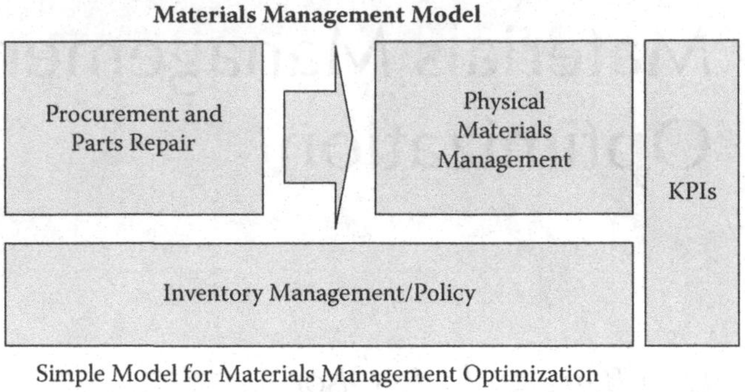

Materials Management Model

Simple Model for Materials Management Optimization

FIGURE 6.1   Materials management model.

many and perhaps many in diverse regions or countries? Other considerations and complexities can include whether we expect to manage rotable parts—also known as rotating assets (parts identified to be repaired and returned to inventory once repaired), parts warranties, or vendor-owned inventories.

To organize some of the areas we expect to focus on in this chapter we will use the simple, high-level model shown in Figure 6.1. This model shows that an optimized macro view of materials management should have at a minimum a focus on the following:

- Key performance indicators for material management to track how this infrastructure supports the maintenance of the asset functionality effectively
- Inventory management policy that supports all of the materials management areas and can ensure that the key performance metrics are tracked and tracking to plan
- Physical materials management set of activities that manage the receiving, physical storing, and distributing of parts
- Procurement of parts for inventory and to support work orders as well as the repair of rotable parts designated as serialized reparable parts that go to inventory once repaired

Availability of maintenance parts (or spare/service parts) on a timely basis is critical to the successful execution of a maintenance plan. They are sourced, placed, managed, and used to support the sustainability and life cycle of the expected functions of valued assets. Overall equipment effectiveness (OEE, defined in Chapter 4) will suffer if critical spare parts are not available for either planned or unplanned maintenance. In a successful maintenance operation, spare parts are available as and where required to maintain the function the asset is expected to perform (e.g., equipment used in a critical manufacturing process or a power generator in a plant).

In this chapter, we will discuss fundamentals and some of the complexities of sourcing and delivering MRO materials. Before that, we introduce the dynamics of the materials management life cycle and how it supports the asset life cycle. Figure 6.1 displays the high-level macro view of a materials management set of processes. Working from left to right on this diagram, we will introduce some of the metrics that have been successfully leveraged in an optimized leading materials management organization.

Materials management begins with understanding the demand for the materials that will be required for MRO for the planning period. Once the specifications, quantities, and timing are known, procurement can find the best suppliers based on multiple factors that include service, quality, and total cost. If suppliers could bundle all the necessary MRO materials for the specific maintenance task and deliver them from their shelves directly to the job when needed, there would be no need for local inventory. In practice, the diversity of materials required and irregular demand usually justify local inventory, whether in central or satellite warehouses, in depots near certain equipment, or in service vehicles. The number of materials, multiple stocking locations, and irregular demand present significant complexity for inventory management. Improvement in MRO management is a balancing act between the availability of parts (service) and the cost of making them available.

Success in materials management optimization requires data, including material specifications, bills of materials for specific maintenance tasks, historical usage, inventory integrity counts by location, lead times, order quantities, and logistics costs. Of equal importance, the consequences of a material not being available when needed must also be known. A rational process to manage and mitigate business risks related to spare parts availability must involve operations, maintenance, and finance. For many organizations, the challenge of assembling the relevant data and making sound risk-based decisions for thousands of materials each year, one material at a time, can be daunting. Fortunately, these decisions can be made easier by information management systems, by grouping materials for similar treatment, and by judicious application of the Pareto principle. For these organizations, getting procurement and inventory control for the limited number of items that represent the majority of value is the obvious way to get started.

While the path to improvement in MRO is a journey, it becomes easier when it merges with that for maintenance excellence in general. When equipment is delivering the expected functions and capability for its projected service life and is requiring maintenance only at expected intervals, there is generally enough lead time to have suppliers bundle all the necessary MRO materials and deliver them from their shelves directly to the job when needed. Conversely, when equipment breaks often and without warning, spares and spare materials may have to be kept close at hand. Excellent asset performance reduces the demand for spare parts and the requirement for a substantial local inventory.

For organizations currently carrying large MRO inventories to support maintenance of assets that are failing in service regularly, improving materials management can seem difficult or impossible. All elements of effective change management, the people side of things, are required to supplement the improvement of processes for strategic procurement and inventory management in support of maintenance excellence, including a robust and well-communicated case for action, participation in analysis and solution design by all affected stakeholders, performance measurement and intentional performance management, and visible and steadfast leadership from executives and senior managers.

In many organizations, a lot of money is spent on materials management, and it's often poorly controlled. Usually, there isn't enough consideration given to how a part will be procured or replenished, the best inventory placement, how to manage surplus, or how to deliver parts so that the technicians can be confident of reliable parts delivery. In addition, inventory reserves for scrap and inventory surplus management are often misunderstood and underfunded. As a result, inventory is often not optimized, and service levels from suppliers and from inventory is low.

The fundamental rationale for storing tens of thousands of part numbers or stock-keeping units (SKUs) on site is to reduce the mean time to repair for critical equipment. The farther an operation is from parts and supplies distribution centers, the more safety stock could be required and the more critical inventory optimization becomes. The position of an organization on the "Innocence to Excellence" scale shown in Figure 6.2 is a visual representation of how much the inventory is rationalized, inventory control is optimized, and stores purchases are strategically sourced. The lower the position on the scale, the more opportunity there is to reduce spending on MRO and to improve OEE and service to users—the main job of maintenance, after all.

While manufacturing inventory is often kept at locations determined by use in the manufacturing process, service parts inventory is more likely to be stocked for proximity to the asset base it is supporting. Maintenance parts inventory can be kept in multiple stockrooms across the company, in service vehicles, in depots, as well as at

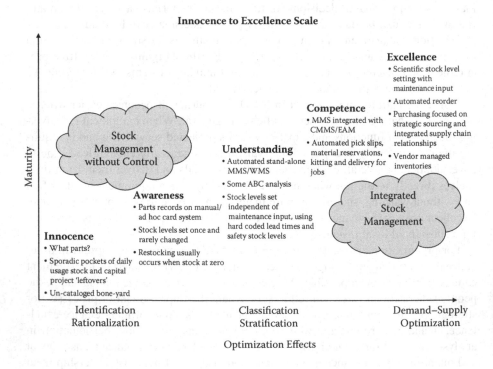

FIGURE 6.2 Innocence to excellence scale for spare parts delivery.

the asset locations—placement in these locations is an optimization issue, balancing proximity (service) with cost.

In a manufacturing environment, inventory planning is based on sales forecasts and orders. One tactic to optimize costs in a manufacturing environment is to use these forecasts to manage a *just-in-time* product inventory. Maintenance parts are stocked for often unpredictable equipment failures driven by risk-averse management approaches. Stocking for OEE in a supply chain is often referred to as the *just-in-case supply chain*.

6.2 ASSET MANAGEMENT LIFE-CYCLE

In maintenance, the availability of a spare part by itself does not completely fulfill a customer's request. A craftsperson or tradesperson is required to perform the maintenance action and install the part. The spare part, the technician with the right qualifications, and the necessary technical documentation all need to be brought together ("rendezvous") to satisfy the maintenance demand efficiently.

The asset criticality, configuration, component makeup (bill of materials), and planned service strategies all contribute to a leading parts inventory planning strategy. To accomplish this effectively, the materials management organization should understand all the attributes of the maintenance service strategy and expected maintenance tasks (planned and unplanned). This planning approach applies before the asset is installed and during its operation and includes understanding when the asset is to be decommissioned so that surplus spare parts can be sold or scrapped as appropriate.

Compared with a typical manufacturer's supply chain, the (spare parts) materials management challenges are unique in many ways. Effective inventory strategy and placement require an aligned understanding of the assets they support. A simple list of some of the areas of focus includes the following:

- The service life-cycle requirements of the asset/systems to be maintained
- The service demand being scheduled or unscheduled (planned or unplanned)
- The criticality of the asset/systems to be maintained
- The downtime tolerance
- The geographical location of the asset/systems to be maintained
- The qualification of the workforce required for maintenance
- The possibility of repair
- The location of repair
- The asset configuration and its associated bill of materials (BOMs)
- The parts that make up the BOM
- The requirement for reverse logistics

The ideal time in the asset management life cycle to develop a sourcing and stocking strategy is right at the asset planning stage (Figure 6.3). If we know what the functional expectations of the asset are and the maintenance characteristics of the asset's components, we can develop a maintenance strategy, or a set of mitigating tasks that will have parts associated with them. This can be done through a basic

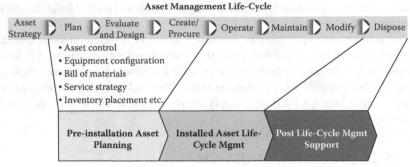

Materials Management Life-Cycle

FIGURE 6.3 Asset management life cycle.

reliability-centered maintenance (RCM) initiative during the asset planning phase of its life cycle (see Chapter 8 on reliability). With this information we can determine a basic parts requirements list and an understanding of how the part may be needed (i.e., corrective maintenance vs. preventive maintenance) and where we should stock parts along with the quantity to fulfill the expected maintenance tasks. At the other end of the life cycle, if we know specific assets will be decommissioned, we can dispose of its spare parts at the same time. For maintenance organizations that have existing assets, an RCM initiative can provide a similar result at any time; however, the best time to do this assessment is during the asset planning phase.

6.3 PERFORMANCE MEASUREMENT AND MANAGEMENT

At a high level, key performance indicators (KPIs) that would serve as true indicators of a leading maintenance parts operation could include the following:

- Parts availability
- Parts acquisition time
- Systems availability
- Distribution quality
- Parts quality
- Parts costs
- Inventory turnover (annual usage vs. average annual inventory)
- Inventory vs. asset value (value supported assets produce in one year vs. average annual inventory)
- Inventory reserves

Parts availability would ideally be measured for every craft or planner's request for parts. Leading organizations would measure how long the craft waited for parts or the time the asset availability was affected due to waiting for parts. Systems must be "available" for a craft to order a part; to confirm prior orders are reserved, picked, or staged; or at a minimum to confirm that the parts are in stock and available to be used. The warehouse would be measured on inventory integrity and distribution

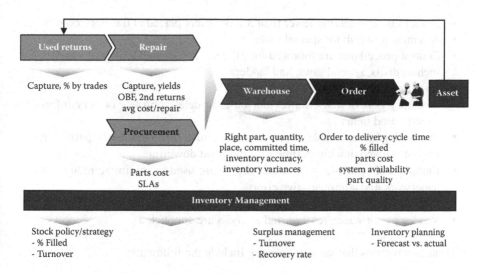

FIGURE 6.4   MRO process sub-KPIs.

quality (right part, right quantity, right place, and within the committed time). The overall process could be measured on the cost of the part versus "new" or "street value" and the confirmation that the part actually performed as expected (parts quality).

These metrics refer to the health of the overall maintenance parts process. Comprehensive performance measurement typically requires submetrics for each of the supporting processes to ensure that the overall process will meet its objectives.

Many metrics can be developed to manage the typical maintenance parts process. Figure 6.4 provides a small example of additional measurements that have been used to support the ultimate leading metrics previously provided. An example of KPIs for the repair of returned defective parts and rotables would be the following:

- Confirmation that all reparable parts are captured in their process
- Monitoring the repair yields of the received defective part
- "Out-of-box failure" (OBF) would monitor the new defective rate reported by a craft for a refurbished part or rotable
- Logging parts that have been returned for repair for the same symptom a second time in the past year
- Cost of repair versus the cost of a new purchase

6.3.1   EXAMPLE OF SOME KPIS FOR SPARE PARTS SUPPLY CHAINS

Initial Spare Parts

- Turnover/asset type
- ISP list quality

Additional characteristics of a successful material management and procurement organization include the following:

- Stock outs representing fewer than 3% of orders placed at the storeroom
- A central tool crib for special tools
- Control procedures are followed for all company-owned tools and supplies such as drills, special saws, and ladders
- Inventory cycle counts are conducted
- Inventory is reviewed on a regular basis to delete obsolete or very infrequently used items
- Purchasing/stores is able to source and acquire rush emergency parts that are not stocked quickly in time to avoid plant downtime
- Blanket and system contracts and orders are used to minimize redundant paperwork and administrative effort
- Stores catalog is up to date and is readily available for use
- Vendor performance reviews and analysis are conducted

Some key metrics that support delivery include the following:

- Inventory variance and accuracy
- Right part and quantity shipped to the right place within the committed time (distribution quality)
- Dock to stock cycle times
- Order to delivery cycle times
- Percentage of lines filled (parts availability)
- Ease of use to order (system availability)
- Quality of the part (not damaged by handlers or shipping)

6.4 PHYSICAL MATERIALS MANAGEMENT

The maintenance parts need to be ordered by the tradesperson, they need to be purchased, and the inventory needs to be managed. In this section we discuss briefing some key points around the physical logistics of materials management and handling.

Traditionally, tradespeople have often turned to stores not only to collect that part but also to search for the "right" part. This has been due largely to a poorly maintained (or completely lacking) catalog or parts book. The stockroom has often been the first place the trades interface with the asset management process. From the trade's point of view, the idea of service levels from a stockroom is simple. When we come to the parts counter to order and receive our parts, the criteria for materials management quality service is easy to understand: "We want the part when we want it."

Six Rs define complete optimization of the sourcing and delivery of spare parts and materials to maintain, repair, and operate equipment assets:

- Right parts
- Right quality
- Right quantity
- Right place

- Right time
- Right price

Clearly, this cannot be achieved without a well-defined and executed inventory management and procurement set of processes; however, the physical logistics process also needs to be well defined and executed, and the parts have to be physically handled to complete the six Rs. For example, a critical step to being able to deliver a part when requested is to ensure that the part is receipted correctly into the inventory system when shipped to and received at the dock. Receiving is pivotal to inventory accuracy and control. Inventory integrity will never be right if the part receipted is incorrectly acknowledged.

There are several key events in receiving:

- Invoices are paid by matching item quantities and attributes to the purchase order and then confirming with the accounting systems.
- The item as ordered is confirmed.
- The item is marshaled for end use (inspection, storage, delivery).
- All variances or inconsistencies are recorded and monitored and used to track vendor performance.

A major consideration is how much to centralize receiving. You must determine whether trained personnel should handle all receipts at a designated location or whether end users should be in charge of receipts of noninventoried purchases. While either decision has merits, you must review internal factors:

- What is the training/motivation level of the staff? Any employee who receives goods for the organization must implicitly agree to be rigorous, prompt, and accurate.
- Who will actually perform the task? Should the engineer at site receive the goods or the warehouse person? Who should be fiscally responsible for the task?
- Does centralized receiving of all goods increase the lead time to the end user? Is this a receiving function or an internal communication or systems limitation?

Another example is that best-practice warehouses or storerooms must have security that is appropriate for the type of item stored in terms of its deterioration or risk of theft. Best-practices warehouses are intended to provide maintenance personnel high levels of service; for example, they may have procedures in place to alert maintenance personnel regarding receipt of their materials. Maintaining inventory involves identification, storage, and auditing. Inventory audit is the process of confirming the absolute inventory accuracy. Proving inventory accuracy is the equivalent of an operations quality control program. You must instill and maintain user confidence that whatever the system says is in the warehouse is actually there.

Service level (the frequency of fulfilled orders that is tolerated) is a measure of inventory placement, procurement service levels, and inventory integrity and control performance. Reduced lead times, inventory accuracy, and the ability to find the right part all contribute to parts availability service levels. Establishing different service levels for each commodity reduces costs while ensuring the availability of critical items. Fulfillment misses can be broken down into two categories:

- Not stocked indicates parts that were not intended to be stocked at the local stockroom.
- Out of stock indicates parts that are normally stocked in the local stockroom but are currently not available in the stockroom.

For inventory control service levels, traditional organizations annually "count everything in a weekend" and then, as a result, adjust stock balances and value. This method is inefficient, often inaccurate, and usually done under strict time and resource constraints.

Perpetual stock count or cycle counting is a better alternative. You set up numerous counts, typically less than 200 items, to be counted at regular intervals. Over a year, all inventory (at least A and B class) is counted at least once. The benefits of cycle counting are as follows:

- Counts are more accurate since the number of items is relatively small and variances can be easily researched.
- Stock identity is validated against the description and changes noted, reducing duplicates, increasing parts recognition efficiency, and helping standardize the catalog.
- Stock location is confirmed.
- Accuracy levels (dollar value and quantity variances) can be used as a performance measure.

6.5 UNDERSTANDING INVENTORY MANAGEMENT DYNAMICS

An effective plan to ensure that MRO parts will be available when needed ideally includes a holistic understanding of the parts service requirement dynamics of each asset and the materials management infrastructure that you have to support this need. When looking at the simple materials management model introduced earlier in this chapter (Figure 6.1), the responsibility to ensure that this infrastructure will support the business need falls into the inventory management and inventory policy area.

The process typically begins with an assessment of what maintenance tasks will be done during the asset management planning period and what parts will be required. For a single part (e.g., a gasket), materials management would like answers to the following questions (and others):

- What is the associated task (e.g., overhaul a pump)?
- When will the task be done? Is the timing predictable with confidence and precision?
- Where should the part be stored?
- What is the lead time for delivery to the job site?
- What is the reliability of delivery within the required timeframe?

- What is the economic order quantity for the designated stockroom?
- Is the cost of shipping significant?
- Have we established a supplier of choice for the material (e.g., gasket)?
- Can the material be bundled by the supplier with some or all of the other parts required to complete the task?

With that information, procurement decisions, including when to order, and inventory decisions, such as stocking location and quantities, can be made rationally with lead time to optimize service and cost. In practice, this information is usually not available on a comprehensive basis. However, by taking a systematic approach, using the information listed on high-cost items first, the total expenditure for MRO parts can be managed effectively.

Demand for many MRO materials is random. Except for parts used for preventive maintenance (notably for time-based replacement and overhaul), most of the items in the MRO stores are used on an irregular basis. In contrast, the consumption rate of tires per automobile manufactured or catalyst used per barrel of oil is highly predictable. For MRO goods, this means that it can be more costly than effective to use automated reorder points and a high economic order quantity (EOQ) according to traditional materials analysis. An exception would be for an extremely high install base and therefore high potential for the same randomly failing component; using EOQ theory for these parts would be a practical and leading exercise. For MRO materials that have random or "lumpy" demand, manual intervention in decisions about stocking quantities and location is appropriate for those with high value or high potential for causing downtime or other unacceptable risks.

6.5.1 LOCATION AND DELIVERY

Deciding what to stock where can become very complicated for many maintenance organizations. Figure 6.5 suggests that in some cases the organization's many stockroom locations may offer both the opportunity and challenge of multiple choices for where to best stock the part. As Figure 6.6 shows, the choices range from the suppliers' shelves to with the equipment or asset itself.

As one option, leaving materials on suppliers' shelves (and on their books) until required, which is the effect of ordering directly to job, offers several significant advantages. If an order is placed to supply parts for a scheduled maintenance task, the specifications and quantity are generally well known, so there is less of a tendency to order too much, potentially building inventory with the surplus, and less requirement for returns and substitutions. If the order can be kitted, meaning all the parts required are assembled by the supplier or are in a reserved receiving area on site, the need to place them physically in bins and later retrieve them can be obviated. Kitting usually makes maintenance execution more efficient, since little or no time is lost looking or waiting for the "one last (necessary) part." Just-in-time ordering reduces the average value of inventory, a key indicator of materials management effectiveness, provided service levels meet or exceed requirements.

Most organizations decide to maintain inventory on site, often in a central warehouse. Spare parts inventory is usually expensive to purchase initially and expensive

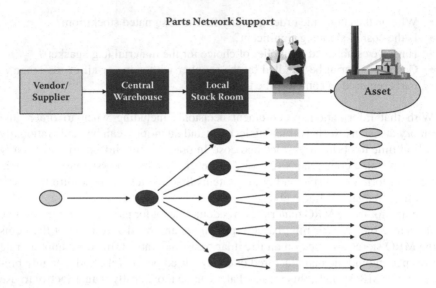

FIGURE 6.5 Example of an MRO inventory network.

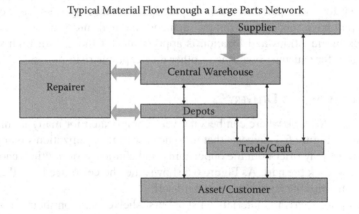

FIGURE 6.6 MRO inventory network.

to maintain—the carrying cost of inventory typically ranges from 20% to 30% of book value per year. Many organizations, especially capital-intensive industries such as mining or chemicals manufacturing, find that they have stocked items that are seldom required—items that sit unused in the warehouse for years. Obsolescence, including electronic components that are past their "use before" date, and spare parts for equipment that has itself been removed from service can further reduce the effectiveness of inventory spending. Despite these cost factors, having inventory locally available may be justified if demand is unpredictable and the consequences of a stock out are significant. Some parts and materials are also used frequently, in repair or maintenance of several assets, for which significant efficiencies in ordering and handling are possible. When these items are low unit value (e.g., gloves, studs, and nuts) they should be ordered in economic quantities and handled as efficiently as possible.

Satellite warehouses, tool cribs, service vehicles, or a craftsperson's workstation may be the location of choice for similar reasons. Gains in the efficiency of maintenance work execution can outweigh the incremental costs of placing and replenishing inventory of spare parts in these places.

Generally, adding locations to the MRO parts network will add cost that must be justified by improved service and, ultimately, enhanced maintenance efficiency. Failure to account for potential downtime, waiting time, travel time, and inconvenience may lead an organization to rely too much on just-in-time ordering or on a central warehouse. Off-system or "squirrel inventories" will proliferate if needed parts cannot be obtained in a timely and convenient basis.

One of the key areas in the MRO supply-chain focus shown in Figure 6.6 is inventory management. Important elements of inventory management include the following:

- Demand management
- Service-specific forecasting algorithms
- Part criticality
- Inventory effectiveness
- (Multiple replenishment) inventory echelon management
- Automated replenishment plans
- Inventory surplus management

The best organizations determine demand characteristics and develop a way to leverage automated replenishment to ensure that they have the parts where they need them. Leading organizations will work with the maintenance planners to confirm when planned maintenance is scheduled and look to leverage their suppliers' inventory rather than their own so that they can keep inventory carrying costs down and service levels up for the random demand and typically urgent needs. This may result in an inventory stockroom network and tailored inventory placement within the network. In addition, leading MRO operations have service-level agreements (SLAs) with their suppliers to ship planned parts requests directly to the work order or craft so that the stockrooms quite often carry only minimal inventory for planned work. Figure 6.6 provides an example of an inventory network diagram for a large organization with multiple tiers of maintenance parts.

System solutions exist to provide inventory planning based on the true cost of stocking each part, including not only what it costs to stock a part but also the costs associated with not having a part when it is needed (i.e., stock out). Such planning would optimize placement of warehouses and field stocking locations, locally by country, or globally. It could define where sites need to be to support critical asset availability and where existing sites are no longer needed. This type of planning would optimize the target stock level for each part number and site and provides intelligent logic to maintain the target stock levels through replenishment and rebalancing of inventory.

Advanced scenario modeling tools exist to provide multiple "what-if" responses, allowing maintenance organizations to fine-tune their service levels on a company-wide scale. Inventory planning system solutions can include real-time, Web-based interfaces that provide an up-to-the-minute snapshot of parts availability and materials requirements. Within these tools, simple alphabetical inventory categorization can be used to help manage inventory placement and stocking strategies. Figure 6.7 provides examples of how some organizations have interpreted MRO parts within an alphabetical or three-level context to work for them. Once defined, all of these strategies can be automated in a system managed replenishment program.

Stacking Strategies	ABC	Active vs. Insurance Stock	Criticality Stocking	Optimization
Level 1	• Top 80% used items by $ value • Less than 10% of stocked items	• Top used parts (typically 2 usages in 3 months) • Less than 10% of stocked items	• Level 1 is supported through either of ABC or Active Level 1 process • Less than 10% of stocked items	• Low value parts that are deemed to be used in asset life-cycle • Balanced with stocking and expediting costs
Level 2	• Next 15% used items by $ value • Less than 20% of stocked items	• New part in past year or at least one usage in past year • Less than 20% of stocked items	• High criticality, high usage parts not covered in level 1 • Can be 20% of parts stocked	• Cost effective stocking of parts expected to be used once a cycle (i.e., year) • Can be 80% of stocked items
Level 3	• Bottom 5% used items by $ value • Can be less than 70% of stocked items	• Parts deemed to be stocked "Just in case" • Can be 70% of stocked items	• Lower criticality parts and lower usage parts • Can be 70% of stocked items	• High value low usage parts stocked, often at a consolidation center
Comments	• Often manual initial stock process with min/max support	• Often scientific intial stock process with min/max support	• Can have many levels of criticality and echelon support depending on the support network	• Considers all costs/ impacts in stocking optimization calculations, by network location

Note: Parts deemed as "critical" or required for "insurance" typically part of level 3.

FIGURE 6.7 MRP strategies for MRO.

6.5.2 Inventory Optimization Assessment Process

The objective is to optimize your stocking decisions by balancing two conflicting cost drivers: stocking materials to minimize stock-out costs versus reducing ownership costs.
Using this basic method, you do the following:

- Identify all MRO (e.g., procurement, repair, returns) sources.
- Identify goods that need to be stocked.
- Develop new and efficient ways of dealing with goods that should not be stocked.

The three steps of inventory optimization are analysis, evaluation, and optimization.

6.5.3 Analysis: Identification and Rationalization

The first step in optimizing inventories is to analyze current inventory sources and MRO practices. This will help you develop fundamental processes to create a strategy to reduce costs and optimize stocking decisions.

Inventory also includes items that are not held in a warehouse. Often, materials are purchased directly and stored on the shop floor or in designated end-user laydowns (local convenient storage areas in the plant). These inventories are kept for numerous reasons, most often distrust with inventory management, but it's an inefficient practice. It can mean poor stock visibility, inappropriate charges, no assured adherence to specifications or loss protection, and excessive on-hand quantities. All of this is costly to the organization.

You want to consolidate this inventory with all other types into a centralized inventory information source and manage it accordingly. This will help to reduce the real MRO spend and ensure material is available.

The basic tasks performed at this point are as follows.

6.5.3.1 Task 1

Identify the entire inventory within the organization, including items that are "off the books":

- List each item's supply and usage date by referring to the inventory information sources.
- Identify all inventory held outside the warehouse, including satellite shops, scrap yards, lay-down areas, lockers, and squirreled inventory.
- Construct matrices that segment the inventory according to such criteria as value, transaction frequency, criticality, and likelihood of being stolen, using tools that perform alphabetical analysis (Figure 6.7).
- Ensure that all inventory items are uniquely identified and adequately described and that all components are currently in use.
- Assess common functions and uses across operations, and identify and eliminate duplicate part numbers and sources.

6.5.3.2   Task 2

Develop a strategy for rationalizing inventories:

* Identify volume and location of material held as inventory.
* Establish a suitable stock control system, based on inventory size and distribution, transaction volumes, and the integration you need for procurement and work management systems.

6.5.4   EVALUATION: CLASSIFICATION AND STRATIFICATION

Make stocking decisions first on a commodity level and then, where warranted, at the individual item level. This ensures not only that the right amount of inventory is available but also that the right types of inventory are stocked and controlled.

To make the appropriate stocking decision, partition the inventory into segments and apply stocking and sourcing strategies to each unit (Figure 6.8). You will be able to estimate cost savings using performance measures or industry KPIs, as described in Chapter 3. Also do the following:

* Establish transactional stock control and auditing processes to get a clear stock position and set a control framework for each item.
* Determine stock-level requirements (using statistical stock models such as EOQ, if demand is predictable, or input from the end user if demand is highly variable, seasonal, or varies other than with time) to set optimum levels for individual items.
* Determine how service levels will be calculated and applied.

It should be noted that a more detailed version of an alphabetical or three-level inventory system for maintenance parts (the "M" in MRO) is shown in Figure 5.7.

From Figure 6.8, you can see that several factors apply to each cell on the matrix, representing activity level (usage volume and frequency) and type (why the item was purchased) for each commodity.

If the item you are evaluating is "A" class (the 10–20% of the inventory that accounts for 80% of the value or spend) and was purchased for operations, you could use the following strategies:

* Purchasing strategies: contracts
* Inventory strategies: medium to large quantities, medium amount of stock, management (if stocked)
* Planning horizon: none
* Impact to downtime: none to maintenance; may be critical to daily operation of organization

6.5.5   INVENTORY CATEGORIZATION AND OPTIMIZATION

You now have an optimal stocking decision for each commodity and maximum availability (service levels) at minimum cost.

		M	R	Ov	Op
Inventory Classification	**A**	• Contracts • Large quantities highly managed • Short planning horizon • Low impact	• Small purchase/ contracts • Small quantities, highly managed • Short/medium PH • Medium– critical impact	• One-off purchases • Very small quantities, highly managed • Very long specific PH • Highly critical impact	• Contracts • Med–large quantities, medium managed • Not in PH • Low impact to maintenance— may be critical to operation
	B	• Contracts/VMS • Large quantities, highly managed • Short planning horizon • Low impact to downtime	• VMS • Small quantities, medium managed • Low impact	• One-off/VMS • Rarely purchased, highly managed • Specialized PH • Medium to critical	• VMS/contracts • Large quantities, not managed • Not in planning horizon • No impact
	C	• VMS • Large quantities, not managed • No planning horizon • Little impact	• Not often found	• Not often found	• One-off purchases • Small quantities, not managed • Low planning horizon • Low impact

Purchasing strategies (VMS—vendor managed solutions)
Inventory strategies
Planning horizon (PH)
Impact to downtime

MRO: Maintenance Repair and Operations Inventory
Traditionally, the "O" in MRO meant Overhaul
The above matrix refers to both options

FIGURE 6.8 Inventory matrix for MRO.

The basic tasks at this step are given in the following sections.

6.5.5.1 Task 1

Determine the stocking decision for each commodity. There are several stocking choices, depending on whether the item is

- Regular inventory
- Unique (highly managed) inventory
- Vendor-managed inventory (consignment, vendor-managed at site)
- Vendor-held inventory (the vendor becomes a remote "warehouse" of the organization)
- Stockless (the item is cataloged and its source identified but not stocked)
- No stock (purchase as required)

6.5.5.2   Task 2

Increase maintenance's ability to find the right part for the job. This has become easier now that you have a standard user catalog. The next step is to link each part with the equipment where it is used, by assigning parts in the bill of materials (BOM), simplifying the process of identifying material to complete tasks. Accurate and complete BOMs reduce duplicates and are a straightforward, reliable tool for requisitioning materials. You can also use them to accurately determine stock levels. There are two types of BOMs:

- *Where-used BOM:* lists all of the repair parts and installed quantities for a piece of equipment or one of its components.
- *Job BOM:* lists all of the parts and consumables required for a particular repair job.

6.5.5.3   Task 3

Develop management processes for difficult-to-manage or unique stocks:

- Develop an inventory recovery (surplus stock) management system, including a decision matrix to help you retain or dispose of surplus stock. This is essential to ensure that surplus stock is managed as an asset to the business. Include a divestment model in this matrix to appraise the immediate disposal return against the probability of future repurchase. Include a plan to identify and value all nonstocked inventories.
- Formalize the repairable stock management process. Establish accounting practices and policies to effectively control repairable components and stocked materials (e.g., motors, mechanical seals, pump pullout units, transformers).

Automating inventory management is a leading practice. If we understand the influences of why we stock the part, this can be set into an algorithm and run on a daily, weekly, or monthly basis. Some computerized maintenance management system (CMMS) and enterprise resource planning (ERP) systems do this in a small way. Frankly, most do not yet do it in a way with which the organization is comfortable, allowing the algorithm to run without an inventory analyst review of each suggested inventory addition. A true leading practice would have this set in a way that inventory and service levels would be optimized and run without line-by-line analysis. Solutions are now starting to be available to manage this specifically for maintenance parts. Material requirements planning (MRP) can be used for highly active or predictable activity, but tailored applications are required for managing less active or insurance parts.

One way to approach this inventory management auto replenishment challenge is to outsource the process to organizations that provide "black box" inventory management services and simply analyze the recommendations and execute inventory adds as appropriate. This can be a bit labor intensive and assumes that the third party understands your business. A second way would be to run an application within your own organization and to monitor the recommendations and adjust the algorithm as needed. This takes the focus from the line-by-line analysis to the algorithm analysis and places the control and responsibility directly with inventory management.

Strategic financial and customer service modeler solutions now exist for inventory management. They can be tailored to provide automated analysis to identify short-term optimized inventory and service strategies while modeling longer-term, continuous improvement benefits. They can sense and quickly respond with adjustments to inventory parameters throughout the life of the product. This can allow users to have greater control of life-cycle decisions by updating factors like the SKU service targets as needed. See Chapter 18 on futures in maintenance management for more detail in this area.

6.6 PROCUREMENT AND PARTS REPAIR

Tody parts procurement has multiple source categories. At a high level it can be sourcing from the original equipment manufacturer or supplier or from a parts repair and refurbishment source. The parts repair source is often a mix of vendor and in-house services that repair the defective part and then return it to stock as if it were as good as a new part. In these cases the MRO supply chain is a "two-way" supply chain. As parts are replaced in the field, many of them are returned to be repaired or refurbished. In many industries, such as aircraft maintenance, the field service is simply the physical swap of a field replacement unit (FRU), and the used part is a rotated part (or "rotable") that will be returned to the bench for repair or overhaul. To support the management of used parts, a "reverse logistics" process must be established to ensure that the used defective part is tracked so that it can be repaired and the opportunity to establish a repaired part back to inventory is maintained. One significant benefit of doing this is that typically the cost of repair is significantly less than the cost of new, and this allows lower-cost quality parts into the inventory.

Managing used parts returns and repairing used parts is a new and growing part of a maintenance parts procurement strategy. For more traditional parts sourcing, procurement can be summarized as three main activities:

- Strategic commodity sourcing
- Supplier relationship management
- Transaction management

The inventory management group typically helps understand what commodities we want to procure when, but procurement will help us refine these requirements. For example, *strategic commodity sourcing* will require a better understanding of the commodities we need and analyze our past and planned spending in these areas. In this area, procurement would assess the market dynamics for these same commodities, would coordinate the request for information (RFI) and request for proposal (RFP) activities, and would develop a suitable commodities strategy for the organization. *Supplier relationship management* would empower procurement to engage and manage selected suppliers; would manage contract specifications, pricing, and compliance; and would assess supplier performance to established service-level agreements. By using the Internet, *transaction management* can execute in an automated interface between the two business, leveraging business-to-business transactions and allowing procurement to focus more on the first two activities of procurement. The Internet now helps to manage parts catalogs, purchase orders, order requisitions, order confirmations, and receipt confirmations. The procurement group can play less of a hands-on role, and the physical logistics folks can complete the cycle with receipt validation.

6.7   CONCLUSIONS

How the materials management area is organized and how the responsibilities are divided can vary from organization to organization. In some, the materials management group reports up through to procurement and finance. In others, it reports up through the maintenance and operations group.

Regardless of where it reports, the mission should be focused on serving the company's expected support role for the assets that drive the value for the organization's reason for existing in the first place. At times, a maintenance organization can presume that operations is king and drive too much inventory or that procurement and finance are king and drive down the dreaded inventory levels and drive up inventory turns. Optimal inventory in a leading operation drives a balanced inventory that understands its role and mission and drives an inventory and service-level balance that complements the value the company's assets are there to optimize. An example of a world-class parts operation is described in some detail in the IBM example provided at the end of this chapter.

Implementing an optimization methodology is essential for you to achieve the benefits identified in the cost/benefit analysis. However, as with any improvement, unless the policies, practices, and culture that support the organization change to sustain optimization, the benefits will be short-lived.

You must carefully address the "What's in it for me" issue with employees. They must understand their importance in sustaining change and supporting new initiatives. You will need to conduct training and education sessions, but, most importantly, the message must be top-driven. Senior management must communicate its support throughout the organization. The message must be repeated over time to encourage continuous improvement. Performance audits should be conducted to ensure that optimization is always achieved.

Once you have evaluated your organization's maturity level and settled on long-term goals and expectations, you can apply the appropriate optimization methodologies:

- First, analyze your current inventory and materials sources. Create a sustainable inventory strategy.
- Second, evaluate and adopt inventory control techniques to make optimized stocking decisions, and apply performance measurements.
- Third, optimize demand and supply management that emphasizes strategic sourcing and long-term planning reliability.
- Fourth, review procurement strategies for operations material, and implement strategies for better supplier management and spending control.

- Last, realize that change is sustainable only if the employees of the organization are prepared for it. Communication driven from the top down and adequate training are absolutely essential to optimize materials management.

Example: IBM Canada Replacement Parts Operation

EXAMPLE OF A LEADING MAINTENANCE PARTS OPERATION

IBM Canada's parts operation is responsible for supporting all of the IBM-manufactured and original equipment manufacturer (OEM)-supported products that are either on a warranty agreement or a maintenance agreement within Canada. Its client base is primarily corporate but can also be the small business and home user. IBM strives to service its client's equipment through the warranty period as well as through its planned functional life cycle. In other words, when a major client determines that it will purchase a set of equipment to perform certain functions for a specific period of time, IBM will contract to support this equipment through the planned installed life cycle. An example of this would be a major bank that elects to set up thousands of teller terminals across Canada using monitors, printers, keyboards, base personal computers, and network equipment with identical/similar distributed software support. The bank may plan to purchase a specific configuration of equipment and then support this new footprint for 6–10 years even though the original manufacturer may have created the equipment with a 3-year life-cycle plan.

The challenge for a maintenance organization such as IBM is to provide a timely forecast and repair service so that the functionality of the client's equipment can be sustained and the client will experience a high level of satisfaction for the service provided, resulting in the customer wanting to renew service contracts when they come due. To meet the service-level expectations of its clients, IBM has placed parts stations in 20 key cities across Canada; two of these locations (Toronto and Montreal) operate 24 hours a day, seven days a week, with the central distribution center located in the Greater Toronto area. However, just having the stockrooms is not enough; the locations need to be able to effectively determine what part to stock where, prior to the installation of the supported assets, while the asset is installed and during the sunset phase of the client footprint. In addition, they need to be able to coordinate the delivery of new parts and return of used parts within the technician's work order process so that both the part and the technician's time are effectively managed.

Within Canada, IBM supports more than 2,000 products and up to potentially 500,000 different part numbers. The actual number of parts stocked in Canada is 60,000; however IBM can system pass on an order for any part through to another parts location in North America, a supplier, or an internal plant and escalate the delivery from these referred locations. Their technicians use a radio-frequency–supported personal terminal to inquire about parts availability, to order parts, or to receive orders and shipping status "live." This order process allows for an internal target of 15 minutes from the time the part is ordered to the time it is picked, packed, and handed to a courier for local emergency delivery. The technicians have a map of their city that declares how long they should expect to wait to receive a part that is in stock. Typically, stocked parts are received within an hour of being ordered in any of the 20 cities in Canada that have an IBM parts stockroom. This high-quality order fulfillment process allows technicians a comfort level that their order will arrive in an acceptable time frame and also supports the concept that they do not need to hide inventory or do other unnatural acts to support the assets and clients to which they are assigned. In the Toronto area, this represents over 300 deliveries of emergency parts each business day. To enhance delivery, each parts station has couriers available on site during operating hours to complete the process and also can send a referred part on the next flight out or obtain a car or air charter if necessary with 15 minutes' notice.

Managing used parts returns and their refurbishment or disposition is an important component of IBM's parts management philosophy. Initial parts are procured through suppliers or other IBM plants; however, the bulk of the used parts (more than 70% in terms of value) are returned to be refurbished as new and returned to stock. This is a significant cost savings in terms of parts unit cost and procurement cost. In addition, refurbished parts have been statistically proven to be more reliable than new parts. This is particularly true with electronic components.

IBM CANADA'S PARTS DISTRIBUTION PROCESS

Along with the typical inventory metrics such as inventory turnover and inventory reserves management, the organization focuses on five main areas of excellence to achieve its key goals (Figure 6.9):

- Parts availability/parts acquisition time
- Parts costs
- Systems availability
- Distribution quality
- Parts quality

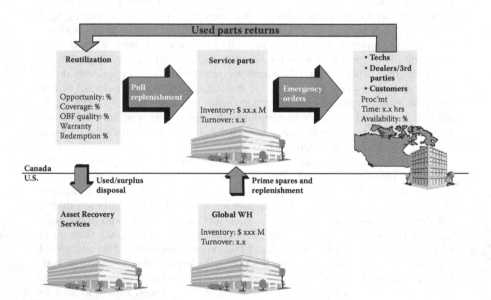

FIGURE 6.9   High-level parts flow diagram.

The organization accomplishes this with a clear focus on delivery (as previously discussed), on the elements of leading inventory management, and on the cost of sourcing each part.

Inventory management centrally manages and controls all the planned adds and deletes of the actual spare parts inventory across Canada as well as the inventory stocking policy set in the system for automated replenishment. It creates year-over-year plans for inventory and reports on the system's monitored key performance indicators of inventory and the overall spare parts materials management operation.

For new asset/product additions, the organization works with maintenance to develop an initial spare parts strategy to support the asset and to monitor the performance of the sparing strategy over the life of the asset. For existing inventory items, the safety stocking levels and economic reorder quantities are set based on asset criticality, supplier lead times, and expected service levels. A spare parts master is managed to ensure standardization of parts identification and harmonization of the parts stocked so that multiple pieces of the same part are not in the spares' network multiple times for the same asset. The organization leverages systems support to flag critical stock situations and to alert an inventory planner that a critical situation needs to be reviewed for expediting.

Like most inventory systems, IBM uses a dynamic min/max and economic order quantity solution that is complemented with an understanding of the required safety stock to buffer supplier lead times. In addition, it has a dynamic initial spare parts strategy to complement the process and also to leverage system managed date parameters such as the date the part was first added or last used as part of its replenishment algorithm.

The algorithm (or "inventory policy") determines target inventory levels for each part at each stocking location and the appropriate replenishment policy for each part at each location (replenishment planning). Some parts are identified to be tactically supported from a second stocking tier (echelon) rather than the stock location closest to the asset. Expensive items or slow movers of noncritical parts often are effectively held at central locations to reduce safety stock carrying cost. The planner in the inventory management group forecasts or simulates inventory requirements at both the location and piece level and is able to predict the service levels these pieces or locations generate. The planner could effectively, through parts planning, set up declining levels of support from the central warehouse to a local warehouse and through to a remote storeroom leveraging an echelon support structure. This could allow (or target) 55% parts availability support at the remote storeroom, 85% parts availability support at the local warehouse, and, if desired, 95% support at the central warehouse. The organization has the ability to run "what if" scenarios to determine the best course of action for a change in sparing strategy or asset mix. In addition, it is capable of developing an "end-of-life" stocking strategy to respond to supplier "end-of-life" or "end-of-support" notifications.

EXAMPLE OF A LEADING-PRACTICE ECHELON HIERARCHY

The planner in the inventory management group forecasts inventory surplus management tactics including surplus redistribution within Canadian operation. Needless to say, systems usability and availability are essential to the success of this inventory management set of processes (Figure 6.10).

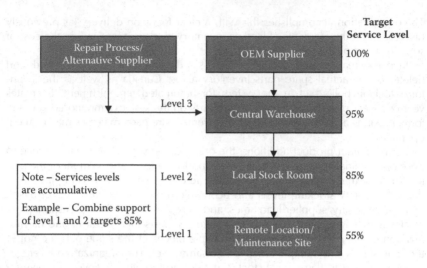

FIGURE 6.10   Service level example by stock location echelon.

The unit cost sourcing focus for IBM Canada has been primarily around supply and repair elements. This includes inbound inventory processes such as formal procurement and the management of used parts returns through to parts repair. Within IBM Canada, the purchasing process supports strong relationships with suppliers and partners.

The procurement processes includes the following:

- Ensuring purchasing strategies align with corporate culture
- Having standardized naming/numbering conventions for items purchased and suppliers used
- Analyzing commodity spending on a regular basis to confirm best sourcing strategy and alignment to corporate culture
- Having a defined and standard set of RFI/request for quotation (RFQ) processes for commodity types or business environment
- Linking supplier targets clearly to total materials management objectives
- Leveraging systems/technology to automate and, as a result, to minimize the need for procurement workload in the procurement transaction process
- Monitoring supplier performance and reviewing in regularly held joint supplier/company meetings
- Leveraging communications and technology in an integrated fashion to optimize inventory within the enterprise

IBM's repair practices look to optimize the scope of parts that can be repaired, based on the forecasted usage of parts, and to treat each part procured as a potential opportunity lost for savings in parts costs from repair. Parts are usually repaired to a "like new" state and returned to inventory at a value not fully burdened as a new part (i.e., cost of repair plus 25% of a new part can be a total cost that is less than 50% of the cost of a new part). Credits generated from the repair action typically become a credit to expense against the original parts usage. For IBM, more than 70% of all parts used (dollar value) are repaired and refurbished and returned to stock as new, and they generate a credit to the original expense. This action lowers the average cost of each part and the overall value of the inventory as well as creates the opportunity to lower overall parts usage costs.

This process also tracks warranty returns to the OEM. In this case the credit to expense is the full value to the credit given by the OEM, and the piece is returned to stock (when a replacement part is provided) at the full value.

IBM does capacity planning against its potential parts it can repair and its capacity to fulfill the repair. It has online access to the OEM for engineering documents to assist in repair strategies and is able to farm out the repair to another organization when it is financially viable to do so.

The organization has forecast visibility to the expected parts volumes to assist in its planning and scheduling of resources (e.g., parts, people, setup). Repair lead times can be integrated into the replenishment systems of the inventory network it supplies.

The inventory management system within IBM Canada also supports the following:

- Content management and substitution management
- Used parts return process
- Used parts sort process
- Reuse strategies
- Working with procurement to have vendor-managed inventory for the component parts used in repair
- Limited supplier SLAs
- Warranty identification process
- Supplier warranty redemption process management

Section II

Managing Equipment Reliability

7 Assessing and Managing Risk

Siegfried F. Sanders
Original by J. Kaderavek and G. Walker

CONTENTS

The risks inherent in asset management are coming under ever greater scrutiny. Because of increasing competition, how well companies manage assets becomes an additional market differentiator as companies strive to create value and thrive in a global marketplace. There are four basic groups of business assets: financial, human, intellectual, and physical. For many years, successful businesses have managed the first three well, but in recent years businesses recognized that managing physical assets is the next improvement opportunity. The term *asset management* has emerged to describe the process of managing a business's physical assets. In the broad sense, asset management is managing physical assets from the cradle to the grave. Over this life cycle, maintenance is responsible for managing risk during the physical assets' productive life. This chapter will explore the issue of risk management in maintenance and describe a number of effective methods to help assess and manage risk.

What is maintenance risk management? Webster's[*] defines the noun *risk* as "the chance of injury, damage, or loss." As a verb it means "to expose to the chance of injury, damage, or loss." Thus, risk management is both identifying the chance and reducing the exposure of "injury, damage, or loss." The reader will learn the nature of asset risk and risk management processes, including a proven method for identifying critical equipment. Managing asset risk includes the function of the asset itself, the safety of workers, and prevention of negative effects on the environment.

Risk management includes both identifying risk and taking action to reduce unacceptable levels of risk. A number of methods systematically identify and develop hazard and risk scenarios. Two comprehensive methods are (1) failure modes, effects,

---

[*] Webster's New World College Dictionary, 4th ed. IDG Books Worldwide, 2001.

and criticality analysis (FMECA) and (2) hazard and operability studies (HAZOPS). Both methods identify hazards during the asset's life cycle and provide guidance for reducing operational, personal, and environmental risks. HAZOPS has been in use in the chemical industry since the early 1980s.

Also included is a list of some relevant national and international standards and regulations for determining asset reliability and risk assessment.

7.1 INTRODUCTION

Asset management in maintenance is about making decisions. It involves determining the optimum maintenance policy to adopt, including preventive maintenance activities, spare parts to keep, worker skills to maintain, tools to provide, and ultimately decisions on repair or replacement. While managers try to base these decisions on the best data available and a rational understanding of the issues and trade-offs involved, there is always some uncertainty involved, and with uncertainty comes risk. It is the nature of the maintenance manager's job, therefore, to manage risk.

The term *risk* refers to an event where the outcome is uncertain and the consequences are generally undesirable. Strictly speaking, buying a lottery ticket is a risk, since it is unknown in advance, even with published odds, whether it will win or how much. Every action or inaction has some level of risk, since every event may end in an undesirable consequence. For example, jumping off the top of a tall building seems very risky, since it is almost certain that this will lead to instant death, but there have been instances of people falling from great heights and surviving. The chance of winning the lottery jackpot and surviving a fall from a tall building may be similar, but certainly more people buy lottery tickets than jump from tall buildings.

Risk is the product of probability and consequence. Therefore, two apparently different situations, one with a high probability and low consequence (e.g., tripping on an uneven floor and being injured) and one with a low probability and high consequence (e.g., an aircraft crashing and killing everyone) can actually have similar risk values. For example, in maintenance, an air system leak may have the same risk as a shaft failure on an air compressor. This is because there is a high probability of a joint failure in a compressed air system, but the resulting minor air leak has small cost, safety, and environmental consequences. On the other hand, there is a low probability that the compressor shaft will fail, but the resulting loss of plant air would cause a major process disruption as well as repair expenses.

It is only human to focus more on high-consequence events, even when they are unlikely, than those with little impact. Even experienced maintenance professionals have this bias. In fact, one of the most significant benefits of a total productive maintenance (TPM) program is that it addresses conditions such as leaks and normal adjustments, which are of minor consequence but happen frequently and still deserve attention.

The rest of this chapter will explore ways to analyze and deal with different risks in a consistent and rational manner.

7.2   MANAGING MAINTENANCE RISK

In any operation, there is always some degree of risk. All activities expose people or organizations to a potential loss of something of value. In maintenance, the impact is typically on equipment failure, human safety, or damage to the environment.

Risk involves three issues:

- The frequency of the loss
- The consequences and extent of the loss
- The perception of the loss to the ultimate interested party

A major equipment failure represents an extreme need to manage maintenance risk. The production downtime could delay product delivery to the customer and cost the business loss of sales or even market share. There could be further losses if the equipment failure threatened the safety of employees or adversely affected the environment.

A critical, high-profile failure could also create the impression that the business is out of control and tarnish its reputation in the marketplace. An example of this was the Perrier water incident in 1990. Perrier had a reputation for purity and promoting health. A minor maintenance failure led to traces of Benzene contaminating the product and damaging Perrier's reputation. It was a huge expense to recall and destroy millions of bottles of product from countries across Europe and from the United States. More important, though, was the damage to the company's reputation, causing the company to launch a high-profile and expensive public relations campaign to reassure customers that the product was still safe. Other significant examples of failure to manage risk that damaged either the firm or their supplier include the following:

- Union carbide, toxic release of methyl isocyanate, Bhopal, India, 1989[*]: 15,000 deaths, $3 billion legal claims.
- BP oil refinery explosion, Texas City, Texas, 2005.[†]
- American Airlines Flight 191, Chicago, Illinois, 1979.[‡]
- Prudhoe Bay oil field shutdown, Alaska, August 7, 2006[§]: 400,000 barrel per day lost production for several weeks.

The previous examples illustrate what the consequences can be of making a wrong maintenance decision. The consequences can be many: lessened plant reliability and availability; reduced product availability; decreased product quality; and increased total operating costs, as well as potential environmental damage, loss of life, and legal claims.

How is the maintenance manager to reduce the risk? One accepted premise is that increasing preventive maintenance results in less downtime and greater production, thus reducing risk and lowering overall maintenance cost. This is not completely

[*] http://en.wikipedia.org/wiki/Bhopal_Disaster, accessed November 7, 2006.
[†] http://www.bp.com/liveassets/bp_internet/us/bp_us_english/STAGING/local_assets/downloads/t/final_report.pdf, accessed November 7, 2006.
[‡] http://en.wikipedia.org/wiki/American_Airlines_Flight_191, accessed November 7, 2006.
[§] http://money.cnn.com/2006/08/07/news/international/oil_alaska/index.htm?cnn=yes, accessed November 7, 2006.

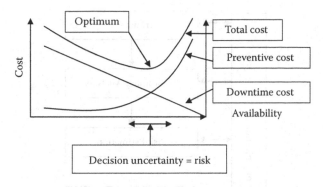

FIGURE 7.1 Preventive maintenance optimization curve.

true; Figure 7.1 shows that there is an optimum point where the combined preventive and downtime costs are at a minimum. This should determine maintenance policy and the amount of preventive work that will lead to the lowest total cost.

While this is a valuable concept, one can never know the exact trade-off between additional preventive maintenance and its impact on downtime. The information on which risk decisions are based will never be very certain or able to accurately foretell the future. Uncertainty is inevitable, so setting the maintenance policy is a question of managing risk and balancing risk versus cost.

Good data can improve the quality of decisions and confidence about the optimum maintenance point. Most risk management decision processes start by using historical data to predict future events. For equipment, this starts with manufacturers' recommended maintenance practices. Generally, that is the only guide available to predict the future, based on the premise that history predicts future events. Over time, you can use local history to modify risk management decisions, but also over time, conditions change so assumptions made in the past may no longer apply. Thus, the prudent risk manager must continuously manage this uncertainty and its resulting risk.

Generally, there is little acknowledgment that conditions change. Systematically planning for risk, including effects of changing conditions, can dramatically improve decision making. To make the process and resulting decisions more credible, there must be adequate support for the data used to make estimates, analysis methods, and unknown factors associated with the data. In addition, because of changing conditions, regularly revisit risk assessments to ensure that original conditions and assumptions still apply.

The objective of risk management is to identify significant risks and take appropriate actions to minimize them as much as is reasonably possible. To get to this point, you must balance risk control strategies, their effectiveness and cost, and the needs, issues, and concerns of stakeholders. Communication among stakeholders throughout the process is critical.

To assist in the risk management process, numerous tools and standards can be of help. However, there is also a trade-off between the amount of effort put into the analysis and its possible benefit. Getting this wrong can lead to some ridiculous situations, for example, a chemical company that used a full hazard study for a microwave oven in the mess room. The investigators demanded additional interlocks, regular

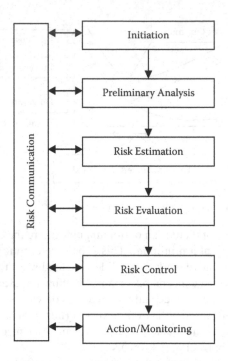

FIGURE 7.2   Risk management process.

condition, and radiation monitoring—even though home kitchens used the identical appliance! Avoid this extreme solution by seeking help to determine an appropriate risk management process.[*]

There are six general steps in the risk management process, as shown in Figure 7.2:

1. Initiation: define the problem and associated risks, form the risk management team, and identify stakeholders.
2. Preliminary analysis: identify hazards and risk scenarios and collect data.
3. Risk estimation: estimate probability of occurrence and consequences.
4. Risk evaluation: estimate benefits, cost, and stakeholders' acceptance of risk.
5. Risk control: identify risk control options and obtain stakeholders' acceptance of controls and any residual risk.
6. Action/monitoring: develop plan to implement risk management decision and to monitor its effectiveness.

To make effective decisions about the risk management process, the risk team and the stakeholders need to communicate frequently. There must be open dialogue to validate each step of the team's hypothesis and to ensure that all stakeholders are involved. Since the process to evaluate risk can be time-consuming

---

[*] CAN/CSA-Q850-97 "Risk Management: Guideline for Decision Makers."

and expensive, it is important to invest the time only on the assets and processes identified as most critical.

7.3 IDENTIFYING CRITICAL EQUIPMENT

To reduce risk there is generally far more to analyze than there are time or resources available. In addition, as described in the microwave oven example, not all equipment deserves the same degree of analysis. To determine where the risk management effort should go, it is crucial to understand equipment criticality (i.e., how critical the asset is to the business). Generally, you should determine the answer by the consequences if it fails.

The elements that make up criticality include safety, health, environment, and financial consequences. Quantify both the financial and nonfinancial impacts to provide a common base for analysis. Commercial criticality analysis (CCA) is valuable to focus and prioritize both day-to-day maintenance and ongoing improvements that manage risk.

For example, an oil terminal used CCA to rank its 40 main systems for both safety hazards and the cost per hour of downtime. Maintenance then used CCA ranking information to decide the priority of work. CCA also helped to determine which units justified flying in spares when a breakdown occurred and for which units cheaper, normal delivery was adequate. In addition, the CCA study helped to refocus the maintenance organization, which had been providing 100% skills coverage, 24 hours a day. Significant savings resulted, without any significant additional risk, by switching to 24-hour coverage only for the critical units.

In addition, when maintenance started a reliability-centered maintenance (RCM) study, CCA findings ensured that maintenance focused on areas of maximum impact. To achieve the most from RCM, it is important to assess the long-term risks in asset maintenance. Therefore, it is vital to consider risks to asset capability and possible shortcomings in information records and analysis methods.

Assessing criticality is specific to each individual facility. Even within the same industry, what is important to one business may not be important to another. Issues such as equipment age, performance, design (a major consideration), technologies using hazardous chemicals, geological issues (typically for mining and oil and gas industries), supplier relationships, product time to market (product cycle), finished goods inventory policy, information technology (IT) infrastructure, and varying national health and safety regulations all influence equipment criticality decisions.

Generally, all businesses need to consider the following asset criticality measures:

- Asset performance (reliability, availability, and maintainability)
- Cost (e.g., direct maintenance and engineering costs, indirect costs of lost production)
- Safety (e.g., lost time incidents, lost time accidents, disabling injuries, fatalities)
- Environment (number of environmental incidences, cost of environmental cleanup, environmental compliance)

7.4 CASE STUDY: A MINERAL PROCESSING PLANT

Now we examine the first two criticality measures with a real case study involving a mineral processing plant. Management wanted a maintenance strategy with clear equipment performance and cost measures and a site-wide improvement program.

7.4.1 EQUIPMENT PERFORMANCE

Rule 1: Talk to the people who know the plant.

Initially, production statistics provided equipment downtime, which was clear and accurate to a process line level. However, breaking the picture down further required meetings with maintenance and production crews to determine what actually occurred. For a first analysis, the main concern was equipment causing whole plant downtime. The downtime statistics are shown in Figure 7.3.

Presenting the data using Pareto analysis, a simple and ranked presentation style, quickly shows that the first three equipment types contributed 85% of total plant downtime. Judging from this, conveyors, pumps, and the clarifier were obvious targets for a program for equipment performance improvement.

The production and maintenance personnel checked the data (see Rule 1). One incident caused the clarifier downtime when a piece of mobile equipment struck it, causing substantial damage. This was an unusual event, not related to normal operating and maintenance, and thus was eliminated from the operating risk analysis. After investigation, the plant implemented several remedial actions that should prevent this incident from recurring. Conveyors, pumps, and screens then became the top three improvement targets. Even though eliminating the clarifier failure was justified, it is wise to remember that not all risk comes from normal operating and

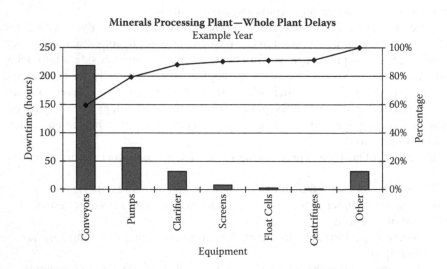

FIGURE 7.3 Pareto analysis of downtime caused by equipment failures.

maintenance activities. External events such as a car striking a power pole, civil unrest, or terrorist activity can still lead to significant risk.

The next step was to identify, agree on, and implement conventional reliability, availability, and maintainability (RAM) measures.

7.4.2   Costs

Maintenance collected breakdown and failure work order cost data from the existing work order system. The costs were organized by target equipment types and subtypes. Even though this analysis concentrated on lost time and repair costs, maintenance could consider other factors such as capacity, quality, safety, or environmental effects. The results of the chemical plant analysis are shown in Figure 7.4.

To increase reliability and to reduce maintenance cost, the facility concentrated on the three highest-cost pump, screen, and conveyor types. This illustrates the 80/20 rule that 20% of your effort should reduce 80% of the cost. Although this was a very systematic analysis of the failures, remember Rule 1: ask the operators and maintenance people. It is very common that if you just ask the maintenance people they can tell you the top three maintenance failures using anecdotal evidence. Don't discount this input to your analysis process. This information can confirm or expand your statistical analysis to determining highest cost failures.

Using this simple analysis, management identified critical equipment by the following:

• Poor overall performance (most downtime)
• Highest failure costs

The facility was able to reduce its risk by using the data to launch a strategic reliability improvement program. The organization has since reduced its risk, has saved over $1 million annually, and has significantly improved equipment performance simply by concentrating on the worst of the "bad actors."

7.5   SAFETY AND ENVIRONMENTAL RISK: DUTY OF CARE

Maintenance has a huge impact on safety and the environment, the other two major areas of risk management. But isn't maintenance just supposed to keep the plant running? As you will see in this section, the general responsibilities of maintenance engineers can be extensive.

Why should a business be concerned with safety? The legal reasons date to a 1932 ruling by Lord Atkin of the British Privy Council that the Stevenson soft drink company was liable for an injury sustained by Scottish widow Mary O'Donahue after drinking a contaminated soft drink. The ramifications of this ruling include the modern "duty of care" legislation that is the basis of most national occupational health and safety (OHS) legislation. The ruling ensures that all employers have a duty of care to their employees, that employees have a duty to each other, and that employees have a duty to their employer. The standard of care and reasonableness in an organization should follow the national standard. An increase in the number and

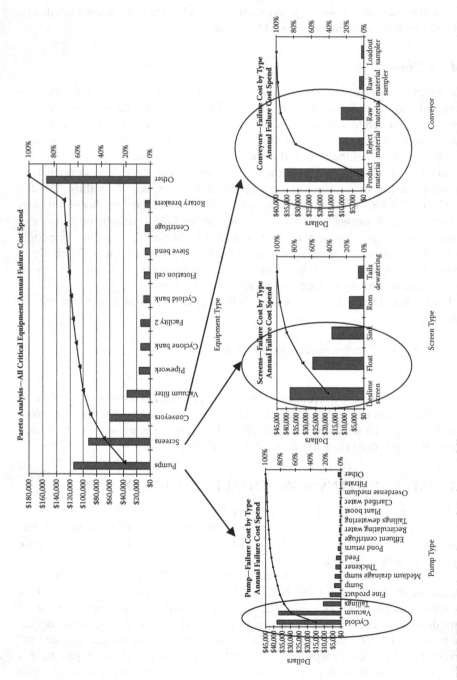

FIGURE 7.4   Pareto analysis of cost of equipment failures.

strength of these regulations is expected. This duty of care also extends to protecting the public and the environment. Thus, managing risks that may affect safety or the environment are part of the responsibility of every employee, including maintenance. Maintenance through its action or inaction can have a significant impact on managing risk. The examples mentioned earlier illustrate this fact.

7.6 MANAGING LONGER-TERM RISKS

Over the operating phase of the life cycle of an asset, maintenance needs to maximize reliability and minimize risk. However, certain conditions can increase risk in the long-term.

To achieve asset reliability, you should use failure and reliability analytical methods to develop effective maintenance plans. Failure modes and effects analysis (FMEA), or the closely related FMECA, and reliability-centered maintenance (RCM) use multidisciplinary teams to develop maintenance plans for existing assets. Although these techniques are invaluable, there are potential long-term risks:

- *Slow degradation failures*: These are often difficult to predict and model, especially for equipment early in its life cycle. The probability of fatigue-related failures increases with every operating cycle of the asset. It can be difficult, even impossible, to predict these failures using conventional reliability methods. One railway freight provider, aware that low-cycle fatigue creates increasing failure rates and costs in its wagon fleet, conducted an accelerated reliability test on several wagons. The aim was to model the failures ahead of time to develop appropriate condition and time-based tactics. Some wagons were fatigue tested to 10 times their current lives.
- *Incomplete execution of reliability methods:* If all plausible failure modes are not identified, then the analysis is not complete, and neither are the maintenance plans. As a result, the asset will display unpredicted failure modes. Review maintenance plans periodically to evaluate the effectiveness of the reliability improvement.
- *Change in operating environment*: Changes in rate of effort (e.g., operating hours/year), physical environment (e.g., different geological composition for mining businesses), and operating procedures and techniques can all trigger new failure modes not predicted by the original analysis. Examples are changes in airplane route or truck service changing from highway transport to city delivery.
- *Change in maintenance environment:* Changes in maintenance tactics (e.g., servicing intervals), maintenance personnel quantity and quality (number, skill sets, and experience), and support and condition monitoring equipment can all lead to unanticipated failures.
- *Modifications and capital upgrades:* Modifications to existing or new assets can also affect existing assets and result in unpredicted failures.
- *Change in plant operations:* This places different demands on installed assets; thus, the criticality of assets may either increase or decrease.

7.7 IDENTIFYING HAZARDS AND USING RISK SCENARIOS

Risk management starts by identifying hazards. Start by analyzing past events, incidents, or lost performance relating to assets, people, and the environment. This could be information from the assessed facility or from events at other facilities. However, as already mentioned, there are limitations with only using past history. Selecting an experienced team of engineers and process managers is also essential for a thorough understanding of the whole process or system and potential hazards or failures.

Develop risk scenarios by identifying hazards and then evaluating the loss, both direct and consequential. A risk scenario is therefore a sequence of events with an associated probability and consequences. There are a variety of approaches and methods to identify and analyze risks. Some are observation and experienced judgment based while others are systematic analysis. Two analytical methods typically applied in the asset management are FMEA or FMECA and hazard and operability studies (HAZOPS).

7.8 FMEA AND FMECA

The failure modes and effects analysis identifies potential system failures and their effects. Criticality analysis (CA) ranks failure severity and probability of occurrence. When performing both steps, FMEA and CA, the result is failure mode, effects, and criticality analysis.

There are two primary ways of doing FMEA. One is the hardware approach, which lists the effects on the system. The other is functional, based on the premise that every item in the system is designed to perform a number of functions that can be classified as outputs. For functional FMEA, list outputs and analyze them to determine their system effects. Variations in design complexity and available data usually dictate the analysis method to use. When detailed design information is available, the hardware is generally studied. Use the functional approach when in conceptual design stages.

7.8.1 FMEA OBJECTIVES

FMEA and FMECA are an integral part of the design process and should be updated regularly to reflect design evolution or changes. FMEA provides inputs to product reviews at various levels of development. Also, FMEA information can minimize risk by defining special test considerations, quality inspection points, preventive maintenance actions, operating constraints, and other pertinent information and activities. This may include identifying additional redundancy, alarms, failure detection mechanisms, design simplification, and derating. FMEA can also be used to do the following:

- Compare various design alternatives and configurations.
- Confirm the system's ability to meet its design reliability criteria or requirements.
- Provide input data to establish corrective action priorities and trade-off studies.

7.8.2   FMEA AND CA METHODOLOGY

FMECA methodology consists of two phases: FMEA and CA. To perform FMEA

- Define the system and its performance requirements.
- Define the assumptions and ground rules to be used in the analysis.
- List all individual components or the various functions at the required indenture level for the analysis.
- Develop a block diagram or other simple model of the system.
- Devise an analysis worksheet to provide failure and effects information for each component, together with other relevant information.

CA ranks each potential failure identified by FMEA, according to its combined severity and probability of occurring. The criticality analysis may be performed qualitatively or quantitatively.

FMEA is very versatile and useful for risk analysis. If CA is included, it will be easy to rank failures for their severity and probability of occurrence. Then the best corrective actions can be determined and prioritized. The FMEA worksheet should be structured so that the analysis's detailed information is tailored to fit the situation. The FMEA method, however, has a major shortcoming, since it requires a great deal of time and effort, making it expensive.

The analysis becomes even more complex if the effects of multiple (i.e., two or more simultaneous) failures are taken into account. It is easy to overlook both human and external system interactions. Often, too much time and effort are spent analyzing failures that have a negligible effect on the system performance or safety. To help with the analysis, a number of computer packages automate the FMECA process. For it to be accurate and effective, even if using a computer, FMECA should be performed by people intimately familiar with the design or process being analyzed.

In conclusion, a rigorous FMECA analysis is a highly effective method to detail risk scenarios. The risk management team and stakeholders will develop a good understanding of risk levels and controls.

7.9   HAZOPS

HAZOPS was developed in the process industry to identify failure, safety, and environmental hazards. It evolved in the chemical industry in the 1970s, particularly under Trevor Kletz, an employee of the U.K.-based chemical company ICI. HAZOPS* is actually the fourth in a series of six hazard studies covering the life cycle of new plant development, from initial plant concept to commissioning and effective operation. The early studies consider product-manufacturing hazards, such as a material's toxicity or flammability, and the pressures and temperatures required. The later studies check that the plant was built according to the requirements of the earlier studies and that operating conditions comply.

---

* http://www.netregs.gov.uk/static/documents/Business/comah-1785585.

Hazard study 4, the hazard and operability study, examines in painstaking detail the hazards inherent in the plant's design and any deviations. This is particularly relevant to asset maintenance. A distressingly high proportion of serious and fatal chemical accidents occur when normal conditions in the plant are temporarily disturbed. Refer to the Bhopal and Texas City risk management failures mentioned earlier. Maintenance is important as both an input and an output of the HAZOP—as an input because maintenance of plant assets may require changes to normal plant operating conditions and as an output in numerous ways. The study could conclude, for instance, that maintenance be done only under certain conditions or that some plant items must meet specified standards of performance or reliability. The HAZOP helps determine the importance of various equipment and the need for either a CCA or an FMECA/RCM analysis.

7.10   METHOD

Small teams carry out a hazard and operability study. Typically, the team requires the following:

- An operations expert, familiar with the way the plant works (which, in the chemical industry, is probably a chemical engineer)
- An expert on material hazards (typically, in the chemical industry, a chemist)
- An expert in the way the plant equipment itself may behave (likely the plant engineer)
- A trained facilitator, usually a safety expert

Call in other experts as required to provide advice on such things as electrical safety and corrosion.

The essential starting point for the HAZOP is an accurate plant diagram (in the process industry, a piping and instrumentation, or P&I, diagram) and processing instructions that spell out how the product is manufactured in the plant.

Start the study by reviewing the intrinsic manufacturing hazards and standards. This includes briefing the team on the flammability, toxicity, and corrosion characteristics of all the materials and process characteristics such as high temperatures or pressures, potential runaway reactions, and associated hazards. In addition to managing the inherent risks of normal operations, the present rise in international terrorism means that asset managers must now consider how a terrorist attack may cause a failure of processes. This is especially important for processes containing dangerous materials that can be released and cause mass casualties.[*]

After doing a desktop study of plant processes to determine potential areas of high risk, the team must then physically inspect the risky manufacturing processes, considering the parts of the plant identified as potential high risk, down to individual items, even to each section of pipe work. For every process considered, follow a sequence of prompts. First, address normal plant operation to understand each hazard and how it is controlled. Then consider deviations from normal operations.

---

[*] http://www.esmagazine.com/CDA/Articles/Feature_Article/115bad5f395fd010VgnVCM100000f932 a8c0, accessed December 10, 2006.

TABLE 7.1
Hazard Checklist Parameters

Hazard	Parameter Change	Parameter
Corrosion	more of	pressure
Erosion	less of	temperature
Abrasion	none of	flow
Cracking	reverse of	Ph
Melting	other than	quantity
Brittleness	more of	concentration
Distortion		mixture
Leakage		bombing
Perforation		impact
Rupture		

Maintain a hazard checklist of such things as those in Table 7.1; also consider what changes could occur in each of the relevant parameters. Use the topics in the table to ask questions such as "Can a *leak* occur because there is a *reverse* in the normal *flow* of the process?"

Perform the analysis, documenting any hazardous results and the appropriate actions to reduce the risk. This could include plant redesign or changes to operating and maintenance procedures. For example, in a facility that relies on automatic safety control systems, more checking, or calibration may be required to ensure the instruments are operating correctly or creating a purchasing specification for replacement parts may be required. Follow this approach until every part of the plant and the process has been considered and documented.

In summary, HAZOP provides plenty of benefits. The plant risk assessment will be rigorous, comprehensive, and done to high safety standards. Besides reducing risk, this process will also reveal other opportunities for improvement.

But there are some limitations to keep in mind. HAZOP can be time-consuming, typically taking several full days for the team to go through a single process in a complex plant. This level of thoroughness is costly, tedious, and resource consuming. Always consider the cost/benefit ratio before initiating a major effort such as HAZOP.

7.11 STANDARDS AND REGULATIONS

7.11.1 INTERNATIONAL STANDARDS RELATED TO RISK MANAGEMENT

A number of international and national standards and regulations are available for a risk management program. The following section lists the number and name of the standards and provides a brief overview of their content. The standards cover basic reliability definitions, reliability management, risk management, environmental and safety regulations, cost of quality, and software reliability concepts. These standards provide excellent further reading and will provide guidance for asset managers.

7.11.2 CAN/CSA-Q631-97 RAM Definitions

7.11.2.1 Scope

This standard lists terms and basic definitions, primarily intended to describe reliability, availability, maintainability, and maintenance support. The terminology is that of engineering but is also adapted to mathematical modeling techniques.

7.11.2.2 Application

RAM addresses general concepts related to reliability, availability, and maintainability—how items perform over time and under stated conditions and concern all the life-cycle phases (concept and definition, design and development, manufacturing, installation, operation and maintenance, and disposal).

Note the following regarding this terminology:

* Reliability, availability, and maintainability are defined as qualitative "abilities" of an item.
* Maintenance support is defined as the qualitative ability of an organization.
* Such general abilities can be quantified by suitable "random variable" conditions, such as "time to failure" and "time to repair."
* You can apply mathematical operations to these random variables using relations and models. The results are called "measures."
* The significance of variables and measures depends on the amount of data collected, the statistical treatment, and the technical assumptions made in each particular circumstance.

In RAM, one key ability measured is uptime. How to define and measure uptime has been a hot topic of debate within the industry. Management spends considerable time grappling with interdepartmental issues—for example, determining operations versus maintenance caused downtime, equipment handover time, logistics, and administrative delays. Often, it is hard to reach a clear agreement on these issues. Figure 7.5, based on CAN/CSA-Q631-97 "Reliability, Availability, and Maintainability (RAM) Definitions," shows one method to define the time available for maintenance and operations. As clear as the diagram is, there is room for disagreement on the definition of the terms and for actually assigning events to one of the categories of downtime. In any individual operation, these terms will have slightly different meanings, and it is management's responsibility to agree to a common definition. This is more important than assigning responsibility; it is crucial to having good data to manage risk. Define and measure downtime correctly; otherwise, risk analysis will use uncertain data resulting in more risk.

RAM is also excellent in establishing the performance of standard asset management measures across an organization. This would include individual equipment, process line, and systems, as well as the maintenance organization's ability to support that performance. The clear definition and detail make these measures easily communicated to stakeholders. This reduces the risk of "miscommunication" and

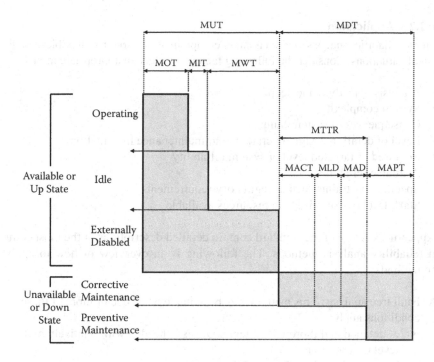

Operating Time
MUT: Mean Up Time
MOT: Mean Operating Time
MIT: Mean Idle Time
MWT: Mean Wait Time

Maintenance Time
MDT: Mean Down Time
MTTR: Mean Time To Repair
MACT: Mean Active Corrective Time (Wrench Time)
MLD: Mean Logistic Delay Time

FIGURE 7.5 Equipment availability diagram.

"miscomprehension" that plagues most managers when establishing an "apples for apples" performance comparison.

7.11.3 CAN/CSA-Q636-93 GUIDELINES AND REQUIREMENTS FOR RELIABILITY ANALYSIS METHODS—QUALITY MANAGEMENT

7.11.3.1 Scope

This standard guides asset managers in selecting and applying reliability analysis methods. Its purpose is to do the following:

1. Describe some of the most common reliability analysis methods that represent international standard methods.
2. Guide you in selecting analysis methods, depending on technology and how the system or product is used.
3. Establish how results will be documented.

7.11.3.2 Application

No single reliability analysis method is either comprehensive enough or flexible enough to suit all situations. Consider the following factors to select an appropriate model:

- Analysis objectives and scope
- System complexity
- Consequences of system failure
- Level of detail in design, operation, and maintenance information
- Required or targeted level of system reliability
- Available reliability data
- Specific constraints such as regulatory requirements
- Staff, level of expertise, and resources available

Appendices A to F of the standard contain detailed descriptions of the most common reliability analysis methods. The following is an overview of how to apply these methods:

A. Fault tree analysis: This method may be suitable when one or more of these conditions apply:
   - A detailed and thorough system analysis is needed with relatively high level of resolution.
   - There are severe safety and economic consequences of a system or component failure.
   - The reliability requirements are stringent (e.g., system unavailability ≤0.001 units?).
   - Considerable staff and resources, including computer facilities, are available.
B. Reliability block diagram: Consider this method if one or more of these conditions apply:
   - Either a rudimentary system study or a higher hierarchical level is needed (although the method may be used at any level or resolution).
   - The system is relatively simple.
   - The analysis needs to be simple and straightforward, even if some detail is lacking.
   - Reliability data can be obtained at a block level, but data for a more detailed analysis are either not available or not warranted.
   - The reliability requirements are not very stringent.
   - There are limited staff and resources.
C. Markov analysis: This method may be best if one or more of these conditions apply:
   - Multistates or multiple failure modes of the components will be modeled.
   - The system is too complex to be analyzed by simple techniques such as a reliability block diagram (which may be too difficult to construct or to solve).

- The system has special characteristics, such as the following:
  - A component can't fail if some other specified component has already failed.
  - You can't repair a component until a certain time.
  - Components don't undergo routine maintenance if others in the system have already failed.

D. FMEA: This may be suitable when one or more of these factors apply:
- Ranking the failure modes' relative importance is required.
- All possible failure modes along with their effect on system performance must be detailed.
- The system components aren't dependent on each other to any important degree.
- The prime concern is single component failures.
- Considerable staff and resources are available.

E. Parts count: Consider this method if one or more of these conditions apply:
- Only a very preliminary or rudimentary conservative analysis will be performed.
- The system design has little or no redundancy.
- There are limited staff and resources.
- The system being analyzed is in a very early design stage.
- Detailed information on components such as part ratings, part stresses, duty cycles, and operating conditions is not available.

F. Stress analysis: You may prefer stress analysis if one or more of the following conditions apply:
- A more accurate analysis than the parts count method is desired.
- Considerable staff and resources, including computer facilities, are available.
- The system being analyzed is in an advanced design stage.
- Access to detailed information on components such as parts ratings, part stresses, duty cycles, and operating conditions is available.

7.11.4 CAN/CSA-Q850-97 RISK MANAGEMENT— GUIDELINES FOR DECISION MAKERS

7.11.4.1 Scope

This standard helps to effectively manage all types of risks, including injury or damage to health, property, the environment, or something else of value. The standard describes a process for acquiring, analyzing, evaluating, and communicating information for decision making.

Note: The Canadian Standards Association has a separate standard to address risk analysis (CSA Standard CAN/CSA-Q634) and environmental risk assessment (CSA Standard Z763)

7.11.4.2   Application

This standard provides a comprehensive decision process to identify, analyze, evaluate, and control all types of risks, including health and safety. Owing to cost constraints, risk management priorities must be set, which this standard encourages.

7.11.5   AS/NZS 4360-1999 RISK MANAGEMENT

7.11.5.1   Scope

This standard helps to establish and implement risk management, including context, identification, analysis, evaluation, treatment, communication, and ongoing monitoring of risks.

7.11.5.2   Application

Risk management is an integral part of good management practice. It is an iterative process consisting of steps that, in sequence, continually improve decision making. Risk management is as much about identifying opportunities as avoiding or mitigating losses.

This standard may be applied at all stages of an activity, function, project, or asset. Maximum benefits will be gained by starting the process at the beginning. It is usual to carry out different studies at various stages of the project.

The standard details how to establish and sustain a risk management process that is simple yet effective. Here is an overview:

- *Establish the context:* Establish the strategic, organizational, and risk management context in which the rest of the process will take place. Define criteria to evaluate risk and the structure of the analysis.
- *Identify risks:* Identify what, why, and how problems can arise, as the basis for further analysis.
- *Analyze risks:* Determine the existing controls and their effect on potential risks. Consider the range of possible consequences and the probability of occurrence. By combining consequence and probability, risk can be estimated and measured against preestablished criteria.
- *Evaluate risks:* Compare estimated risk levels with preestablished criteria. They can then be ranked to identify management priorities. For low-risk scenarios, no action may be required.
- *Treat risks:* Accept and monitor low-priority risks. For higher-priority risks, develop a specific management plan with sufficient funding to implement risk reduction measures.
- *Monitor and review:* Monitor and review how the risk management system performs, looking for ways to continuously improve.
- *Communicate and consult:* Communicate and consult with internal and external stakeholders about the overall process and at each individual stage.

Figure 7.6 shows the steps in a risk management program.

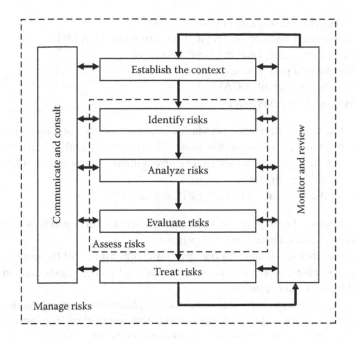

FIGURE 7.6 Risk management process overview.

Appendix B of the standard details the steps to develop and implement a risk management program:

- Step 1: Support of senior management.
- Step 2: Develop the organizational policy.
- Step 3: Communicate the policy.
- Step 4: Manage risks at organizational level.
- Step 5: Manage risks at the program, project, and team levels.
- Step 6: Monitor and review.

CAN/CSA-Q850-97 describes the risk communication process, including stakeholder analysis, documentation, problem definition, and general communications, while AS/NZS 4360-1999 has a well-developed and articulated risk management process model and provides a useful summary for implementing a risk management program. If both standards are available, the risk manager can use both to decide an appropriate process for a particular maintenance environment.

7.11.6 ISO 14000 ENVIRONMENTAL MANAGEMENT SYSTEMS

ISO 14000 is a series of international, voluntary environmental management standards. Given that environmental damage is one of the key factors to be considered in risk management, it is included here. Developed under ISO Technical Committee 207, the 14000 series of standards addresses the following aspects of environmental management:

- Environmental management systems (EMS)
- Environmental auditing and related investigations (EA&RI)
- Environmental labels and declarations (EL)
- Environmental performance evaluation (EPE)
- Life-cycle assessment (LCA)
- Terms and definitions (T&D)

The ISO series of standards provide a common framework for organizations worldwide to manage environmental issues. They broadly and effectively improve environmental management, which in turn strengthens international trade and overall environmental performance.

The key elements of an ISO 14001 EMS are as follows:

- *Environmental policy* includes the environmental policy and how to pursue it via objectives, targets, and programs.
- *Planning* includes analyzing the environmental aspects of the organization (e.g., processes, products, and services as well as the goods and services used by the organization).
- *Implementation and operation* includes implementing and organizing processes to control and improve operations that are critical from an environmental perspective (e.g., both products and services).
- *Checking and corrective action* includes monitoring, measuring, and recording characteristics and activities that can significantly impact the environment.
- *Management review* includes top management review of the EMS to ensure it continues to be suitable and effective.
- *Continual improvement* is a key component of the environmental management system. It completes the cycle: plan, implement, check, review, and improve continually.

ISO 14000 standards and related documents can be obtained from National Standards Association (ISO Member Body), which is usually a country's primary ISO sales agent. In countries where the national standards association is not an ISO member body, ISO 14000 documents can be obtained directly from the ISO Central Secretariat.

7.11.7   BS 6143-1990 GUIDE TO THE ECONOMICS OF QUALITY

7.11.7.1   Scope
BS 6143 has two parts:

- Part 1—Process cost model: Using this model, process measurement and ownership are key. Quality costing can be applied to any process or service. The quality cost categories simplify classification by making clear the cost of conformance and nonconformance. The method involved is process modeling, and there are guidelines for various techniques. In addition, the process control model is compatible with total quality management.

- Part 2—Prevention, appraisal, and failure model: This is a revised version of traditional product quality costing in manufacturing industries. With recent improvements, this approach has become more effective, though it may be combined with the process cost model.

Using this standard will help to determine the cost of preventing defects, appraisals, and internal and external failures as well as quality-related cost systems for effective business management.

7.11.7.2   Application

For asset managers unfamiliar with this standard, it deals with a manufacturing cost structure that can readily be applied to direct maintenance charges. Costs are defined as follows:

- Prevention cost is the cost of any action to investigate, prevent, or reduce the risk of nonconformity or defect.
- Appraisal cost is the cost of evaluating quality requirement achievements, such as verification and control performed at any stage of the quality loop.
- Internal failure cost is the cost of nonconformities or defects at any stage of the quality loop, including, for example, scrap, rework, retest, reinspection, and redesign.
- External failure cost is the cost of nonconformities or defects after delivery to a customer/user. This can include claims against warranty, replacement, and consequential losses as well as evaluating penalties.
- Identifying cost data means that quality-related costs should be identified and monitored. It is essential that the way data are classified is relevant and consistent with other accounting practices within the company. Otherwise, it will be difficult to compare costing periods or related activities.
- Quality-related costs are a subset of business expenses, and it is useful to maintain a subsidiary ledger or memorandum account to track them. By using account codes within cost centers, the quality cost of individual activities can be better monitored. Allocating costs is important to prevent failures and should not be done solely by an accountant. The analyst may need technical advice as well.
- Quality costs alone do not provide managers with enough perspective to compare them with other operating costs or to identify critical problem areas. To understand how significant a quality cost actually is, compare it with other regularly reported organizational costs.

7.11.8   EPA 40 CFR 68 CHEMICAL ACCIDENT PREVENTION PROVISIONS

7.11.8.1   Scope

This regulation requires U.S. facilities with more than a threshold quantity of certain chemicals to develop and publish a risk management plan to mitigate the effects to the surrounding public of an accidental release, fire, or explosion. Section G of this regulation provides the basis for a risk manager to develop a risk management program even for those facilities not covered by this regulation.

7.11.8.2 Application

Facilities must analyze risk at least every five years using approved methods of analysis including the following:

- What-if
- Checklist
- What-if/checklist
- HAZOP
- FMEA
- Fault tree analysis

The process hazard analysis must address the hazards of the process, past incidents, controls to prevent failures, consequences and health effects of failure, location of hazards, and human and safety factors.

A team with expertise in engineering and process operations must perform the process hazard analysis. The team must include at least one employee who has experience and knowledge specific to the evaluated process and one member knowledgeable in the specific process hazard analysis methodology being used.

Most importantly, though, there must be a system to promptly address and resolve the team's findings and recommendations. Documentation of action taken is vital for compliance. This follows the simple rule:

- Say what you will do.
- Do what you say.
- Document that you did it.

For facilities covered by this regulation, review the hazard analysis at least every five years after the completion of the initial process hazard analysis.

7.11.9 OSHA 29 CFR 1910.119 PROCESS SAFETY MANAGEMENT OF HIGHLY HAZARDOUS CHEMICALS

7.11.9.1 Scope

The regulation requires U.S. facilities with more than a threshold quantity of certain chemicals to develop and publish a process safety management (PSM) plan to protect the employees of a covered facility from the effects of failures of systems containing highly hazardous chemicals.

7.11.9.2 Application

The application of PSM is very similar to the U.S. Environmental Protection Agency's (EPA's) 40 CFR 68 risk management plan. There is even a provision in 40 CFR 68 that allows the Occupational Safety and Health Administration (OSHA) PSM plans to qualify for inclusion in 40 CFR 68 risk management plans. Both regulations have similar requirements for the following:

- Qualified personnel are requied to use recognized hazard analysis methods to develop the PSM plan.
- An action plan is required to act on the findings and recommendations.
- Documentation of findings, plans, and actions taken is required.
- Review of plans at least every five years is required.

7.11.10   ANSI/AIAA R-013-1992 SOFTWARE RELIABILITY

7.11.10.1   Scope

Software reliability engineering (SRE) is an emerging discipline that applies statistical techniques to data collected during system development and operation. The purpose is to specify, predict, estimate, and assess how reliable software-based systems are. This is a recommended practice for defining software reliability engineering, which is becoming much more important, even to industrial plants, since more process equipment and instrumentation includes software for control and maintenance.

7.11.10.2   Application

The techniques and methods in this standard have been successfully applied to software projects by industry practitioners to do the following:

- Determine whether a specific software process is likely to produce code, which satisfies a given software reliability requirement.
- Determine the need for software maintenance by predicting the failure rate during operation.
- Provide a metric to evaluate process improvement.
- Assist software safety certification.
- Determine whether to release a software system or to stop testing it.
- Estimate when the next software system failure will occur.
- Identify elements in the system that most need redesign to improve reliability.
- Measure how reliably the software system operates to make changes where necessary.

7.11.10.3   Basic Concepts

There are at least two significant differences between hardware and software reliability. First, software does not fatigue, wear out, or burn out. Second, because software instructions within computer memories are accessible, any line of code can contain a fault that could produce a failure. The failure rate over time of a software system is generally decreasing due to fault identification and removal. Software failures are unlikely to reoccur after they are identified and removed (Figure 7.7).

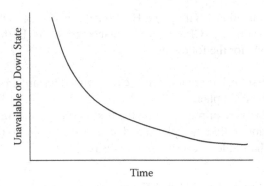

FIGURE 7.7 Software reliability measurement curve.

7.11.10.4 Procedure

Following is an 11-step generic procedure for estimating software reliability. Tailor this to the specific project and the current life-cycle phase. Not all steps will be used in every application, but the structure provides a convenient and easily remembered standard approach. The following steps are a checklist for reliability programs:

1. Identify application.
2. Specify the requirement.
3. Allocate the requirement.
4. Define failure: a project-specific failure definition is usually negotiated by the testers, developers, and users. It is agreed upon before the test begins. What is most important is that the definition be consistent over the life of the project.
5. Characterize the operational environment, including three aspects: system configuration, system evolution, and system operating profile.
6. In modeling software reliability, keep in mind that systems frequently evolve during testing. New code and components can be added.
7. Select tests: software reliability engineering often involves operations and collecting failure data. Operations should reflect how the system will actually be used. The standard includes an appendix of information to help determine failure rates.
8. Select models: included are various reliability models. We recommend that you compare several models before making a selection.
9. Collect data: to make an effective reliability program, learn from previous lessons. This does not mean you need to keep every bit of information about the program as it evolves. Also, clearly define data collection objectives. When a lot of data are required, it is going to affect the people involved. Cost and schedule can suffer, too.
10. Two additional points to keep in mind when collecting data: (1) motivate the data collectors; and (2) review the collected data promptly. If this advice is not followed, quality will suffer.

11. Estimate parameters: there are three techniques in the standard to determine model parameters—method of moments, least squares, and maximum likelihood.
12. Validate the mode: to properly validate a model, first address the assumptions about it. This can be done effectively by choosing appropriate failure data items and relating specific failures to particular intervals or changes in the life cycle.
13. Perform analysis: after the data have been collected and the model parameters have been estimated, perform the appropriate analysis. The objective may be to estimate the software's current reliability, the number of faults remaining in the code, or when testing will be complete.

Be careful about combining a software reliability value with a system reliability calculation. The risk analysis may require a system reliability figure, while execution time is the basis for software reliability. In that case, it must be converted to calendar time to be combined with hardware reliabilities. After converting to common units, one will be able to calculate the system reliability value.

7.12 CONCLUSION

This chapter has shown that managing asset risk is required to remain competitive in the changing global marketplace. Companies that ignore asset risk management place themselves in a position where every day becomes a roll of the dice. Will a major failure occur that can cost millions, lose customers, cause fatalities, or damage the environment? Management must decide that asset risk management is important. Maintenance professionals should embrace risk management because risk management improves maintenance effectiveness.

After concluding this chapter, the reader should have a better understanding of both risk and risk management. This includes how to define risk as well as several proven methods to manage risk. These include methods such as FMECA and HAZOPS. There are accepted standards to help the risk manager with the process to identify and manage risk.

Management needs to accept asset risk management as a vital maintenance process in managing a company. Without an adequate risk management process, it is very possible that the next edition of this book may include another company in the list of major failures. Do not let it be yours.

8 Reliability by Design
Reliability-Centered Maintenance

Don Barry
Original by James Picknell

CONTENTS

In an evolving technological world, the growing mechanization and automation as well as the increased focus on cost, productivity, and risk demand precise asset management strategies and maintenance tactics. Reliability-centered maintenance (RCM) is well established as the preeminent technique for establishing the best maintenance task to be done in a scheduled maintenance program. In this chapter we introduce RCM, describe it in detail, and explore its history. We discuss who should be using RCM and why. RCM is increasingly important as society becomes less tolerant of risk and more interested in holding real people responsible for business failures (e.g., safety, environmental, or operational) and as productivity demands increase. RCM can be deployed to effectively improve plant availability and reliability, product quality, return on (equipment) asset, as well as equipment life. When well executed, RCM can be the effective tool for proactively defining maintenance to ensure safe and environmentally friendly plants. A demonstrated effective maintenance program can also help plants qualify for lower commercial insurance risk.

The entire RCM process is described, along with important factors to consider as you work through an RCM analysis. The chapter includes a flow diagram for a suggested process that complies with the Society of Automotive Engineers (SAE) standard for RCM programs and a simplified decision logic diagram for selecting appropriate and effective maintenance tactics. We explain the deliverables and how to get them from the vast amount of data usually produced by the process. The scope of RCM projects is also described to give you a feel for the effort involved. We thought it would be helpful also to include an effective RCM implementation that shows team composition, size, time required, effort required, and tools that are available to make the task easier.

Of course, RCM has been used in several environments and in numerous ways. Some are only slight variations of the thorough process, others are less rigorous, and some are downright dangerous. These methods are discussed along with their advantages and disadvantages. We also examine why RCM programs fail and how to recognize and avoid those problems. As responsible maintenance and engineering professionals, we all want to improve our organization's effectiveness. You will learn how to gradually introduce RCM, successfully, even in the most unreceptive environments.

This chapter is likely to generate some controversy and discussion. That is just what we intend. In law, the concept of justice and legal realities are sometimes in conflict. Similarly, you will find that striking a balance between what is right and what is achievable is often a difficult challenge.

8.1   INTRODUCTION

As mentioned, RCM is the preeminent method for establishing the best maintenance task to be done in a scheduled maintenance program. For years it has been demonstrated to be highly effective in numerous industries—civil and military aviation, military ship and naval weapon systems, electric utilities, and the chemical industry. It's mandated in civil aircraft and often, as well, by government agencies procuring military systems. Increasingly, RCM is selected by companies when reliability is important for safety or environmental reasons or simply to keep the plant running at maximum capacity.

The published SAE Standard JA1011,[1] "Evaluation Criteria for Reliability-Centered Maintenance (RCM) Processes," outlines the criteria a process must meet to be called RCM. This new standard determines, through seven specific questions, whether a process is RCM, although it doesn't specify the process itself. In this chapter, we describe a process designed to satisfy the SAE criteria. Plus, we've included several variations that don't necessarily answer all seven questions but are still called RCM. Consult the SAE standard for a comprehensive understanding of the complete RCM criteria.

In our increasingly litigious society, we are more and more likely to be sued for accidents that at one time would have been accepted as being out of our control. Today, the courts take a harsh stand with those who haven't done all they could to eliminate risks. There are many examples in recent decades of disastrous accidents that could have been avoided, such as the carnage at Bhopal, the Challenger explosion, and the tendency of the original Ford Pinto gas tanks to explode when rear-ended.

The incident at Bhopal triggered sweeping changes in the chemical industry. New laws were established, such as the Emergency Planning and Community Right to Know Act passed by the U.S. Congress in 1986 and the Chemical Manufacturers Association's "Responsible Care" program. Following the Pinto case and others of a similar nature, consumer goods manufacturers are being held to ever more stringent safety standards.

Despite its wide acceptance, RCM has been criticized as being too expensive just to solve the relatively simple problem of determining what maintenance to do. These criticisms often come from those in industries where equipment reliability and environmental compliance and safety are not major concerns. Sometimes, as well, they result from failing to manage the RCM project properly as opposed to flaws within the RCM process itself. Alternative methods to RCM are covered later in this chapter, although we don't recommend that you use them. A full description of their risks is included.

We all want to get the maximum from our transportation and production systems, infrastructure, and plants. They're very expensive to design and build, and downtime is costly. Some downtime is needed, of course, to sustain operations and for logical breaks in production runs or transportation schedules. RCM helps eliminate unnecessary downtime, saving valuable time and money.

RCM generates the tasks for a scheduled maintenance program that logically anticipate specific failure modes. It can also effectively do the following:

- Detect failures early enough for them to be corrected quickly and with little disruption.
- Eliminate the cause of some failures before they happen.
- Eliminate the cause of some failures through design changes.
- Identify those failures that can safely be allowed to happen

This chapter provides an overview of the different types of RCM:

- Aircraft versus military versus industrial
- Functional versus hardware
- Classical (thorough) versus streamlined and lite versions

We describe the basic RCM process step by step. This includes a brief overview of critical equipment and failure modes, effects, and criticality analysis (FMECA), which are covered thoroughly in Chapter 7, "Assessing and Managing Risk."

8.2   WHAT IS RCM?

RCM is a logical, technical process that determines which maintenance tasks will ensure a reliable, "as-designed" system, under specified operating conditions, in a specified operating environment. Each of the various reference documents describing RCM applies its own definition or description. We refer readers to SAE JA1011[1] for a definitive set of RCM criteria.

RCM takes you from start to finish, with well-defined steps arranged in a sequence. It is also iterative: it can be carried out a few different ways until initial completion. RCM determines how to improve the maintenance plan, based on experience and optimizing techniques. As a technical process, RCM delves into the depths of how things work and what can go wrong with them. Using RCM decision logic, you select maintenance interventions or tasks to reduce the number of failures, to detect and forecast when one will be severe enough to warrant action, to eliminate it altogether, or to accept it and run until failure.

The goal of RCM is to make each system as reliable as it was designed to be. Each component within a system has its own unique combination of failure modes and failure rates. Each combination of components is also unique, and failure in one component can cause others to fail. Each "system" operates in its own environment, consisting of location, altitude, depth, atmosphere, pressure, temperature, humidity, salinity, exposure to process fluids or products, speed, and acceleration. Depending on these conditions, certain failures can dominate. For example, a level switch in a lube oil tank will suffer less from corrosion than if it were in a salt water tank. An aircraft operating in a temperate maritime climate is likely to corrode more than it would in an arid desert. The environment and operating conditions can have significant influence on what failures will dominate the system.

It is this impact of operating environment on a system's performance and failure modes that makes RCM so valuable. Technical manuals often recommend a maintenance program for equipment and systems, and, sometimes, they include the effects of the operating environment. For example, car manuals specify different lubricants and antifreeze densities that vary with ambient operating temperature. However, they don't usually address the wear and tear of such things as driving style (aggressive vs. timid) or how the vehicle is used (taxi or fleet vs. weekly drives to visit grandchildren). In industry, manuals are not often tailored to any particular operating environment. An instrument air compressor installed at a subarctic location may have the same manual and dew point specifications as one installed in a humid tropical climate. RCM specifically addresses the environment experienced by the fleet, facility, or plant.

8.3   WHY USE RCM?

RCM works and is cost-effective. RCM has been around for about 30 years, since the late 1960s, beginning with studies of airliner failures carried out by Nowlan and Heap.[2] United Airlines wanted to reduce the amount of maintenance for what was then the new generation of larger wide-bodied aircraft. Previously, aircraft maintenance was based on experience, and, because of obvious safety concerns, it was quite conservative. As aircraft grew larger, with more parts and therefore more things to go wrong, maintenance requirements similarly grew, eating into flying time needed to generate revenue. In the extreme, achieving safety could have become too expensive to make flying economical. But, thanks to RCM and to United's willingness to try a new approach, the airline industry has been able to develop almost entirely proactive maintenance. This resulted in increased flying hours, with a drastically improved safety record.

In fact, aircraft safety has been consistently improving since RCM was introduced. In addition, RCM has reduced the number of maintenance man hours needed for new aircraft per flight. Why? RCM identifies functional failures that can be caught through monitoring before they occur. It then reveals which failures require some sort of usage or time-based intervention, develops failure finding tests, and indicates whether system redesign is needed. Finally, it flags failures that can be left to occur because they cause only minor problems. Where aircraft are concerned, this is a very small number indeed. Frequent fliers seldom experience delays for mechanical or maintenance-related problems, and airlines are usually able to meet their flight schedules.

RCM has also been used successfully outside the aircraft industry. Those managing military capital equipment projects, impressed by the airlines' highly reliable equipment performance, often mandate the use of RCM. In one ship-building project, the total maintenance workload for the crew was cut by almost 50%, compared with other similarly sized ships. At the same time, the ship's service availability improved 60% to 70%. The amount of downtime needed for maintenance was greatly reduced. In that project, the cost of performing RCM was high, in the millions of dollars, but the payback, in hundreds of millions, justified it.

The mining industry usually operates in remote locations far from sources of parts, materials, and replacement labor. Consequently, miners, like the navy and airline industry, want high reliability, that is, minimum downtime and maximum productivity from the equipment. RCM has been a huge benefit. It's made fleets of haul trucks and other equipment more available while reducing maintenance costs for parts and labor and planned maintenance downtime.

In process industries, RCM has been successful in chemical plants; oil refineries; gas plants; remote compressor and pumping stations; mineral refining and smelting; steel, aluminum, pulp, and paper mills; tissue converting operations; food and beverage processing; and breweries. RCM can be applied anywhere that high reliability and availability are important.

A recently published study on asset management strategies revealed that improved reliability was the top motivator for improving their maintenance processes over just a focus on costs or asset uptime. However, when asked in the same survey where they were in their approach around proactively planning tasks for maintenance, less than 30% indicated that they had actually engaged in RCM.

Figure 8.1 shows that almost 90% of companies polled by Aberdeen Group,[3] Dec. 2006, reported that preventive maintenance (PM) is their most commonly used proactive maintenance strategy. While predictive maintenance and RCM have also proven to produce desirable results, many companies have yet to leverage this as part of their living program in maintenance. This suggests that more than 70% of

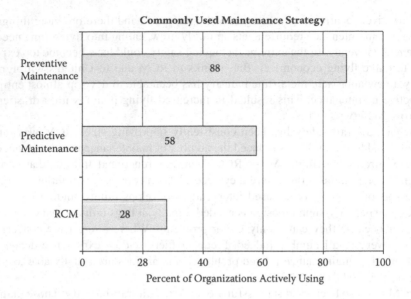

FIGURE 8.1   Currently adopted proactive maintenance strategies.

companies have yet to reap the real benefits of RCM. Many companies do too little of the right maintenance or too much of the wrong maintenance (i.e., preventive maintenance) and leave money and risk on the table when it comes to leveraging their assets.

Today, RCM enables maintenance and operations to respond quickly and positively to the business dynamics. It can be used to support new production technologies such as digital electronics, pneumatics, and hydraulics. RCM recognizes that successful maintenance demands a complete understanding among the operators, maintainers, and design engineers.

Properly executed RCM will eliminate maintenance tasks that provide no value in terms of equipment's expected functionality. It will generate a comprehensive understanding of the required maintenance tasks and the frequency and resources (skills, tools, and spare parts) required to perform them. It will direct the team to identify real opportunities and tasks to improve the following:

- Safety and environmental integrity
- Plant operating performance
- Maintenance cost-effectiveness
- The length of an asset's life cycle

Teams that focus on RCM typically have a greater understanding of how each area of their plant contributes to the value of their business; they become focused and motivated. The resulting database of maintenance requirements provides an audit trail of assessment findings and recommendations. It helps a plant adapt to change should its asset's operating context change or personnel change. It defines the maintenance that should be done or not done and provides a framework for helping a plant work through all of these issues.

Commercial insurance performs a critical role in the world economy. Managing risk requires an in-depth knowledge of systems and processes used by a plant. As insurance companies look to manage a plant's risk, many perceive RCM to be onerous and resource intensive; however, the payback can be in a few months or even weeks depending on the application and the effectiveness of the team working the RCM process. In short, leading maintenance organizations do an RCM analysis, implement RCM as a prime influence on which maintenance task will be placed into their computerized maintenance management system (CMMS) and workflow, and leverage RCM as part of their living program.

8.4   WHO SHOULD USE RCM?

RCM should be used by any plant, fleet, or building where productivity is crucial. That includes companies that can sell everything they produce, where uptime and high equipment reliability and predictability are very important. It also includes anyone producing to meet tight delivery schedules, such as just-in-time parts delivery to automotive manufacturers, where equipment availability is critical. Availability means that physical assets (e.g., equipment, plant, fleet) are there when needed: the higher the equipment availability, the more productive the assets. Availability is measured by dividing the time assets are available by the total time needed for them to run.

If a failure causes damages, there's a growing trend in our increasingly litigious society for those affected to sue. Practically everyone, then, can benefit from RCM. There is no better way of ensuring that the right maintenance is being done to avoid or mitigate failures.

$$Ao = Uptime/Total\ Time$$

Uptime is simply *Total Time* minus *Downtime*. Downtime for unplanned outages is the total time to repair the failures, or mean time to repair (MTTR). Reliability takes into account the number of unplanned downtime incidents you suffer. Reliability is, strictly speaking, a probability. A commonly used interpretation is that large values of mean time between failures (MTBF) indicate more highly reliable systems. Generally, plants, fleets, and buildings benefit from greater MTBF because it means fewer disruptions.

RCM is important in achieving maximum reliability—longest MTBF. For most systems, MTBF is long, and, typically, repair time (MTTR) is short. Reducing MTTR requires a high level of maintainability. Adding the two (MTTR + MTBF) gives the total time that a system would be available if it never broke down. Availability is the portion of this total time that the system is actually in working order and available to do its job.

Availability can be rewritten as

$$Ao = MTBF/(MTBF + MTTR)$$

Mathematically, we can see that maximizing MTBF and minimizing MTTR will increase *Ao*.

Generally, an operation is better off with fewer downtime incidents. If downtime is seriously threatening the manufacturing process, delivery schedule, and overall productivity, additional measures clearly are needed. Reliability is paramount. RCM will help you maximize MTBF while keeping MTTR low.

8.5 THE RCM PROCESS

RCM has seven basic steps to meet the criteria of the published SAE standard:

1. Prioritize and identify the equipment and system to be analyzed.
2. Determine its functions and asset operating context.
3. Determine what constitutes failure of those functions.
4. Identify what causes those functional failures.
5. Identify their impacts or effects.
6. Use RCM logic to select appropriate maintenance tactics.
7. Document the final maintenance program, and refine it as operating experience is gained.

Steps 2 through 5 constitute the failure modes and effects analysis (FMEA) portion of RCM. FMEA is discussed in greater detail in Chapter 7. Some practitioners limit their FMEA analysis only to failures that have occurred, ignoring those that can be effectively prevented. But FMEA, as used in the RCM context, must consider all possible failures—the ones that have occurred, can currently be prevented, and haven't yet happened (Figure 8.2).

In the first step of RCM, you decide what to analyze. A plant usually contains various processes, systems, and equipment. Each does something different, and some are more critical to the operation than others. Some equipment may be essential for environmental or safety reasons but have little or no direct impact on production. For example, if a wastewater effluent treatment system that prevents untreated water being discharged from the plant goes down, it doesn't stop production. The consequences can still be great, though, if environmental regulations are flouted. The plant could be closed and the owners fined or jailed.

Start by establishing criteria to determine what is important to the operation. Then, use them to decide which equipment or systems are most important, demanding the greatest attention. There are many possible criteria, including the following:

- Personnel safety
- Environmental compliance
- Production capacity
- Production quality
- Production cost (e.g., maintenance costs)
- Public image

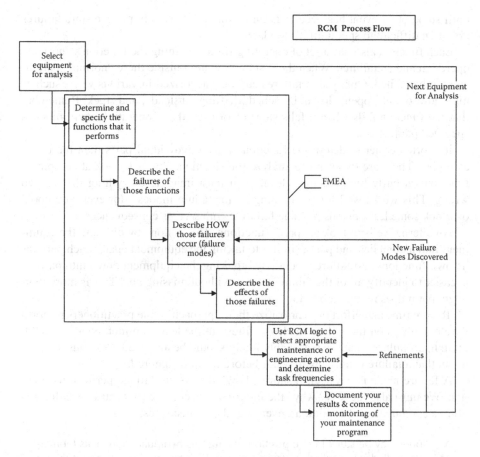

FIGURE 8.2 Functions are more easily identified as the hardware detail increases.

When a failure occurs, the effect on each of these criteria can vary from "no impact" or "minor impact" to "increased risk" or "major impact." Each criterion and how it's affected can be weighted. For example, safety usually rates higher than production capacity. Likewise, a major impact is weighted higher than no impact. For each system or equipment being considered for RCM analysis, imagine a "worst-case" failure, and then determine its impact on each criterion. Multiply the weights of each criterion, and add them together to arrive at a "criticality score." Items with the highest score are most important and should be analyzed first.

There are both active and passive functions in each system. Active functions are usually the obvious ones, for which we name our equipment. For example, a motor control center controls the operation of various motors. Some systems also have less obvious secondary or even protective functions. A chemical process loop and a furnace, for instance, both have a secondary containment function. They may also include protective functions such as thermal insulating or chemical corrosion resistance. Keep in mind that some systems, such as safety systems, do not become active until some other event has occurred. Unfortunately, normally passive state failures are often difficult to spot until it's too late.

Each function also has a set of operating limits, defining the function's "normal" operation and its failures. When the system operates outside these "normal" parameters, it has failed. Our system failures can be categorized in various ways, such as high, low, on, off, open, closed, breached, drifting, unsteady, and stuck. Remember that the function fails when it falls short of or exceeds its operating environment's specified parameters.

It is often easier to determine functions for the individual parts than the entire assembly. There are two ways to analyze the situation. One is to look at equipment functions, a fairly high assembly level. You must imagine everything that can go wrong. This works well for pinpointing major failure modes. However, you could overlook some less obvious possible failures, with serious consequences.

An alternative is to look at "part" functions. This is done by dividing the equipment into assemblies and parts, similar to taking the equipment apart. Each part has its own functions and failure modes. By breaking the equipment down into parts, it is easier to identify all of the failure modes, without missing any. This is more thorough, but it does require a bit more work.

To save time and effort or to prioritize their approach, some practitioners perform Pareto analysis on the failure modes to filter out the least common ones. In RCM, though, all failures that are reasonably likely should be analyzed. You must be confident that a failure is unlikely to occur before it can be ignored.

A failure mode is physical. It shows "how" the system fails to perform its function. We must also identify "why" the failure occurred. The root cause of failures is often a combination of conditions, events, and circumstances.

A cylinder may be stuck in one position because its hydraulic fluid lacks lubrication. The cylinder has failed to stroke or provide linear motion. "How" it fails is the loss of lubricant properties that keep the sliding surfaces apart. There are many possibilities, though, for "why." For instance, the cause could be a problem with the fluid, either using the wrong one, leaking, dirt, or surface corrosion due to moisture. Each of these can be addressed by checking, changing, or conditioning the fluid.

Not all failures are equal. They can have varying effects on the rest of the system, plant, and operating environment. The previously described cylinder's failure could be severe if, by actuating a sluice valve or weir in a treatment plant, excessive effluent flows into a river. Or the effect could be as minor as failing to release a "dead-man" brake on a forklift truck stacking pallets in a warehouse. In one case, it's an environmental disaster; in the other, it's only a maintenance nuisance. But if an actuating cylinder on the brake fails in the same forklift, while it's in operation, there could be serious injury.

By knowing the consequences of each failure we can determine what to do: whether it can be prevented, can be predicted, can be avoided altogether through

FIGURE 8.3 Normally, a fire sprinkler is dormant. We know of failure only when it's too late!

periodic intervention, can be eliminated through redesign, or requires no action. We can use RCM logic to choose the appropriate response.

RCM helps to classify failures as hidden or obvious and to classify whether they have safety, environmental, production, or maintenance impacts. These classifications lead the RCM practitioner to default actions if appropriate predictive or preventive measures cannot be found. For example, a fire sprinkler cannot be detected or predicted while it is in normal operation (dormant), but by designing in redundancy of sprinklers we can mitigate the consequences of failure (Figure 8.3). More severe consequences typically require more extensive mitigating actions. A version of the decision logic is depicted in Figure 8.4.

Most systems failures involving complex mechanical, electrical, and hydraulic components will fail randomly. You can't confidently predict them. Still, many are detectable before the functional failure takes place. For example, if a booster pump fails to refill a reservoir providing operating head to a municipal water system, the

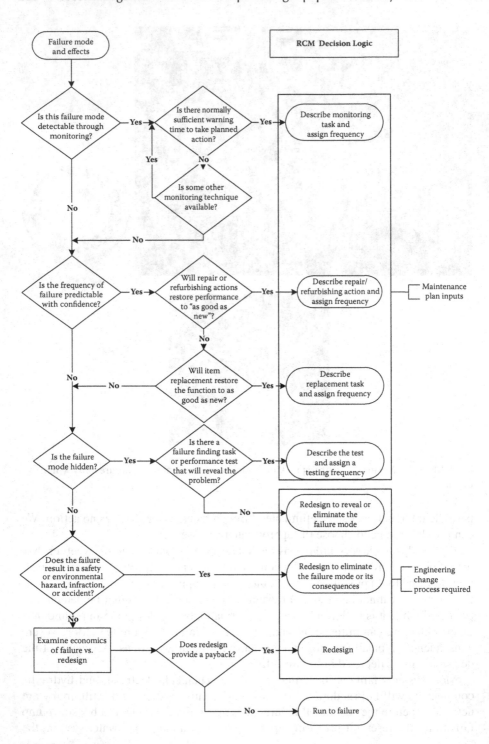

FIGURE 8.4 RCM decision logic.

system doesn't have to break down. If you watch for the problem, it can be detected before municipal water pressure is lost. That is the essence of condition monitoring. We look for the failure that has already happened but hasn't progressed to the point of degrading the system. Finding failures in this early stage helps protect overall functional performance.

Since most failures are random, RCM logic first asks if it's possible to detect the problem in time to keep the system running. If the answer is "yes," condition monitoring is needed. You must monitor often enough to detect deterioration, with enough time to act before the function is lost. For example, in the case of the booster pump, check its performance once a day if you know that it takes a day to repair it and two days for the reservoir to drain. That provides a buffer of at least 24 hours after detecting and solving the problem before the reservoir system is adversely affected.

If you can't detect the problem in time to prevent failure, RCM logic asks if it's possible to reduce the impact by repairing it. Some failures are quite predictable even if they can't be detected early enough. For example, we can safely predict brake wear, belt wear, tire wear, erosion. These failures may be difficult to detect through condition monitoring in time to avoid functional failure or may be so predictable that monitoring for the obvious isn't warranted. Why shut down equipment monthly to monitor for belt wear if you know it isn't likely to appear for two years? You could monitor every year, but, in some cases, it's more logical to simply replace the belts automatically every two years. There is the risk, of course, that a belt could either fail earlier or still be operating well when it's replaced. If you have sufficient failure history you may be able to perform Weibull analysis to determine whether the failures are random or predictable.

If the previous approach isn't practical, you may have to replace all the equipment. Usually, this makes sense only in critical situations because it typically requires an expensive sparing policy. The cost of lost production would have to be more than entirely replacing equipment and storing the spares.

Because safety and protective systems are normally inactive, you may not be able to monitor for deterioration. If the failure is also random, it may not make sense to replace the component on a timed basis because the new part might fail as soon as it has been installed. We simply can't tell, again because the equipment won't reveal the failure until switched to its active mode. In these cases, some sort of testing may be possible. In our earlier classification of failures we sorted these hidden ones from the rest. If condition monitoring and usage or time-based intervention aren't practical, you can use RCM logic to explore functional failure finding tests. These tests can activate the device and reveal whether it's working. If such a test isn't possible, redesign the component or system to eliminate the hidden failure. Otherwise, there could be severe safety or protective consequences, which obviously are unacceptable.

For nonhidden problems that can't be prevented either through early prediction or usage or time-based replacements, you can either redesign or accept the consequences. For safety or environmental cases, where the consequences are unacceptable, redesign is the best decision. For production-related cases, whether to redesign or run to failure may depend on the cost of the consequences. If it's likely that production will completely shut down for a long time, it would be wise to redesign. If the production loss is negligible, run to failure is appropriate. If there aren't any production consequences but there are maintenance costs, the same applies. In these cases, the decision is based on economics—the cost of redesign versus the cost of failure (e.g., lost production, repair costs, overtime).

Task frequency is often difficult to determine. It's invaluable when using RCM to know the failure history, but, unfortunately, it isn't always available. There isn't any operating history at all, for instance, for a system still being designed. You're up against the same problem with older systems where records haven't been kept. An option is to use generic failure rates from commercial or private databases. Failures, though, don't happen exactly when predicted: some will be random, some will become more frequent late in the "life" of the equipment, and so on. Allow some leeway. Recognize also that the database information may be faulty or incomplete. Be cautious, and thoroughly research and deal with each failure mode in its own right.

Once RCM is completed, you need to group similar tasks and frequencies so that the maintenance plan applies in an actual working environment. You can use a slotting technique to simplify the job. This consists of a predetermined set of frequencies such as daily, weekly, monthly, every shift, quarterly, semiannually, and annually or by units produced, distances traveled, or number of operating cycles. Choose frequency slots closest to those that maintenance and operating history show are best suited to each failure mode. Later, these slots can be used to group tasks with similar characteristics together in a workable maintenance plan.

After you run the failure modes through the previous logic, consolidate the tasks into a maintenance plan. This is the final "product" of RCM. Then, those maintaining and operating the system must continually strive to improve the product. The original task frequencies may be overly conservative or too long. If too many preventable failures occur, it shows that proactive maintenance isn't frequent enough. If there are only small failures or the preventive costs are higher than before, maintenance frequencies may be too high. This is where optimizing task frequency is important.

The output of RCM is a maintenance requirements document, which describes the condition monitoring, time- or usage-based intervention, failure finding tasks, and the redesign and run-to-failure decisions. It is not a "plan" in the true sense. It doesn't contain typical maintenance planning information like task duration, tools and test equipment, parts and materials requirements, trades requirements, and a detailed sequence of procedures to follow. These are logical next steps after RCM has been performed.

The maintenance requirements document is not, however, just a series of lists transferred from the RCM worksheets or database into a report. RCM takes you through a rigorous process that identifies and addresses individual failure modes:

- For each plant, there are numerous systems.
- For each system, there can be various equipment.
- For each piece of equipment, there are several functions.
- For each function, there may be several failures.
- For each failure, there may be many failure modes with varying effects and consequences.
- For each failure mode, there is a task, which may have assigned parts to be considered.

In a complex system, there can be thousands of tasks. To get a feel for the size of the output, consider a typical process plant that has spares for only about 50% of its components. Each may have several failure modes. The plant probably carries some 15,000 to 20,000 individual part numbers (stock-keeping units) in its inventory. That means there could be around 40,000 equipment components with one or more failure modes. If we analyzed the entire plant, the maintenance requirements document would be huge.

Fortunately, a limited number of condition monitoring techniques are available—just over 50. These cover most random failures, especially in complex mechanical, electrical, and hydraulic systems. Tasks can be grouped by technique (e.g., vibration analysis), location (e.g., the machine room), and sublocation on a route. Hundreds of individual failure modes can be organized this way, reducing the number of output tasks for detailed maintenance planning. It's important to watch the specified frequencies of the grouped tasks. Often, we don't precisely determine frequencies, at least initially. It may make sense to include tasks with frequencies like "every four days" or "weekly" but not "monthly" or "semiannually." This is where the slotting technique described earlier comes in handy.

Time- or usage-based tasks are also easy to group together. You can assemble all the replacement or refurbishment tasks for a single piece of equipment by frequency into a single overhaul task. Similarly, multiple overhauls in a single area of a plant may be grouped into one shutdown plan.

Another way to organize the outputs is by who does them. Tasks assigned to operators are often performed using the senses of touch, sight, smell, or sound. These are often grouped logically into daily, shift, or inspection rounds checklists. In the end there should be a complete listing that tells what maintenance to do and when. The planner determines what is needed to execute the work.

8.6 WHAT DOES IT TAKE TO DO RCM?

RCM very thoroughly examines plants and equipment. It involves detailed knowledge of how equipment and systems operate, what's in the equipment, how it can fail to work, and the impact on the process, plant, and its environment.

You must practice RCM frequently to become proficient and make the most of its benefits. To implement RCM, do the following:

- Select a team of practitioners.
- Train them in RCM.
- Teach other "stakeholders" in the plant operation and maintenance what RCM is and what it can achieve.
- Select a pilot project to demonstrate success and improve upon the team's proficiency.
- Roll out the process to other areas of the plant.

Typically it is best to demonstrate the success of RCM through a pilot. Before the RCM team begins the analysis, determine the plant baseline reliability and availability measures, as well as proactive maintenance program coverage and compliance. These measures will be used later to compare what has been changed and how successfully.

8.6.1   THE TEAM, SKILLS, KNOWLEDGE, AND OTHER RESOURCES NEEDED

A multidisciplinary team is essential, bringing in specialists when needed. The team needs to know the day-to-day operations of the plant and equipment along with detailed knowledge of the equipment itself. This dictates at least one operator and one maintainer. They must be hands-on and practical, be willing and able to learn the RCM process, and be motivated to make it a success. The team must also be versed in plant operations, supplied usually by a senior operations person, such as a supervisor who has risen through the ranks. The team needs to know planning, scheduling, and overall maintenance operations and capabilities to ensure that the tasks are truly doable in the plant. You may have to contract out some of the work to qualified service providers, especially infrequent yet critical equipment monitoring. This expertise can be provided by someone with a supervisory maintenance background.

Finally, detailed equipment design knowledge is important. The maintainers will know how and why the equipment is put together, but they may have a hard time quantifying the reasons or fully understand the engineering principles. An engineer or senior technician or technologist from maintenance or production, usually with a strong background in either mechanical or electrical, is needed on the team.

RCM is very much a learning process for its practitioners. Facilitating five team members is about optimum to fulfill the requirements. Too many people slow the progress. Too few people equal too much time spent trying to understand the systems and equipment. The team will need help to get started. Someone in-house may already have done RCM, but more likely you will have to look outside the company. Training is usually followed up with a pilot project, producing a real product.

RCM is thorough, which means it can be time-consuming. Training the team usually takes three days to a week but can last as long as a month, depending on the approach. Training of other stakeholders can take from a couple of hours to a day or two, depending on their degree of interest and need to know. Senior executives and plant managers should be involved, so they know what to expect and what support is needed. Operations and maintenance management members must understand the time demands on their staff and what to expect in return. Finally, operators and maintainers must also be informed and involved. Since their coworkers are on the RCM team, they are probably shouldering additional work.

The pilot project time can vary widely, depending on the complexity of the equipment or system selected for analysis. A good rule of thumb is to allow the team a month for pilot analysis to ensure they learn RCM thoroughly and are comfortable using it. On average, each failure mode takes about half an hour to analyze. Determining the functional failure to task frequency can take from 6 to 10 minutes per failure mode. Using our previous process plant example, a very thorough analysis of all systems comprising at least 40,000 items (many with more than one failure mode) would entail over 20,000 man hours (that's nearly 10 person years for an entire plant). When divided by five team members, the analysis could take up to two years. It's a big job.

8.7   IS RCM AFFORDABLE?

8.7.1   WHAT TO EXPECT TO PAY FOR TRAINING, SOFTWARE, CONSULTING SUPPORT, AND STAFF TIME

In the previous example, RCM requires a lot of effort. That effort comes at a price. A total of 10 man years at an average of, say, $70,000 per person, adds up to $700,000 for staff time alone. The training for the team and others will require a couple of weeks from a third party, at consultant rates. The consultant should also be retained for the entire pilot project—that's another month. Even though consulting rates are steep, running into thousands of dollars per day, it's worth it. The price is small when you consider that lives can be saved and environmental catastrophe or major production outages can be avoided. Experienced RCM experts in the field have seen RCM prevent major safety and production calamities.

Software is available to help manipulate the vast amount of analysis data you generate and record. There are several databases to step you through the RCM process and store results. Some of the software costs only a few thousand dollars for a single user license. Some of it is specifically RCM training, while some comes as part of large computerized maintenance management systems. Prices for these high-end systems that include RCM are typically hundreds of thousands of dollars. If you're working with RCM consultants, you'll find that many have their preferred software tools. A word of caution about RCM systems: there are many versions available. Some were created before the SAE standard was published, and some have been marketed since. They should be examined to see if they comply.

To decide on task frequencies, you must know the plant's failure history, which is generally available through the maintenance management system. If not, you can obtain failure rates from databases. Another option for documented failure rates is to inquire from the current experience in your existing plant. Chances are they have a good idea how often a failure has happened. Likely your parts keeper will know how often a part was ordered for a specific piece of equipment. However, to build queries and run reports you may need help. There are external reliability databases you can use, but they're often difficult to find and charge a user or license fee.

8.8   RCM VARIETIES

RCM comes in different varieties, depending on the application, including the following:

- Aerospace (commercial airlines—described in MSG-3)[5]
- Military (various for naval and combat aircraft—described in numerous U.S. Military Standards)[6]
- Commercial as described by Smith[7] and Moubray[8]
- Streamlined versions, some of which don't meet the SAE JA1011 criteria and are no longer called RCM

In the aerospace variation, there are two basic types (besides the terminology). In the first, any structural components are thoroughly analyzed for stress, often using finite element modeling techniques. You identify weaknesses in the airframe structure that must be regularly inspected and undergo nondestructive testing. In the second case, once you decide on a maintenance plan, you follow the RCM logic questions about time-based methods, such as condition-based monitoring (CBM). This is a conservative approach, justified by serious safety concerns if an aircraft fails.

The military standards describe the same processes, with examples from military applications, using military equipment terminology. The commercial versions are what we have described in this book.

8.9 CLASSICAL RCM, STREAMLINED, AND ALTERNATIVE TECHNIQUES

8.9.1 CLASSICAL RCM

Classical RCM is sometimes used to describe the original process, laid out by Smith and Moubray and referenced in the SAE standard. Classical RCM has proven highly successful in numerous industries, particularly at the following:

- Reducing overall maintenance effort and costs
- Improving system and equipment performance to achieve design reliability
- Eliminating planned to-be-installed redundancy and reducing capital investment

Many cost and effort reductions have occurred in industries that were:

- Overmaintaining (e.g., civil and military aircraft, naval ships, nuclear power plants)
- Not maintaining, with low reliability (e.g., thermal power plants, mining–haul trucks, water utilities)
- Overly conservative in design practices (e.g., former public utilities that must now survive in deregulated environments, oil and gas/petrochemical)

Despite these successes, many companies fail trying to implement RCM.
There are many reasons for failure, including the following:

- There is a lack of management support and leadership.
- There is a lack of vision about what RCM can accomplish (i.e., the RCM team and the rest of the plant don't really know what it's for and what it will do for them).
- There is no clearly stated reason for doing RCM (i.e., it becomes another "program of the month").
- There are not enough resources to run the program, especially in "lean manufacturing."

- There is a clash between RCM's proactive approach and a traditional, highly reactive plant culture (e.g., RCM team members find themselves being pulled from their work to react to day-to-day crises).
- The company gives up before RCM is completed.
- There is a lack of focus on the priority of critical equipment to be assessed.
- There are continued errors in the process and results that don't stand up to practical "sanity checks" by rigorous maintainers. This is often due to a lack of full understanding of FMEA, criticality, RCM logic, condition-based monitoring techniques, the distinction between condition-based maintenance (CBM) and time-based maintenance (TBM), and reluctance to accept run-to-failure conclusions.
- There is a lack of available information about the equipment or systems being analyzed, which isn't necessarily significant but often stops people cold.
- There is criticism that RCM-generated tasks seem the same as those already in long use in the preventive maintenance (PM) program. It can be seen as a big exercise that is merely proving what is already being done.
- There is a lack of measurable success early in the RCM program. This is usually because the team hasn't established a starting set of measures, an overall goal, and ongoing monitoring.
- Results don't happen quickly enough, even if measures are used. The impact of doing the right type of PM often isn't apparent immediately. Typically, results are seen in 12 to 18 months.
- There is no compelling reason to maintain the momentum or even start the program (e.g., no legislated requirement, the plant is running well, the company is making money in spite of the lack of RCM).
- The program runs out of funding.
- The organization lacks the ability to implement RCM results (e.g., no system that can trigger PM work orders on a predetermined basis).

This list is a blueprint for how to ensure failure. One criticism of RCM is that it's "the $1 million solution to the $100,000 problem." This complaint is unfounded. It's how you manage RCM that's usually at fault, not the process itself. All of the previous reasons for failure can be traced back to management flaws.

There are many solutions to the problems we've outlined. One which works well is using an outside consultant. A knowledgeable facilitator can help get you through the process and maintain momentum. Often, however, companies stop pursuing change as soon as the consultant leaves. The best chance of success is using help from outside in the early stages of your RCM program. Successful consultants recognize the reasons for failure and avoid them.

8.9.2 Streamlined RCM

There are now methods that shortcut the RCM process, which, where appropriate, can be effective. But make sure they're proper and responsible, especially where there are health or safety risks. All risks should be quantified and managed. No matter how effective, shortcuts cannot be considered RCM unless they comply with SAE JA1011. Some of these shortcut RCM methods have become known as streamlined or lite RCM.

In one variation, RCM logic is used to test the validity of an existing PM program or existing failure modes. This approach, though, doesn't recognize what may already be missing from the program, which is the reason for doing RCM in the first place. This is not RCM. For example, if the current PM program extensively uses vibration and thermographic analysis but nothing else, it probably works well identifying problems causing vibrations or heat but not failures such as cracks, fluid reduction, wear, lubricant property degradation, wear metal deposition, surface finish, and dimensional deterioration. Clearly, this program does not cover all possibilities. Applying RCM logic will result, at best, in minor changes to what exists. The benefits may be reduced PM effort and cost, but anything that isn't already covered or any failure mode that has not already been experienced or identified will be missed. This streamlined approach adds minimal value. In fact, it's irresponsible.

8.9.3 CRITICALITY

Criticality as a focus to determine which equipment to focus on first (perhaps as a pilot) or to help determine focus priorities works well when applied to RCM right in the beginning, as mentioned in the RCM process section of this chapter. However, criticality as another RCM variation can be used to weed out failure modes from ever being analyzed. This must be done carefully. One approach is described thoroughly in MIL STD 1629A,[4] but there are several different techniques. Basically, failures are not analyzed if their effects are considered noncritical, if they occur in noncritical parts or equipment, or if they don't exceed the set criticality hurdle rate. Reducing the need for analysis can produce substantial savings.

When criticality is applied to weed out failure modes, there should be relatively little risk of causing a critical problem. Because classical RCM analysis has already largely been performed, using criticality is relatively risk-free. The drawback, though, is the effort and cost expended deciding to do nothing. In the end, there are virtually no savings. You can cut costs in both areas by reducing RCM analysis before most of it is done. Using a criticality hurdle rate indicates many possible equipment failures, without having to document them, which can mean big maintenance savings. The downside is that, even when you look at worst-case scenarios, you run the risk that a critical failure will slip through unnoticed. How well this technique works depends on how much the RCM team knows. The greater their plant knowledge and experience, the less risk can be taken. Often, then, the strongest team members are the plant's best maintainers and operators.

This approach may be the right choice if you confidently know and accept the consequences of failure in production, maintenance, cost, environmental, and human terms. For example, many failures in light manufacturing have relatively little fallout other than lost production time. But in other industries, you could be sued if a failure could have been prevented or was mitigated through RCM. The nuclear power, chemical processing, pharmaceutical, and aircraft industries especially are vulnerable. SAE JA1011 stipulates that the method used to identify failure modes must show what is likely to occur. Of course, the level at which failure modes are identified must be acceptable to the owner or user. There is room for judgment, and, if done properly, this method can meet the SAE criteria that define RCM. Criticality helps you prioritize so that the most important items are addressed first. It cuts the RCM workload that typically comes with large volumes, systems, and equipment, and limited resources to analyze them.

Many companies suffer from the failures we've described, as well as others. Without the force of law, though, RCM standards such as SAE JA1011 are often treated as mere guidelines that don't have to be followed. Sometimes, the people making the decisions aren't familiar with RCM and its benefits. So what do you do if you know that your plant could suffer a failure that can be prevented? You must responsibly and reasonably do whatever it takes to avert a potentially serious situation. Recognize the doors that are open to you, and use them to get started.

Similarly, if you can foresee that an RCM implementation is likely to fail, you must eliminate it. Even if this doesn't always seem practical or easy, consider the consequences of not taking action. Hypothetically, if a company ignores known failures and does nothing, it could be sued, get a lot of negative publicity, and suffer heavy financial loss. The gas tank problem in the early model Ford Pinto, dramatized in the 1991 movie *Class Action* starring Gene Hackman, is an example where the court assigned significant damages. In November 1996, a New Jersey court certified a similar nationwide class action lawsuit against General Motors due to rear brake corrosion. In another hypothetical situation, if you acknowledge a potential problem but discount it as negligible and then experience the failure, you could be blamed for ignoring what was clearly recognized as a risk.

Risk can never be eliminated entirely, but it can be lessened. Even if you can't fully implement RCM, take at least some positive action and reduce risk as much as possible. Simply reviewing an existing PM program using RCM logic will accomplish very little. You'll gain more, where it counts most, by analyzing critical equipment. If you follow that up by moving down the criticality scale, you'll gain even more and eventually successfully complete RCM. If performing RCM is simply too much for your company, consider an alternative approach that you can achieve. You'll at least reduce risk somewhat. If you do nothing at all, you could be branded as irresponsible later on, which can be deadly for both your business and professional reputation.

Your ultimate challenge is to convince the decision makers that RCM is their best course. You need to build credibility by demonstrating clearly that RCM works. Often, however, maintenance practitioners have relatively little influence and control. A maintenance superintendent may be encouraged to be proactive as long as he or she doesn't ask the operators or production staff for help or need upper management approval for additional funding. Sound familiar? RCM needs operator and production help, though, to succeed. Without this support, the best attempts to implement an RCM program can flounder.

What can a maintainer realistically do to demonstrate success and increase influence? Realize first of all that many companies do value at least some degree of proactive maintenance. Even the most reactive may do some sort of PM. They know that an ounce of prevention is worth a pound of cure, even if they're not using it to their best advantage. The climate may already exist for you to present your case. The cost of using RCM logic can be a stumbling block. The bulk of the work—identifying failure modes—is where most of the analysis money is spent. Reduce the cost and you'll generate more interest.

8.9.4 CAPABILITY-DRIVEN RCM

Even if you're under severe spending constraints, you can still make improvements by being proactive. By using what's known as capability-driven RCM, or CD-RCM, you do the following:

- Reverse the logic of the RCM process, starting with the solutions (which are a finite number) and look for appropriate places to apply them. Since RCM progresses from equipment to failure modes, through decision making to a result, the opposite process can pinpoint failures, even if they aren't clearly identified.
- Extend existing condition monitoring techniques to other pieces of equipment. For instance, vibration analysis that works on some equipment can also work elsewhere.
- Look specifically for wear-out failures and simply do time-based replacements.
- Check standby equipment to ensure that it works when needed.

These are examples of proactive maintenance that can make a huge difference. It's crucial, though, to do root-cause failure analysis when a failure does occur, so you can take preventive action for the future. Among the benefits of using CD-RCM, it can be a means of building up to full RCM.

There are some risks involved in CD-RCM, though. Some items may be over-maintained, especially in cases where run to failure has previously been acceptable. Overmaintenance could even cause failures if it disrupts operations and equipment in the process. The risk, though, is relatively small. The bigger problem may be the cost of using maintenance resources that aren't needed.

There is also a risk of missing some failure modes, maintenance actions, and redesign opportunities that could have been predicted or prevented if upfront analysis had been done. While the consequences could be significant, the risk is usually minimal if the techniques used are broad enough. For instance, Nowlan and Heaps found that condition monitoring is effective for airlines because 89% of aircraft failures in their study were not time related.

CD-RCM could be used in a similar way with other complex electromechanical systems using complicated controls and many moving parts. In industrial plants, for instance, looking for wear-out failure modes is like the traditional approach to maintenance. Most failures are influenced by operating time or some other measure. By searching for failure conditions only where parts are in moving contact with each other and with the process materials, there will be fewer items to be examined. The potential failures are generally obvious and easy to spot. With this approach it is possible to overmaintain, especially if the equipment you decide to perform TBM on has many other random failure modes.

In RCM, failure finding is used for hidden failures that either are not detected using CBM or avoided using TBM. One favored method is to run items that are usually "normally off" to test their operation. This is done under controlled conditions so that a failure can be detected without significant problems occurring. There is a failure risk on startup in activating the system or equipment, but there is also control over the consequences because the check is done when the item isn't really needed. Correcting the failure reduces its consequences during "normal" operation, when it would be really needed if the primary equipment or device failed. Doing failure finding tasks without knowing what you're looking for may seem foolish, but it's not. Failures often become evident when the item is operated outside of its normal mode (which is often "off"). Although all "hidden failures" may not be found, many will.

Again, this is not as thorough as a complete RCM analysis, but it's a start in the right direction. By showing successful results, the maintainer may be able to extend his proactive approach to include RCM analysis. CD-RCM is not intended to avoid or shortcut RCM. It is a preliminary step that provides positive results consistent with RCM and its objectives.

To be successful, CD-RCM must do the following:

- Ensure that the PM work order system actually works (i.e., PM work orders can be triggered automatically, and the work orders get issued and carried out as scheduled). If this is not in place, help is needed beyond the scope of this chapter.
- Identify the equipment and asset inventory (this is part of the first step in RCM).
- Identify the available conditioning monitoring techniques that may be used (which is probably limited by plant capabilities).
- Determine the kinds of failures that each of these techniques can reveal.
- Identify the equipment where these failures dominate.
- Decide how often to monitor and make the process part of the PM work order system.
- Identify which equipment has dominant wear-out failure modes,
- Schedule regular replacement of wearing components and others that are disturbed in the process.
- Identify all standby equipment and safety systems (e.g., alarms, shut-down systems, standby redundant equipment, back-up systems), which are normally inactive but needed in special circumstances.
- Determine appropriate tests that will reveal failures detected only when the equipment runs. The tests implement them in the PM work order system.
- Examine failures that are experienced after the maintenance program is put in place to determine their root cause so that appropriate action may be taken to eliminate them or their consequences.

The result of using CD-RCM can be

- Extensive use of CBM techniques like vibration analysis, lubricant/oil analysis, thermographic analysis, visual inspections, and some nondestructive testing
- Limited use of time-based replacements and overhauls
- In plants where redundancy is common, extensive "swinging" of operating equipment from A to B and back, possibly combined with equalization of running hours
- Extensive testing of safety systems
- Systematically capturing and analyzing information about failures that occur to determine the causes and eliminate them in the future

All of these actions will move the organization to be more proactive. As CD-RCM targets proven methods where they make sense, it builds credibility and enhances the likelihood of implementing full-blown RCM.

8.10   HOW TO DECIDE?

8.10.1   SUMMARY OF CONSIDERATIONS AND TRADE-OFFS

RCM is a lot of work (see Figure 8.5). It is also expensive. The results, although impressive, can take time to accomplish. One challenge of promoting a full-blown RCM program is justifying the cost without being able to concretely show what the savings will be.

The cost of not using RCM, however, may be much higher. Some alternatives to RCM are less rigorous and downright dangerous. We believe that RCM is a thorough and complete approach to proactive maintenance that achieves high system reliability. It addresses safety and environmental concerns and identifies hidden failures and

FIGURE 8.5   It can be a lot of work!

appropriate failure finding tasks or checks. It identifies where redesign is appropriate and where run-to-failure is acceptable or even desirable.

Simply reviewing an existing PM program with an RCM approach is not really an option for a responsible manager. Too much can be missed that may be critical, including safety or environmental concerns. It may be the start of a reliability program for someone newly assigned to the task, but it is not RCM.

Streamlined or "lite" RCM may be appropriate for industrial environments where prioritizing criticality is an issue. RCM results can be achieved on a smaller but well-targeted subset of the failure modes on the critical equipment and systems. While this is a form of RCM, care must be taken to ensure that it meets the SAE criteria for RCM.

Where RCM investment is not an immediately achievable option, the final alternative is to build up to it using CD-RCM. This adds a bit of logic to the old approach of applying a new technology everywhere. In CD-RCM, you take stock of what you can do now and make sure you use it as widely as possible. Once success is demonstrated, you can expand upon the program. Eventually, RCM can be used to make the program complete.

8.10.2 RCM DECISION CHECKLIST

Throughout this chapter, you have had to consider certain questions and evaluate alternatives to determine if RCM is needed. We summarize them here for quick reference:

1. Can the plant or operation sell everything it can produce? If the answer is "yes," high reliability is important, and RCM should be considered. Skip to question 5. If the answer is "no," focus on cost-cutting measures.
2. Does the plant experience unacceptable safety or environmental performance? If "yes," RCM is probably needed. Skip to question 5.
3. Is there already an extensive proactive or preventive maintenance program in place? If the answer is "yes," consider RCM if the program costs are unacceptably high. If "no," consider RCM if maintenance costs are high compared with others in the same business.
4. Are maintenance costs high relative to others in your business? If "yes," RCM is right. Proceed to question 5. If not, RCM won't help. Stop here. Consider root-cause failure analysis as your living program. Confirm that all assets have been considered in prioritization. At this point, one or several of the following apply:
   - A need for high reliability
   - Safety or environmental problems
   - An expensive and low-performing PM program
   - No significant PM program and high overall maintenance costs
5. RCM is right. Next, you need to ensure that the organization is ready for it. Is there a controlled maintenance environment where most work is predictable and planned? Does planned work, like PM and predictive maintenance (PdM), generally get done when scheduled? If "yes," the organization passes this basic test of readiness—the maintenance environment is under

control. Proceed to question 6. RCM won't work well if the decided tasks can't be applied in a controlled environment. If this is the case, RCM alone isn't enough. Get the maintenance activities under control first. Stop here.

6. RCM is needed and the organization is ready for it. But senior management support is still likely required for the investment of time and cost in RCM training, piloting, and rollout. If that is not forthcoming, consider the alternatives to full RCM. Proceed to question 7.

7. Can senior management support be obtained for the investment of time and cost in RCM lite training and piloting? This investment will require about one month of team time (five people) plus a consultant for the month. If "yes," consider RCM lite to demonstrate success before attempting to roll RCM out across the entire organization.

8. If "no," you must prove credibility to senior management with a less thorough approach that requires little up-front investment and uses existing capabilities. The remaining alternative here is CD-RCM and a gradual build-up of success and credibility to expand on it.

8.11   MAKING RCM A LIVING PROGRAM

As indicated earlier in this chapter, doing an early prioritization of your assets and piloting RCM will get things started. Working through your prioritized assets with trained subject matter experts will allow you to experience the benefits of determining the most effective maintenance.

The introduction of new assets or new technologies within your operation would be very well served by performing an RCM analysis as part of the design phase of an asset's life cycle. Leveraging this process at this early design cycle stage would do the following:

- Verify design integrity.
- Identify any required modifications.
- Identify risks.
- Develop failure management policies and maintenance tasks.
- Identify required spares to support the maintenance tasks.
- Develop operating strategies.

One of the biggest reasons RCM is perceived to have failed is that little attention was given to implementing the tasks identified after the hard part of deciding on the most appropriate task and frequency, given a specific failure mode and asset operating context. Many organizations keep their RCM data quite separate from their CMMS or workflow process. If an RCM panel determines that time-based maintenance is the best course of action for a specific failure mode and asset operating context, then postponing or canceling that PM should not be an easy option. As with many projects, they should be started with the end in mind. An RCM analysis will help you pick the most effective maintenance. Getting these tasks documented into your process or system needs to be the next logical step. In other words, RCM helps you pick the most effective thing to do in a given failure mode, and a well-set-up CMMS helps you execute the effective task efficiently.

Software companies now exist that will help us take the RCM decisions made and integrate them directly with a CMMS. Many CMMS solutions today can directly accept the RCM data (functional failure, failure mode, and associated task and resources requirements) into their aligned data fields so the implementation of the selected tasks can be executed efficiently. Processes can be set up so that should a task be canceled by an operator or maintenance supervisor, the maintenance planner would have immediate access to review the original RCM assigned task and frequency and would determine the risk of accepting a maintenance postponement or cancellation.

An organization that has completed a full RCM analysis for its key assets and has set up the assigned tasks to be automated in its CMMS would be considered to be a leading maintenance organization. If its full maintenance program is well defined through RCM, if assigned tasks are set up in its CMMS, and if its business performance is better than desired, then it likely does not require further RCM work except when considering new assets or a new operating context.

Low-priority assets that do not directly contribute to the business goals of a company and may not be deemed as needing a full RCM assessment may require different handling. In these cases perhaps a root-cause failure analysis is all that is required to keep its reliability program living. However, if an early assessment indicates that safety, environment, or operational impact is at risk, a leading company should assemble a trained RCM group of experts to do an assessment on these assets and automate the decided tasks and resources into the workflow managed by their CMMS.

ENDNOTES

1. SAE JA 1011. Evaluation Criteria for Reliability-Centered Maintenance (RCM) Processes, Society of Automotive Engineers, Aug 1999.
2. Nowlan, F. S. and Heap, H. Reliability-Centered Maintenance. Report ADIA066-579. National Technical Information Service, Dec 19, 1978.
3. Aberdeen Group, December 2006, "Collaborative Asset Maintenance Strategies—Redefining the Roles of Product Manufacturers and Operators in the Service Chain."
4. MIL-STD 1629A, Notice 2. Procedures for Performing a Failure Mode, Effects and Criticality Analysis. Washington, DC: Department of Defense, 1984.
5. MSG-3. Maintenance Program Development Document. Revision 2. Washington, DC: Air Transport Association, 1993.
6. MIL-STD 2173 (AS), Relability-Centered Maintenance Requirements for Naval Aircraft, Weapons Systems and Support Equipment, US Naval Air Systems Command; NAVAIR 0025403, Guideliens for the Naval Aviation Reliability Centered Maintenance Process, US Naval Air Systems Command; S9081AB-GIE-OI0/MAINT-Reliability-Centered Maintenance Handobok, US Naval Sea Systems Command.
7. Smith, A. M. Reliability Centered Maintenance. New York: McGraw-Hill, 1993.
8. Moubray, J. Reliability-Centered Maintenance. 2nd ed. Oxford: Butterworth-Heinemann, 1997.

9 Reliability by Operator
Total Productive Maintenance

Doug Stretton and Patrice Catoir

CONTENTS

Total productive maintenance (TPM) is a highly powerful philosophy for managing maintenance, operations, and engineering in a plant environment. It harnesses the power of the entire workforce to increase the productivity of the company's physical assets, optimizing man–machine interaction. It is an internal continuous improvement process to meet increasingly difficult market demands to provide mass customization for individual customers. Only with a highly flexible manufacturing process and workforce can a company achieve this.

In this chapter you will learn the fundamental functions of TPM, what they mean, and how they are used and integrated into a comprehensive program. When completely implemented, TPM becomes more than a program to run the plant—it becomes part of the culture. We also explore the implementation issues that you can expect, and we compare a TPM approach with typical legacy environments. This will dispel some of the many myths about TPM. Finally, we link TPM to other optimizing methodologies discussed elsewhere in this book. With their combined effect, you can implement a true continuous improvement environment.

Traditionally, TPM and overall equipment effectiveness (OEE) have been used to increase throughput in environments that are capacity constrained. Unfortunately, very few companies are faced with this challenge in these days of mergers, acquisitions, and rationalizations, and it would appear that OEE is not the right tool for cost-cutting exercises and labor optimization. However, this is far from true: OEE's improvements are playing a key role, since they provide the opportunity to drastically cut costs not only by allowing consolidation on fewer shifts and production lines but also at a plant level for global companies.

9.1 INTRODUCTION: WHAT IS TPM?

TPM is the most recognized continuous improvement philosophy, but it's also the most misunderstood. TPM has the power to radically change your organization and boost overall production performance.

Some people claim that TPM reflects a certain culture and isn't applicable everywhere. That's been proven wrong countless times. Others maintain that TPM is just common sense, but there are plenty of people with common sense who haven't been successful using TPM. Clearly, TPM is much more than this.

The objectives of TPM are to optimize the relationship between human–machine systems and the quality of the working environment. What confuses skeptics is the approach that TPM uses to eliminate the root causes of waste in these areas. TPM recognizes that the roles of engineering, operations, and maintenance are inseparable and codependent. It uses their combined skills to restore deteriorating equipment, to maintain basic equipment and operating standards, to improve design weaknesses, and to prevent human errors. The old paradigm of, "I break, you fix" is replaced with, "Together we succeed." This is a radical change for many manufacturing and process organizations.

TPM is more about changing your workplace culture than about adopting new maintenance techniques. For this reason, it can be agonizingly difficult to implement TPM even though its concepts seem so simple on the surface. As a result, the published TPM methodologies are associated with implementation techniques. While technical change is rapid, though, social change takes time and perseverance. Most cultures need an external stimulus. Modern manufacturing philosophies, specifically just-in-time (JIT) and total quality management (TQM), are market driven; they force an organization to make a cultural change. TPM has grown along with the need for flexible manufacturing that can produce a range of products to meet highly variable customer demands. Once TPM is in place it continues to develop and grow, promoting continuous improvement. A TPM organization drives change from within.

In companies that have developed a thorough understanding of TPM, it stands for *total productive manufacturing*. This recognizes that TPM encompasses more than maintenance concerns, with the common goal of eliminating all waste in manufacturing processes. TPM creates an orderly environment where routines and standards are methodically applied. Combining teamwork, individual participation, and problem-solving tools maximizes your equipment use.

What do you need to develop a TPM culture? Besides the tools of TPM, it requires production work methods, production involvement in minor maintenance activities, and teaming production and maintenance workers. The operator becomes key to machine reliability rather than a major impediment, as many believe.

This concept must be accepted and applied at all levels of your organization, starting from the bottom up and nurtured by top management. The result is an organization committed to the continuous improvement of its working environment and its human–machine interface.

9.2 FUNDAMENTALS OF TPM

TPM has five fundamental functions:

* Autonomous maintenance (AM)
* Equipment improvement
* Quality maintenance
* Maintenance prevention
* Education and training

9.2.1 AUTONOMOUS MAINTENANCE

Many people confuse AM with TPM, but autonomous maintenance is only one part of TPM, though a fundamental one. The confusion arises because, during a TPM implementation, AM directly affects the greatest number of people.

Autonomous maintenance is a technique to get production workers involved in equipment care, working with maintenance to stabilize conditions and to stop accelerated deterioration. You must teach operators about equipment function and failures, including prevention through early detection and treating abnormal conditions. It can create conflict because of past work rules. For it to be successful, operators must see improvements, strong leadership, and control elements delivering satisfactory service levels. See Campbell[1] for a thorough understanding of maintenance leadership and control—the platform for developing TPM.

Often, AM's impact on maintenance is overlooked. In fact, it helps your staff support the operators, make improvements, and solve problems. More time is spent on maintenance diagnostics, prevention, and complex issues. Operators perform routine equipment inspections and cleaning, lubrication, adjustment, inspection, and minor repair (CLAIR) maintenance tasks, which are critical to how the equipment performs.

The first AM task is to have equipment maintainers and operators complete an initial cleaning and equipment rehabilitation. During this time the operators learn the details of their equipment and identify improvement opportunities. They learn that "cleaning is inspection," as described by the Japan Institute of Plant Maintenance.[2] Regular cleaning exposes hidden defects that affect equipment performance. Inspection routines and equipment standards are established. The net effect is that the operator becomes an expert on his or her equipment.

This development of the "operator as expert" is critical to the success of TPM. An "expert" operator can judge abnormal from normal machine conditions and communicate the problem effectively while performing routine maintenance. It is precisely this expert care that maximizes equipment.

9.2.2 EQUIPMENT IMPROVEMENT

A key function of TPM is to focus the organization on a common goal. Since people behave the way they are measured, it is critical to develop a comprehensive performance measure for all employees. The key TPM performance measure is overall equipment effectiveness (OEE), as described by Nakajima.[3]

OEE combines equipment, process, material, and people concerns and helps identify where the most waste occurs. It focuses maintenance, engineering, and production on the key issue of plant output.

Using OEE, you will be able to better identify whether your operation is producing quality product. Simply put, an operation is always either producing on-spec product or is not. OEE forces the organization to address all the reasons for lost production, turning losses into opportunities for improvement. In OEE, the measurement of all equipment activities is in a given period. At any one time, equipment will always perform one of the following: on-spec product, downtime, quality loss, or rate loss.

The size of the pie in Figure 9.1 is the amount of product produced at the ideal rate for a period of calendar time. This is the ideal state where all of the organization's efforts produce on-spec product.

Often, an organization will decide to remove calendar time from the OEE calculation. If your plant operates five days a week, you may want to eliminate the downtime caused by not operating on Saturday and Sunday, but keep in mind that this OEE calculation is really a subset of the OEE for the plant. In Figure 9.1, a decision has been made to reduce the size of the pie.

OEE is calculated by the following formula:

$$\text{OEE} = \text{Availability \%} \times \text{Production Rate \%} \times \text{Quality Rate \%}$$

where

Availability = Production Time/Total Time
Production Rate = Actual Production Rate/Ideal Production Rate
Quality Rate = Actual On-Spec Production/Actual Total Production

Availability is simply the ratio of production time to calendar time. In practice, it is more convenient to measure the downtime and perform some simple mathematics

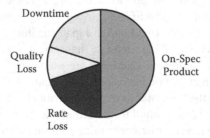

FIGURE 9.1 OEE: time allocation of activities.

to arrive at the production time. Downtime is any time the equipment is not producing. Equipment, systems, or plants may be shut down but available for production. The downtime could be unrelated to the equipment, caused, for example, by a lack of raw materials to process. Count all downtime, including any you have scheduled. Excluding some downtime violates the key principle is that no one should "play" with the numbers.

At one plant, "planned maintenance" was excluded from the availability calculation. If an eight-hour shift had two hours of "planned maintenance," the total time for that shift was said to be six hours. Supervisors reacted by calling almost any maintenance activity "planned." While OEE went up, total output did not. Downtime is downtime, planned or not.

Production rate is the actual product ratio when running to the "ideal" instantaneous production rate. Setting the ideal rate can be difficult because there are different approaches to determine its value. The most common approach is to use either a rate shown to be achievable or the design production rate, whichever is greater. The design rate is a good target if your plant, system, or equipment has never achieved it—common in newer plants. If the plant has been modified over the years and production capacity has increased or if you've used various "tricks" to increase production levels successfully, the demonstrated maximum production rate is useful. In many cases, this corresponds to an upper control limit.

You can measure production rate for continuous operation "on the fly" by just looking at the production speed indicators. Batch processes, however, can be difficult to measure. In a batch process, cycle time, or average rate based on production time output, is used to measure production rate. Quality rate is the ratio of on-spec product to actual production rate. *On-spec* means producing what is needed in a condition that complies with product specifications. Product that does not meet spec may be saleable to some lower spec if customers order large enough quantities. An often forgotten part of the quality rate is rework, which shouldn't affect the production rate. Where rework is fed back through the process (e.g., steel scrap from the hot strip mill is fed back to the basic oxygen furnace at a steel mill), it displaces virgin material that could be processed in its place.

Although the focus of OEE is to identify waste, it also allows for benchmarking of similar operations. Since the calculation is determined based on the efficiency of the asset, and not the output, benchmarking of similar assets is the much easier.

There is a problem when trying to benchmark OEE for assets that have very different characteristics. Since some assets require stoppages are regular intervals, those assets will always have a lower level. For example, in underground mining, equipment must be moved to a safe area during blasting. This time can be minimized but it is always there as a loss. OEE can range from 95% to 98% for a processing operation with a high degree of redundant equipment (where MTTR and MTBF of components is a more effective measure) to 70–80% in a mining operation. Both operations may be considered world class, but the discrepancy in OEE may cause people to believe that the mining operation is poorly operated. This may not be the case, it may just be that the mining operation has more losses that cannot be removed. However, that does not mean that the mine should ignore those losses. Measure all losses, and work to minimize them—that is the spirit of OEE. What is impossible to do today is often routine tomorrow.

There are, however, a few quick rules of thumb for OEE. Where OEE is not measured, you can often observe OEE as low as 40%. Where it is measured, a reasonable OEE to expect is 80% for a standard batch process and 85% for a continuous process. To move OEE beyond these levels requires effort by the organization.

Once OEEs of 80% and up have been achieved consistently, this opens the window to drastically decrease operational costs by consolidating the same throughput on fewer shifts, lines, and, ultimately, plants.

Figure 9.3 illustrates an example where the entire throughput on a non-optimized production system could be consolidated from 5 lines to 3 lines assuming that sufficient compatibility exists between the lines and that OEEs of 80% are achieved consistently. Although it is clear that tremendous efforts must be put in place in order to transform this production system since it's only running at 40–60% OEE, the cost reduction that could be achieved would significantly impact the production system.

As TPM improves OEE and increases the output from existing assets, a choice is often required. Should the organization try to increase sales, or decrease assets?

Keep in mind that knowing the OEE doesn't provide information to improve it. You need to determine what causes each loss and how significant it is. If you know where most waste occurs, you can focus resources to eliminate it through problem solving, root-cause failure analysis (RCFA; see Chapter 2), reliability-centered management (RCM; see Chapter 8), and some very basic equipment care techniques.

9.2.3   QUALITY MAINTENANCE

You have implemented OEE in your key production areas, and you have a wealth of data about your losses. But how do you use that data to improve the OEE? Although the OEE number is the focus, identifying what causes wasted availability and process rate is essential to improving it along with applying quality Pareto analysis. By working on the most significant losses, you make the most effective use of your resources.

There are many forms of waste that TPM can eliminate:

- Lost production time due to breakdowns
- Idling and minor stoppage losses from intermittent material flows
- Setup and adjustment losses (time lost between batches or product runs)
- Capacity losses from operating the process at less than maximum sustainable rates
- Start-up losses from running up slowly or disruptions
- Operating losses through errors
- Yield losses through less than adequate manufacturing processes
- Defects in the products (quality problems)
- Recycling losses to correct quality problems

You can apply OEE at the plant, production line, system, work cell, or equipment level. It can be measured yearly, monthly, weekly, daily, by shift, by hour, by minute,

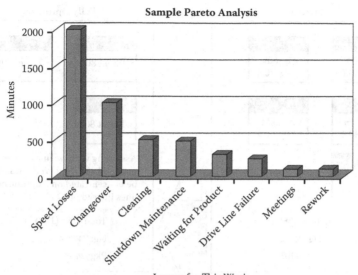

FIGURE 9.2 Sample Pareto analysis.

or instantaneously. The measurement frequency must ensure that both random and systematic events are identified. You must report the data frequently enough to detect trends early on. A 90% OEE target is world class. To successfully get there, first improve availability (largely a maintenance and reliability effort), and then target production and quality rates.

However, as you can see in Figure 9.2, using Pareto analysis, as described by Ishikawa,[4] is the key to improving OEE. Losses caused by operating at less than ideal rate or producing off-spec product can be converted to time. That is, a machine operating at 90% of rated speed for 10 hours has lost the equivalent of 1 hour of production time. Pareto analysis prioritizes the losses so that the organization focuses on the largest piece.

In most organizations, there is a narrow set of measures that zero in on defects or failures. Many organizations monitor mechanical downtime but not availability. In Figure 9.2 the organization would try to correct the "drive line failure." However, completely removing the cause of this failure would be equivalent to a 10% reduction in the amount of product lost to operating at a reduced speed. In the figure, Pareto analysis is critical to prioritized OEE data. Note that the five highest causes in this example are losses often considered normal. In a non-TPM plant, "drive line failure" and "rework" would receive the most management attention.

The solution to many of these losses extends beyond maintenance to include production, engineering, and materials logistics. All elements of the plant's entire supply chain can impact how much quality product is produced. Correcting problems that lead to low availability, production rates, or quality rates can involve maintenance, engineering, and production process or procedural changes. Teamwork is essential to pulling these disciplines together.

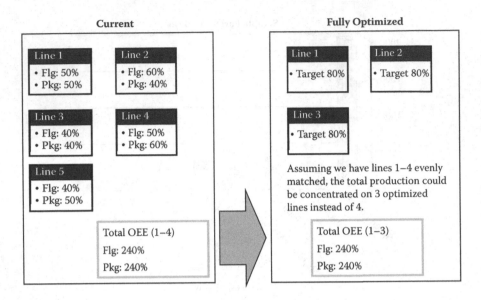

FIGURE 9.3

Key to TPM is the use of teams. Usually, you organize the teams around production areas, lines, or work cells. They comprise production and maintenance workers in a ratio of about 2 to 1. The teams work mostly in their assigned areas to increase equipment familiarity, a sense of "ownership," and cooperation among production and maintenance. Selecting the pilot area is important. It must obviously need change. You want impressive results that you can use later to "sell" the concept to other plant areas. Through teaming, production and maintenance goals are the same because they are specific to the area instead of to a department or function.

9.2.4 MAINTENANCE PREVENTION

You can eliminate a lot of maintenance by studying equipment weaknesses. Many maintenance prevention techniques, from simple visual controls to relatively complex engineering design improvements, can greatly reduce losses. This is very important for effective autonomous maintenance, such as reducing cleaning requirements or increasing ease of adjustment.

Using maintenance prevention effectively converts random breakdown maintenance to routine scheduled maintenance. When operators use lubricants properly, they prevent many unnecessary failures. As the scope of the preventive maintenance program and availability of maintenance tactics increase, problems will be found before failure occurs.

A valuable maintenance prevention tool is an effective computerized maintenance management system (CMMS) or enterprise asset management (EAM) system. The CMMS data help you prevent recurring failures and effectively plan and schedule maintenance tasks.

9.2.5 EDUCATION AND TRAINING

Individual productivity is a function of skills, knowledge, and attitude. The drastic change in the operator's work environment when implementing TPM makes education and training essential.

In many organizations, operators are supposed to follow the supervisor's directions without question. The operator is trained never to deviate from a specific procedure. Often, the training method is the buddy system. The result, however, can be operators who complete the minimum work required to perform the task, without any understanding of their role in the overall operation. The operator may also learn bad habits from his or her buddy.

In TPM, the operator is asked to participate in the decision-making process and constantly question the status quo. This is a basic requirement of continuous improvement. The initial impact, however, can be negative. The operator's first reaction is often that management is "dumping" work that has traditionally been the responsibility of maintenance or plant management. The operator may also worry about not being able to do what's required. It's no wonder that in many plants the mention of TPM immediately mobilizes the union.

You absolutely must educate employees about the benefits of TPM and your business needs while training them to use the tools of TPM. Education is about developing an individual into a whole person, while training provides specific skills. But you must implement and update education and training at the same time. The level of each must increase as the operator learns new concepts and skills.

Before starting TPM, the operators need to learn its philosophy and practices. They must also know about their company. If you are to involve operators in decision making, they must understand the context. The minimum training requirements are as follows:

- An introduction to TPM
- General inspection techniques
- Diagnostic techniques
- Analytical problem-solving techniques
- Selected technical training

You need to appoint a special TPM team to teach operators and other personnel specific problem-solving methods such as Pareto analysis, root-cause failure analysis, and statistical process control for them to take a more proactive role in the organization. Training and education must be ongoing to ensure knowledge transfer and to keep skills totally up to date.

9.3 HOW DO YOU IMPLEMENT TPM?

TPM is implemented in four major phases:

- Establish acceptable equipment operating conditions to stabilize reliability.
- Lengthen asset life.

- Optimize asset conditions.
- Optimize life-cycle cost.

The fundamentals continue and expand as you implement the four phases.

The first phase stabilizes reliability by restoring equipment to its original condition. This is done by cleaning the equipment and correcting any defects. Note major problems, and establish a plan to resolve them. Make sure operators get sufficient training to turn simple equipment cleaning into a thorough inspection to spot machine defects.

The second phase maintains the equipment's basic operating condition. Standards to do so are developed. Begin data collection, and set equipment condition goals. The operators perform minor maintenance activities to eliminate abnormal wear.

In phase three, improve the equipment's operation from its stabilized level. Cross-functional teams should target chronic losses to increase overall machine performance. Review and update standards. Find and analyze opportunities to increase equipment performance and operating standards beyond original capabilities.

Phase four is about optimizing the cost of the asset over its entire life. You achieve this by extending equipment life, increasing performance, and reducing maintenance cost. Keep the machine at its optimal condition. Regularly review processes that set and maintain operating conditions. New equipment will become part of the TPM process. Operator reliability is "built in."

A cross-functional team approach should be used during all phases of TPM implementation. Building effective teams is a prerequisite to entrenching TPM ideas and behaviors. TPM principles and techniques are simple and straightforward. The initial focus will be project management and carefully applying change management. Change is central to TPM. If your organization isn't used to it or has a history of unsuccessful changes, this will be a major hurdle to success. We recommend a pilot project to demonstrate success in one area before tackling TPM throughout the entire plant.

The choice of location for a TPM pilot is critical. The pilot area must clearly need improving and be visible to as many people as possible. Once you establish momentum, apply TPM to other areas. Divide it into manageable portions, and implement one at a time.

Successful TPM requires a transfer of responsibility between management and employees. It depends on a sincere and dedicated management team that pays appropriate attention to change management issues. When implementing TPM, the legacy culture presents the greatest change management problems. See Table 9.1 for samples of typical change management issues.

TPM is implemented gradually over what can be several (three to five) years. Once established, it becomes part of the plant's way of doing things: its culture. Since TPM is about changing the behavior of both workers and managers, it requires patience and positive reinforcement to achieve permanent change. Even very poorly maintained plants, with the right focus and commitment, can be "turned around" through TPM.

TABLE 9.1
Implementing TPM: The Legacy Culture

Legacy Approach	TPM Approach	Change Management Issues
Clear lines of responsibility exist between production and maintenance employees. When a machine breaks down, operators call maintenance.	Employees work together to solve problems. It is recognized that production and maintenance are inseparable and problems need to be solved jointly.	Employees may believe that the goal of TPM is to eliminate maintenance jobs. Most people equate productivity gains to job losses. It is difficult to see TPM's objectives.
Supervisors direct employee actions. Employees do what they are told when they are told.	Self-directed teams develop and execute plans to achieve progressive goals.	Front-line supervisors have difficulty changing to their new "coach" role. Many don't trust their employees.
Management announces a new program to improve operations. Employees are trained. This is commonly referred to as the "flavor of the month."	Management announces that TPM will be implemented. TPM training is conducted, and a TPM team is formed. A pilot site is chosen, and work begins to improve the condition and performance of the pilot site.	At first, few employees recognize that TPM will actually be implemented. When the TPM team shows progress and starts to make improvements, some employees will see the benefits and want to participate. Others will fear the change and reject the process. Over time the naysayers convert or fade away. This is an exceptionally difficult issue where a plant has failed at a TPM implementation in the past.
Increased production level is achieved in one of two ways: employees are pushed to work faster, or equipment must be added.	Reducing losses due to availability, quality rate, and production rate increases production level. The bigger the loss the greater the potential benefit.	Operators believe that OEE is implemented to rate them and make them work harder. In fact, it is the losses that make them work harder. Eventually they begin to see that OEE helps them quantify problems that they have always wanted corrected.
The relationship between the union and management is adversarial. Each tries to beat the other in negotiations. Grievances and disciplinary actions are used as negotiating tools.	Union and management work together to achieve the goals of TPM. Each represents its own interests, but negotiations are considered successful if each side benefits.	TPM is a process that does what most unions have always wanted. It gives employees a voice in their workplace and considers them a valuable resource. However, if the union is not involved from the beginning, its distrust of management's intentions will be a significant hurdle for the TPM team.
Performance measures exist for each department. Maintenance is evaluated on downtime, production output.	OEE is the key measurement. The organization is evaluated on OEE. Pareto analysis prioritizes losses that affect OEE.	Invariably, when OEE tracking starts, big losses are considered "normal" (e.g., changeovers). It is difficult for people to accept that something they have lived with for years can be changed.

9.4 THE CONTINUOUS IMPROVEMENT WORKPLACE

Successfully implementing TPM creates an efficient, flexible, and continually improving organization. The process may be long and arduous, but, once TPM has been accepted, it is as hard to remove as the culture it replaced. This is significant because managers may change, but TPM will continue.

The TPM workplace is efficient because it follows tested procedures that are continually reviewed and upgraded. Change is handled fluidly because effective education and training prepares the workforce to participate in the decision-making process.

TPM embraces other optimizing maintenance management methodologies. RCM and RCFA are often very effective in a TPM environment. RCM affects the preventive and predictive maintenance program, and RCFA improves specific problem areas. TPM affects the working environment in virtually all respects—the way production and maintenance employees work, are organized, use other techniques like RCM an RCFA, solve problems, and implement solutions.

Your competitors may be able to purchase the same equipment, but not the TPM experience. The time required to implement TPM makes it a significant competitive advantage, one that can't be easily copied.

REFERENCES

1. Campbell, J. D. *UpTime: Strategies for Excellence in Maintenance Management.* Productivity Press, 1995.
2. Japan Institute of Plant Maintenance. *Autonomous Maintenance for Operators.* Productivity Press, 1997.
3. Nakajima, S. *Introduction to TPM.* Productivity Press, 1988.
4. Ishikawa, K. *Guide to Quality Control.* Asian Productivity Organization, 1986.

Section III

Optimizing Maintenance Decisions

10 Reliability Management and Maintenance Optimization

Basic Statistics and Economics

Andrew K. S. Jardine
Original by Murray Wiseman

CONTENTS

As global industrial competitiveness increases, showing value, particularly in equipment reliability, is an urgent business requirement. Sophisticated, user-friendly software is integrating the supply chain, forcing maintenance to be even more mission critical. We must respond effectively to incessantly fluctuating market demands. All of this is both empowering and extremely challenging. Mathematical and statistical models are invaluable aides. They can help you increase your plant's reliability and efficiency, at the lowest possible cost.

    This chapter is about the statistical concepts and tools you need to build an effective reliability management and maintenance optimization program. We'll take you from the basic concepts to developing and applying models for analyzing common maintenance situations. Ultimately, you should know how to determine the best course of action or general policy, given a defined situation.

We begin with the relative frequency histogram to discuss the four main reliability-related functions: (1) the probability density function (PDF); (2) the cumulative distribution function (CDF); (3) the reliability function; and (4) the hazard function. These functions are used in the modeling exercises in this and subsequent chapters. We describe several common failure distributions and what we can learn from them to manage maintenance resources. The most useful of these is the Weibull distribution, and you'll learn how to fit that model to a system or component's failure history data.

In this chapter, the words *maintenance, repair, renewal,* and *replacement* are used interchangeably. The methods we discuss assume that maintenance will return equipment to "good-as-new" condition.

10.1   INTRODUCTION: THE PROBLEM OF UNCERTAINTY

Faced with uncertainty, our instinctive, human reaction is often fear and indecision. We would prefer to know when and how things happen. In other words, we would like all problems and their solutions to be *deterministic*. Problems where timing and outcome depend on chance are probabilistic or *stochastic*. Many problems, of course, fall into the latter category. Our goal is to quantify the uncertainties to increase the success of significant maintenance decisions. The methods described in this chapter will help you deal with uncertainty, but our aim is greater than that. We hope to persuade you to treat it as an ally rather than as an unknown foe.

"Failure is the mother of success." "A fall in the pit is a gain in the wit." If your maintenance department uses reliability management, you'll appreciate this folk wisdom. In an enlightened environment, the knowledge gained from failures is converted into productive action. To achieve this requires a sound quantitative approach to maintenance uncertainty. To start, we'll show you an easily understood relative frequency histogram of past failures.

In addition to the relative frequency histogram, we look at the probability density function, the cumulative distribution function, the reliability function, and the hazard function.

10.2   THE RELATIVE FREQUENCY HISTOGRAM

Assume that 48 items purchased at the beginning of the year all fail by November. List the failures in order of their failure ages. Group them, as in Table 10.1, into convenient time segments, in this case by month, and plot the number of failures in each one. The high bars in the center of Figure 10.1 represent the highest (or most probable) failure times: March, April, and May. By adding the number of failures occurring before April (i.e., 14) and dividing by the total number of items (i.e., 48) the *cumulative probability* of the item failing in the first quarter of the year is 14/48. The probability that all of the items will fail before November is 48/48, or 1.

Transforming the numbers of failures by month into probabilities, the relative frequency histogram is converted into a mathematical, and more useful, form called

TABLE 10.1

Failures in Month

Month	Jan	Feb	Mar	Apr	May	Jun	Jul	Aug	Sep	Oct
Failures	2	5	7	8	7	6	5	4	3	1

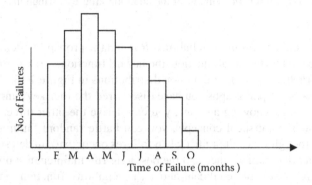

FIGURE 10.1   The relative frequency histogram.

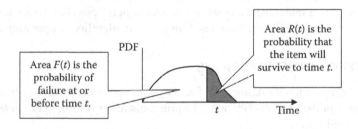

FIGURE 10.2   The probability density function.

the *probability density function*. The data are replotted so that the *area* under the curve, between time 0 and any time *t*, represents the cumulative probability of failure. This is shown in Figure 10.2. (How the PDF plot is calculated from the data and then drawn is discussed more thoroughly in Section 10.6.)

The total area under the curve of the probability density function $f(t)$ is 1, because sooner or later the item will fail. The probability of the component failing at or before time *t* is equal to the area under the curve between time 0 and time *t*. That area is $F(t)$, the cumulative distribution function. It follows that the remaining (shaded) area is the probability that the component will survive to time *t* and is known as the reliability function, $R(t)$. $R(t)$ and $F(t)$ can, themselves, be plotted against time.

The *mean time to failure* (MTTF) is

$$\int_0^\infty tf(t)dt$$

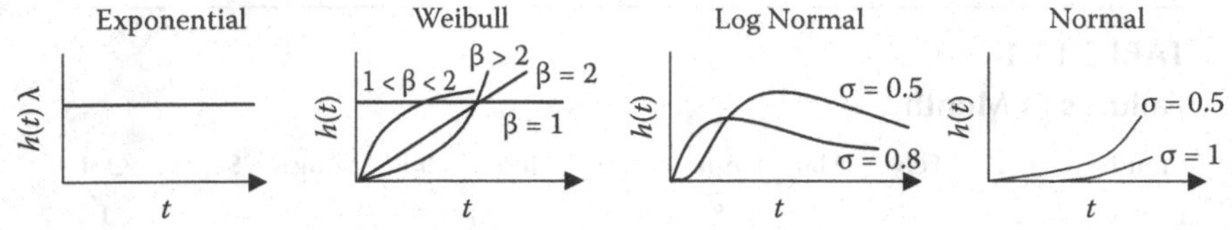

FIGURE 10.3 Hazard function curves for the common failure distributions.

(shown in Appendix 1). From the reliability, $R(t)$ and the probability density function, $f(t)$, we derive the fourth useful function, the hazard function, $h(t) = f(t)/R(t)$, which is represented graphically for four common distributions in Figure 10.3.

In just a few short paragraphs you have discovered the four key functions in reliability engineering. Knowing any one, you can derive the other three. Armed with these fundamental statistical concepts, you can battle random failures throughout your plant. Although we can't predict when failures occur, we can determine the best times for preventive maintenance and the best long-run maintenance policies.

Once you're reasonably confident about the reliability function, you can use it and its related functions, with optimization *models*. Models describe typical maintenance situations by representing them as mathematical equations. That makes it convenient to adjust certain decisions to get the optimum outcome. Optimization reduces long-term maintenance costs to the lowest point possible. Other objectives include the highest reliability, maintainability, and availability of operating assets.

10.3 TYPICAL DISTRIBUTIONS

In the previous section we defined the four key functions once data have been transformed into a probability distribution. The prerequisite step of *fitting* or *modeling* the data is covered next.

How do you find the appropriate reliability function for a real component or system? There are two different approaches to this problem. In one, you estimate the reliability function by curve-fitting the failure data from extensive life testing. In the other, you estimate the parameters (unknown constants) by statistical sampling and numerous statistical confidence tests.[1] We'll take the latter approach.

Fortunately, we know from past failures that probability density functions (and their reliability, cumulative distribution, and hazard functions) of real maintenance data usually fit satisfactorily one of several mathematical equations already familiar to reliability engineers. These include exponential, Weibull, log-normal, and normal distributions. Each failure distribution is a "family" of equations (or graph curves) whose members vary in shape by their differing parameter values. So, their cumulative distribution functions can be fully described by knowing the value of their parameters. For example, the Weibull (two parameter) CDF is

$$F(t) = 1 - e^{-\left(\frac{t}{\eta}\right)^{\beta}}$$

You can estimate the parameters β and η using the methods described in the following sections. Usually, through one or more of these four probability distributions, you can conveniently process failure and replacement data. The modeling process involves manipulating the statistical functions you learned about in section 10.2 (the PDF, CDF, reliability, and hazard functions). The objective is to understand the problem, to forecast failures, and to analyze risk to make better maintenance decisions. Those decisions will impact the times you choose to replace, repair, or overhaul machinery as well as optimize many other maintenance management tasks.

The problem entails the following:

- Collecting good data
- Choosing the appropriate function to represent your situation and estimating the function parameters (e.g., the Weibull parameters β and η)
- Evaluating how much confidence you have in the resulting model

Modern reliability software makes this process easy and fun. What's more, it helps us communicate with management and share the common goal of business— implementing procedures and policies that minimize cost and risk while maintaining, even increasing, product quality and throughput. The most common failure rate or hazard functions are depicted in Figure 10.3. They correspond to the exponential, Weibull, log-normal, and normal distributions.

Real-world data most frequently fit the Weibull distribution. Today, Weibull analysis is the leading method in the world for fitting component life data,[2] but it wasn't always so. While delivering his hallmark paper in 1951, Waloddi Weibull modestly said that his analysis "...may sometimes render good service." That was an incredible understatement, but initial reaction varied from disbelief to outright rejection. Then the U.S. Air Force recognized Weibull's research and provided funding for 24 years. Finally, Weibull received the recognition he deserved.

10.4 THE ROLE OF STATISTICS

Most enter an unacceptable state at some stage in life. One of the challenges of optimizing maintenance decisions is to predict when. Luckily, it's possible to analyze previous performance and to identify when the transition from "good" to "failed" is likely to occur. For example, while your household lamp may be working today, what is the likelihood that it will work tomorrow? Given historical data on the lifetime of similar lamps in a similar operating environment, you can calculate the probability of the lamp still working tomorrow or, equivalently, failing before then. Here's how this is done.

Assume that a component's failure can be described by the normal distribution illustrated as a PDF in Figure 10.4. In the graph you'll notice several interesting and useful facts about this component. First, the figure shows that 65% of the items will fail at some time within 507 to 693 hours, and 99% will fail between 343 hours and 857 hours. Also, since the PDF is constructed so that the total area under the curve adds to 1.0 (or 100%), there is a 50% probability that the component will fail before its mean life of 600 hours.

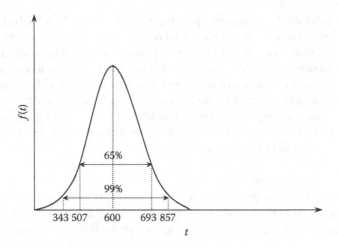

FIGURE 10.4   The normal (Gaussian) distribution.

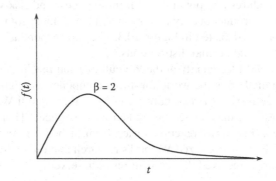

FIGURE 10.5   Skewed distribution.

While some component failure times fit the normal distribution, it is too restrictive for most maintenance situations. For example, the normal distribution is bell-shaped and symmetrical, but many times the data are quite skewed. A few items might fail shortly after installation and use, but most survive for a fairly long time. This is depicted in Figure 10.5—a Weibull distribution whose shape parameter β equals 2.0.

In this case, you can see that the distribution is skewed with a tail to the right. Weibull distribution is popular because it can represent component failures according to the bell-shaped normal distribution, the skewed distribution of Figure 10.5, and many other possibilities. Professor Weibull's equation includes two constants: β (beta) known as the shape parameter, and η (eta) known as the characteristic life. It is this flexible design that has made Weibull distribution such a success.

The hazard function, depicted in Figure 10.6, shows clearly the risk of a component failing as it ages. If the failure times have a beta value greater than 1.0, the risk

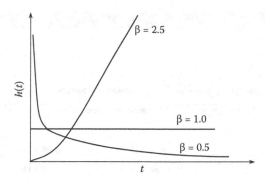

FIGURE 10.6 Weibull hazard function.

increases with age (i.e., it is wearing out). If beta is less than 1.0, the risk declines (e.g., through work hardening or burn-in). If beta takes a value equal to 1.0, the failure isn't affected by age (i.e., failures are purely random, caused by external or unusual stress). That is usually the case when a stone hits a car windshield, severely cracking it, which is just as likely to occur on a new car as an old one. In fact, many failures in industrial, manufacturing, process, and transportation industries are random or stress failures.

You want to optimize the maintenance decision, to better know when to replace a component that can fail. If the hazard function is increasing (i.e., beta is greater than 1.0), you must identify where on the increasing hazard curve the optimal replacement time occurs. You do this through blending together the hazard curve and the costs of preventive maintenance and failure replacement, taking into account component outages for both. Establishing this optimal time is covered in Chapter 11, "Maintenance Optimization Models."

If the hazard function is constant (beta is equal to 1.0) or declining (beta is less than 1.0), your best bet is to let the component run to failure. In other words, preventively replacing such components will not make the system more reliable. The only way to do that is through redesign or installing redundancy. Of course, there will be trade-offs. To make the best maintenance decision, study the component's failure pattern. Is it increasing? If so, establish the best time to replace the component. Is it constant or declining? Then the best action, assuming there aren't other factors, is to replace the component only when it fails. There isn't any advantage, either for reliability or cost, in preventive maintenance.

Earlier, we mentioned the importance of reliability software in maintenance management. You can easily establish a component's beta value using such standard software. OREST[3] software was used in Figure 10.7, where the sample size is 10 with six failure observations and four suspensions, the beta (β) value is 3.91, and the mean time to failure of the item is 4851.13 time units. Additional aspects of the table are covered later in the chapter. Weibull ++,[4] M-Analyst,[5] and Winsmith[6] perform similar functions.

We must stress that, so far, we have been focusing on items termed line replaceable units (LRUs). The maintenance action replaces or renews the item and returns

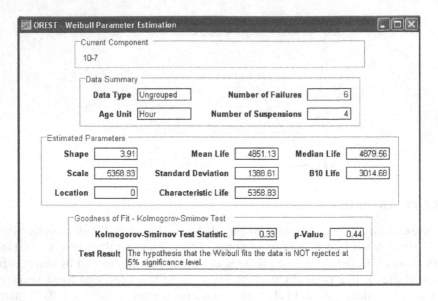

FIGURE 10.7   OREST—Weibull analysis.

the component to a statistically good-as-new condition. For complex systems with multiple components and failure modes, the form of the hazard function is likely to be the bathtub curve in Figure 10.8. In these cases, the three underlying causes of system failure are wear-out, quality, and random. Adding them creates the overall bathtub curve. There are three distinct regions: running-in period, regular operation, and wear-out.

Example

Here is an example illustrating how you can extract useful information from failure data. Assume (using the methods to be discussed later in this chapter) that an electrical component has the exponential cumulative distribution function, $F(t) = 1 - e^{-\lambda t}$, where $\lambda = 0.0000004$ failures per hour.

a. What is the probability that one of these parts fails before 15,000 hours of use?

$$F(t) = 1 - e^{-\lambda t} = 1 - e^{-0.0000004 \times 15000} = 0.006 = 0.6\%$$

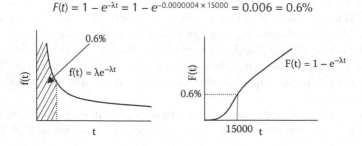

Equipment Life Periods

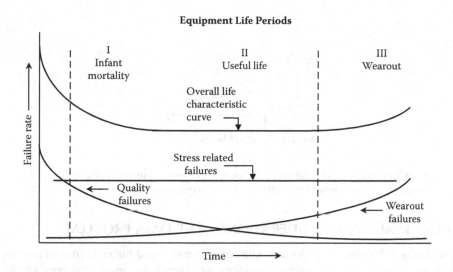

FIGURE 10.8   System hazard function.

b. How long until you get 1% failures? Rearranging the equation for $F(t)$ to solve for t:

$$t = -\ln(1 - F(t))/\lambda = -\ln(1 - 0.01)/0.0000004 = 25{,}126 \text{ hr.}$$

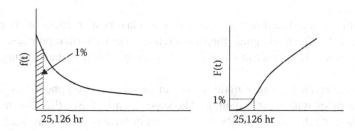

c. What would be the MTTF?

$$MTTF = \int_0^\infty tf(t)dt = \int_0^\infty t\lambda e^{-\lambda t}dt = \frac{1}{\lambda} = 250{,}000 \text{ hr.}$$

d. What would be the median time to failure (the time when half the number will have failed)?

$$F\left(T_{50}\right) = 0.5 = 1 - e^{-\lambda T_{50}}$$

$$T_{50} = \ln 2/\lambda = 0.693/0.0000004 = 1{,}732{,}868 \text{ hr.}$$

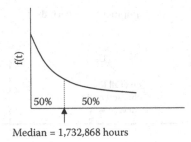

Median = 1,732,868 hours

This is the kind of information you can retrieve using the reliability engineering principles in user-friendly software. Read on to discover how.

10.5 REAL-LIFE CONSIDERATIONS—THE DATA PROBLEM

Ironically, reliability management data can slip unnoticed through your fingers as you relentlessly try to control maintenance costs. Data management can prevent this from happening and provide critical information from past experience to improve your current maintenance management process. It is up to upper-level managers, though, to provide trained maintenance professionals with ample computer and technical tools to collect, filter, and process data.

Without doubt, the first step in any forward-looking activity is to get good information. In fact, this is more important than anything else. History proves that progress is built on experience, but there are countless examples where ignoring the past results in missed opportunities. Unfortunately, many maintenance departments are guilty of this too.

Companies can be benchmarked against world-class best practices by the extent to which their data effectively guide their maintenance decisions and policies. Here are some examples of how you can make decisions using reliability data management:

1. A maintenance planner notices that an in-service component has failed three times within three months. The superintendent uses this information to estimate the failure numbers in the next quarter to make sure there will be enough people available to fix them.
2. When ordering spare parts and scheduling maintenance labor, determine how many gearboxes will be returned to the depot for overhaul for each failure mode in the next year.
3. An effluent treatment system must be shut down and overhauled whenever the contaminant level exceeds a toxic limit for more than 60 seconds in a month. Avoid these production interruptions by estimating the level and frequency of preventive maintenance needed.
4. After a design modification to eliminate a problem, determine how many units must be tested, for how long, to verify that the old failure mode has been eliminated or significantly improved with 90% confidence.
5. A haul truck fleet of transmissions is routinely overhauled at 12,000 hours, as stipulated by the manufacturer. A number of failures occur before the

overhaul. Find out how much the overhaul should be advanced or delayed to reduce average operating costs.

6. The cost in lost production is 10 times more than for preventive replacement of a worn component. From this, determine the optimal replacement frequency.

7. You can find valuable information in the database to help with maintenance decisions. For instance, if you know the fluctuating values of iron and lead from quarterly oil analysis of 35 haul truck transmissions, and their failure times over the past three years, you can determine the optimal preventive replacement age (examined in Chapter 12).

Obviously, it is worth your while to obtain and record life data at the system and, where warranted, component levels. When tradespeople replace a component, for example, a hydraulic pump, they should indicate which specific pump failed. It may be one of several identical pumps on a complex machine that is critical to operations. They should also specify how it failed, such as "leaking" or "insufficient pressure or volume." Because we know how many hours equipment operates, we can track the lifetime of individual critical components. That information will then become a part of the company's valuable intellectual asset—the reliability database.

10.6   WEIBULL ANALYSIS

Weibull analysis supported by powerful software is formidable in the hands of a trained analyst. Many examples and comments are given in the practical guidebook, *The New Weibull Handbook*.[2]

One of the distinct advantages of Weibull analysis is that it can provide accurate failure analysis and forecasts with extremely small samples.[2] Let's look closely at the prime statistical failure investigation tool, the Weibull plot. Failure data are plotted on Weibull probability paper, but, fortunately, modern software[3-6] provides an electronic version. The Weibull plot uses x and y scales, transformed logarithmically so that the Weibull CDF function

$$F(t) = 1 - e^{-\left(\frac{t}{\eta}\right)^{\beta}}$$

takes the form $y = \beta x + $ constant. By using Weibull probability paper, you get a straight line when you plot failure data that fit the Weibull distribution. What's more, the slope of the line will be the Weibull shape parameter β, and the characteristic life η will be the time at which 63.2% of the failures occurred.

10.6.1   WEIBULL ANALYSIS STEPS

Software makes Weibull analysis far more pleasant than it used to be. Conceptually, the software does the following:

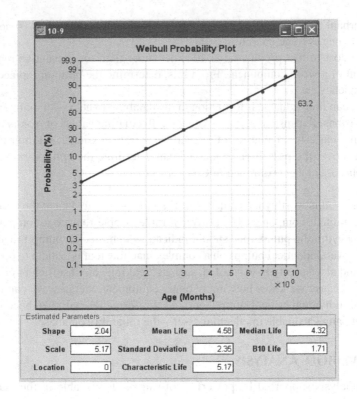

FIGURE 10.9   Weibull plot.

- Groups the data in increasing order of time to failure, as in Section 9.2
- Obtains the *median rank* from tables for each time group (The median rank is explained in Section 10.6.3, and the median rank table for up to 12 samples is provided in Appendix 2.)
- Plots on Weibull probability paper the median rank versus failure time of each observation

The April failures, for example, will have a median rank of 44.83% (Appendix 2), meaning that roughly 44.83% of them occurred up to and including April. The result is a plot such as Figure 10.9. You can see from the graph that the shape parameter beta (β) is 2.04 and the characteristic life eta (η) is 5.17. Furthermore, the mean life is 4.58 months.

TABLE 10.2

Failures and Median Ranks by Munn

Month	Jan	Feb	Mar	Apr	May	Jun	Jul	Aug	Sep	Oct
Failures	2	5	7	8	7	6	5	4	3	1
Med Rank	3.47	13.80	28.28	44.83	59.51	71.72	82.06	90.34	96.53	98.57

10.6.2 ADVANTAGES

Much can be learned from the plot itself, even how the data deviate from the straight line—for example:

- Whether and how closely the data follows the Weibull distribution
- The type of failure (infant mortality, random, or wear-out)
- The component's B_n life (the time at which $n\%$ of a population will have failed)
- Whether there may be competing failures (e.g., from fatigue and abrasion occurring simultaneously)
- Whether some other distribution, such as log-normal, is a better fit
- Whether there may have been predelivery shelf-life degradation
- Whether there is an initial failure free period that needs to be accounted for
- Forecasts for failures at a given time or during a future given period
- Whether there are batch or lot manufacturing defects

For more information about these deductions, see Abernethy.[2]

Surprisingly few data, as you will see, are required to draw accurate and useful conclusions. Even with inadequate data, engineers trained to read Weibull plots can learn a lot from them. The horizontal scale is a measure of life or aging such as start/stop cycles, mileage, operating time, take-off, or mission cycles. The vertical scale is the cumulative percentage failed. The two defining parameters of the Weibull line are the slope, β, and the characteristic life, η. (See Figure 10.3.)

The characteristic life, η, also called the $B_{63.2}$ life, is the age when 63.2% of the units will have failed. Similarly, where the Weibull line and the 10% CDF horizontal intersects is the age at which 10% of the population fails, or the B_{10} life. For more serious or catastrophic failures, the $B_{1.0}$, $B_{0.1}$, or $B_{0.01}$ lives are readily obtained from the Weibull plot.

The slope of the Weibull line, beta, shows which failure class is occurring. This could be infant mortality (decreasing hazard function, $\beta < 1$), random (constant hazard function, $\beta = 1$), or wear-out (increasing hazard function, $\beta > 1$). η, or the $B_{63.2}$ life, is approximately equal to MTTF. (They are equal when $\beta = 1$. That is when Weibull is equivalent to the exponential distribution.)

A significant advantage of the Weibull plot is that even if you don't immediately get a straight line (e.g., in Figure 10.10) when plotting the data, you can still learn something quite useful. If the data points are curved downward (Figure 10.11a), it could mean the time origin is not at zero. This could imply that the component had degraded on the shelf or suffered extended burn in time after it was made but before delivery. On the other hand, it could show that it was physically impossible for the item to fail early on. Any of these reasons could justify shifting the origin and replotting.

Alternatively, the concave downward curve could be saying that the data really fit a log-normal distribution. You can, using software, quickly and conveniently test these hypotheses, replot with a time origin shift, or transform the scale for log-normal probability paper. Once you apply the appropriate origin shift correction (adding or

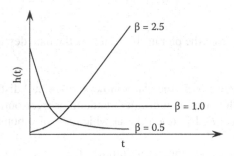

FIGURE 10.10   Hazard function for burn-in, random, and wear-out failures.

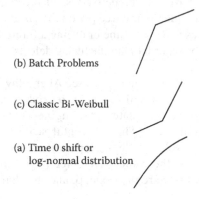

FIGURE 10.11   Various Weibull plotted data.

subtracting a value t_0 to each data point time), the resulting plot will be straight, and you will know a lot more about the failure process.

Further, if the plotted data form a dog leg downward (Figure 10.11b), it could mean that something changed when the failed part was manufactured. The first, steeper sloped leg reflects low time failures. The second, lower sloped line indicates failures from batches without the defect. When there are dual failure modes like this, it's known as a batch effect. When there are many suspensions (parts that have been removed for reasons other than failure, or parts that haven't failed at the end of the sampling period) it can be a clue that a batch problem exists. Failure serial numbers clustered closely along a leg support the theory that there was some manufacturing process change before the specific units were produced. Scheduled maintenance can also produce batch effects.[2]

An upward pointing dog leg bend (Figure 10.11c) indicates multiple failure modes. Investigate the failed parts. If the different failures are plotted separately, treating the other failure as a suspension, two straight lines should be observed. This is the classic bi-Weibull, showing the need for a root-cause failure (RCF) analysis to eliminate, for example, an infant mortality problem. Several dog leg bends distinguish various multiple failure modes.[2]

When the Weibull plot curves concave downward, you may need to use an origin shift, t_0, equivalent to using a three-parameter Weibull:[2]

$$F(t) = 1 - e^{-\left(\frac{t - t_0}{\eta - t_0}\right)^{\beta}}$$

There should be a physical explanation of why failures cannot occur before t_0. For example, bearings, even in an infant mortality state, will continue to operate for some time before failing. Use a large sample size, at least 20 failures. If you know, from earlier Weibull analyses, that the third parameter is appropriate, you might use a smaller sample size, say, 8 to 10.[2]

For Weibull and other statistical modeling methods, the data requirements are straightforward. Maintenance personnel should be able to collect it during routine activities. There are three criteria for good data stipulated by D. R. Cox:[2]

- The time origin must be clear.
- The scale for measuring time must be agreed upon.
- The meaning of failure must be clear.

Although all modern computerized maintenance management systems (CMMSs) and enterprise asset management (EAM) systems can provide this level of data collection, unfortunately they have been mostly underused. To be fair, writing out failure and repair details isn't part of the tradesperson's job description. Many in the field don't yet see that adding meticulous information to the maintenance system makes organization assets function more reliably. Training in this area could yield untapped benefits.

Although the technicians performing the work provide the most useful information, collecting it is the responsibility of everyone in maintenance. Data must be continually monitored for both quality and relevance. Once the data flow is developed, it can be used in several areas, such as developing preventive maintenance programs, predictive maintenance, warranty administration, tracking vendor performance, and improved decision making.[7] As Weibull analysis and other reliability methods become prevalent through advanced software, data will improve. The result? Management and staff will recognize the potential of good data to sharpen their company's competitive edge.

10.6.3 MEDIAN RANKS

Suppose that five components fail at 67, 120, 130, 220, and 290 hours. To plot these data on Weibull probability paper, you need the CDF's corresponding estimates. You must estimate the fraction that is failing before each of the failure ages. You can't simply say that the percentage failed at 120 hours is 2/5, because that would imply that the cumulative probability at 290 hours, a random variable, is 100%. This small sample size doesn't justify such a definitive statement. Taken to the absurd, from a sample size of 1, you certainly couldn't conclude that the time of the single failure reflects 100% of total failures. The most popular approach to estimating the y axis plotting positions is the median rank. Obtain the CDF plotting values from the median ranks table in Appendix 2 or from a reasonable estimate of the median rank, Benard's formula:

$$Median\ Rank = \frac{i - 0.3}{n + 0.4}$$

Determined by either method, this item's cumulative failure probabilities are 0.13, 0.31, 0.5, 0.69, and 0.87, respectively for the first, second, third, fourth, and fifth ordered failure observations. When you use reliability software, you do not have to look up the median ranks in tables or perform manual calculations. The program automatically calculates and applies the median ranks to each observation.

10.6.4   CENSORED DATA OR SUSPENSIONS

It is an unavoidable data analysis problem that, at the time you are observing and analyzing, not all the units will have failed. You know the age of the unfailed units and that they are still working properly but not when they will fail. Also, some units may have been replaced preventively. In those cases, too, you don't know when they would have failed. These units are said to be *suspended* or *right-censored*. Left-censored observations are failures whose starting times are unavailable. While not ideal, statistically you can still use censored data, since you know that the units lasted *at least* this long. Ignoring censored data would seriously underestimate the item's reliability.

Account for censored data by modifying the order numbers before plotting and determining the CDF values (from Benard's formula or from the median ranks table). Assuming item 2 was removed without having failed, the order numbers 1, 2, 3, 4, and 5 in the previous section (10.6.2) would become 1, 2.5, 3.75, and 5.625. The formula used to calculate these modified orders is in Appendix 3. Consequently, the modified median ranks would become those shown in   10.3.

Now you can plot the observations on Weibull probability paper normally.

TABLE 10.3

Modified Orders and Median Ranks for Samples with Suspensions

Hours	Event	Order	Modified Order	Median Rank
67	F	1	1	0.13
120	S	2		
130	F	3	2.25	0.36
220	F	4	3.5	0.59
290	F	5	4.75	0.82

10.6.5 THE THREE-PARAMETER WEIBULL

For the reasons described in Section 10.6.1, including physical considerations, the time of the observations may need to be shifted for the Weibull data to plot to a straight line. By changing the origin, you activate another parameter in the Weibull equation. To make the model work, you must estimate three parameters rather than two. See Appendix 4 for the procedure to estimate the third parameter, also called the location parameter, the guarantee life, or the minimum life parameter.[8] A two-parameter Weibull model is a special case of the more general three-parameter model, with the location parameter equal to zero.

10.6.6 THE FIVE-PARAMETER BI-WEIBULL

As shown in Figure 10.11, there can be more than one behavior (of burn-in, random, and wear-out), reflected by your sample data. The bi-Weibull model can represent two failure types, for example, a stress (random) failure rate phase followed by a wear-out phase, a random phase followed by another random phase, at some higher failure rate, or a burn-in failure followed by random failure. Six common failure patterns are illustrated in Figure 10.12.

A limitation of the Weibull distribution is that it does not cover patterns B and F (Figure 10.12), which often occur.[9] A bi-Weibull distribution is formed by combining two Weibull distributions. The five-parameter Weibull, known as the Hastings bi-Weibull, is implemented in the Relcode[9] package. Its hazard function form is

$$h(t) = \lambda\theta\left(\lambda\gamma\right)^{\theta-1} + \left(\frac{\beta}{\eta}\right)\left(\frac{t-\gamma}{\eta}\right)^{\beta-1}$$

If $\gamma = \theta = 0$, the Hastings bi-Weibull reduces to a Weibull distribution. The bi-Weibull distribution includes the Weibull as a special case but allows two failure phases, so patterns B and F are now covered. Software[9] will fit the most appropriate

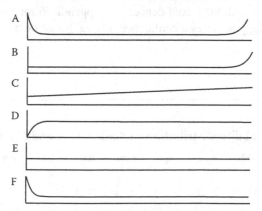

FIGURE 10.12 Failure rate patterns.

of the various models (two-parameter Weibull, three-parameter Weibull, or five-parameter bi-Weibull) to the data. For a random failure phase followed by a wear-out phase, knowing the onset of the wear-out helps decide when to begin preventive repair or replacement.

10.6.7 CONFIDENCE INTERVALS

You need to be confident that any actions you take based on your statistical analysis and observed data modeling will be successful. In Section 10.6 and Figure 10.9, we plotted failure observation times of 48 failures over a period of 10 months against the CDF values estimated by the median rank. The median rank estimates that 50% of the time the true percentage of failures lies above and below it.

Similarly, you can estimate the CDF according to another set of tables, the 95% rank tables. This allows that 95% of the time the percentage of failures will be below this value. The same applies for a 5% rank, where 95% of the time the percentage will be above this value. Table 10.4 includes rows for the 5% and 95% rank values into our original example.

The 5% and 95% rank tables are given in Appendix 5. Figure 10.13 shows the data of Table 10.4 on the Weibull plot. From the Weibull plot of the median, between 95%, and 5% rank lines, you can conclude that the distribution function at the end of May will have a value between 69% to 49%, with a confidence of 90%. This means that, after five months, in 90% of the tests you conduct, between 47% and 17% of the items will fail. Or, the reliability (which is 100% minus the CDF value) of this type of item surviving five months is between 30% and 53% for 90% of the time.

The vertical distance between the 5% and 95% lines represents a 90% confidence interval. By plotting all three lines, you get a confidence interval showing that the cumulative probability of failure will range from a to b with 90% confidence (90% of the time). As a second example, an item's failure probability, up to and including June, is between 60.43% and 81.43% for 90% of the time. Or, knowing that $R(t) = 1 - F(t)$, the item's reliability that it will survive to the end of six months is between 18.57% and 39.57% with 90% confidence. See Appendix 5 for another example of how this methodology achieves a confidence interval for an item's reliability.

TABLE 10.4
Cumulative Probability Distribution Including 95% and 5% Ranks

Month	Jan	Feb	Mar	Apr	May	Jun	Jul	Aug	Sep	Oct
Failures	2	5	7	8	7	6	5	4	3	1
Med rank	3.47	13.80	28.28	44.83	59.31	71.72	82.06	90.34	96.53	98.57
95% rank	9.51	23.17	39.57	56.60	70.41	81.43	89.85	95.91	29.26	99.89
5% rank	6.75	7.05	18.57	33.43	47.52	60.43	71.93	81.94	90.49	13.95

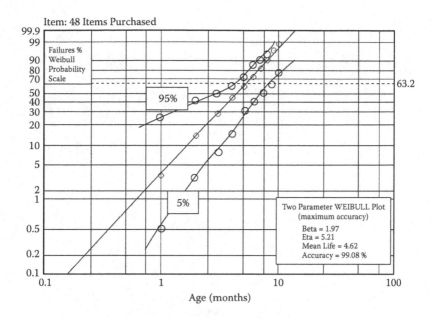

FIGURE 10.13   Weibull Plot Showing 5% and 95% Confidence Limits.

10.6.8   GOODNESS OF FIT

You would, quite naturally, like to have a quantitative measure of how well your model fits the data. How good is the fit? Goodness-of-fit testing provides the answer. Methods such as least squares or maximum likelihood (described in Appendix 6) are used to fit an assumed distribution model to the data and estimate the parameters of its distribution function. That information, and the confidence intervals discussed in the previous section, will help you judge the validity of your model choice (be it Weibull, log-normal, negative exponential, or another) and your estimation method. One of the methods commonly used is the Kolmogorov-Smirnov test described in Appendix 7.

REFERENCES

1. Frankel, E.G., *Systems Reliability and Risk Analysis*, 2d ed., Norwell, MA: Kluwer Academic, 1983.
2. Abernethy, R.B., *The New Weibull Handbook*, 5th ed., Houston: Gulf Publishing, 2006.
3. Optimal Maintenance Decisions Inc., http://www.omdec.com.
4. ReliaSoft, http://www.weibull.com.
5. M-Tech Consulting Engineers, http://www.m-tech.co.za
6. WinSmith for Windows, Fulton Findings, http://www.weibullnews.com.
7. Jardine, A.K.S., and A.H.C. Tsang, *Maintenance, Replacement and Reliability: Theory and Applications*, Boca Raton, FL: CRC Press, 2006.
8. Kapur, K.C., and L.R. Lamberson, *Reliability in Engineering Design*, New York: John Wiley & Sons, 1977.
9. Hastings, N.A.J., *RelCode for Windows*, user's manual.

11 Maintenance Optimization Models

Andrew K. S. Jardine

CONTENTS

Maintenance optimization is all about *getting the best result possible, given one or more assumptions.* In this chapter, we introduce the concept of optimization through a well-understood traveling problem: identifying the best mode of travel, depending on different requirements. We also examine the importance of building mathematical models of maintenance decision problems to help arrive at the best decision.

We look at key maintenance decision areas: component replacement, capital equipment replacement, inspection procedures, and resource requirements. We use optimization models to find the best possible solution for several problem situations.

11.1 WHAT IS OPTIMIZATION ALL ABOUT?

Optimal means the most desirable outcome possible under restricted circumstances. For example, following a reliability-centered maintenance (RCM) study, you could conduct condition monitoring maintenance tactics, time-based maintenance, or time-based discard for specific parts of a machine or system. In this chapter, we introduce maintenance decision optimization. In the next chapter, we discuss detailed models for asset maintenance and replacement decision making.

To understand the concept of optimization, consider this travel routing problem: you have to take an airplane trip, with three stops, before returning home to Chicago. The first destination is London, followed by Moscow and then Hawaii. Before purchasing a ticket, you would weigh a number of options, including airlines, fares, and schedules. You'd make your decision based on factors such as economy, speed, safety, and extras:

- If *economy* is most important, you'd choose the airline with the cheapest ticket. That would be the optimal solution.
- If *speed* was it, you'd consider only the schedules and disregard the other criteria.
- If you wanted to optimize *safety* you'd avoid airlines with a dubious safety record and pick only a well-regarded carrier.
- If you wanted a free hotel room (an *extra*) for three nights in Hawaii, you'd opt for the airline that would provide that benefit.

This list illustrates the concept of optimization. When you optimize in one area—economy, for example—then almost always you get a less desirable (suboptimal) result in one or more of the other criteria.

Sometimes, you have to do a trade-off between two criteria. For example, though speed may be most important to you, the cost of traveling on the fastest schedule could be unacceptable. The solution is somewhere in the middle—providing an acceptable cost (but not the very lowest) and speed (but not the very fastest).

In any optimization situation, including maintenance-decision optimization, you should do the following:

- *Think* about optimization when making maintenance decisions.
- Consider *what* maintenance decision you want to optimize.
- Explore *how* you can do this.

11.2 THINKING OPTIMIZATION

Thinking about optimization means considering trade-offs: the pros and cons. Optimization always has to do with getting the best result where it counts most while consciously accepting less than that elsewhere.

A customer service manager was asked by the vice president of marketing what he thought his main mission was. His answer: To get every order for every customer delivered without fail on the day the customer specified, 100% of the time. To achieve this goal, the inventory of ready-to-ship goods would have to include every color, size, and style in sufficient quantities to ensure that no matter what was called for, it could be shipped. In spite of unusually big orders, a large number of customers randomly wanting the same thing at the same time, or machinery failure, the manager would have to deliver. His inventory would have had an unacceptably high cost.

The manager failed to realize that a delivery performance just slightly less, say 95%, would be better. In fact, it would be a profit-optimization strategy, the best trade-off between the cost of inventory and an acceptable and competitive customer satisfaction level.

11.3 WHAT TO OPTIMIZE

Just as in other areas, you can optimize in maintenance for different criteria—including cost, availability, safety, and profit.

Lowest-cost optimization is often the maintenance goal. The cost of the component or asset, labor, lost production, and perhaps even customer dissatisfaction from delayed deliveries are all considered. Where equipment or component wear-out is a factor, the lowest possible cost is usually achieved by replacing machine parts late enough to get good service out of them but early enough for an acceptable rate of on-the-job failures (to attain a "zero" rate, you'd probably have to replace parts every day).

Availability can be another optimization goal: getting the right balance between taking equipment out of service for preventive maintenance and suffering outages due to breakdowns. If *safety* is most important, you might optimize for the safest possible solution but with an acceptable impact on cost. If you optimize for *profit,* you would take into account not only cost but also the effect on revenues through greater customer satisfaction (better profits) or delayed deliveries (lower profits).

11.4 HOW TO OPTIMIZE

One of the main tools in the scientific approach to management decision making is building an evaluative model, usually mathematical, to assess a variety of alternative decisions. Any model is simply a representation of the system under study. When applying quantitative techniques to management problems, we frequently use a symbolic model. The system's relationships are represented by symbols and properties described by mathematical equations.

To understand this model-building approach, examine the following maintenance stores problem. Although simplified, it illustrates two important aspects of model use: constructing the problem being studied and its solution.

A Stores Problem

A stores controller wants to know how much to order each time the stock level of an item reaches zero. The system is illustrated in Figure 11.1.

The conflict here is that the more items ordered at any time, the more ordering costs will decrease, but holding costs increase, since more stock is kept

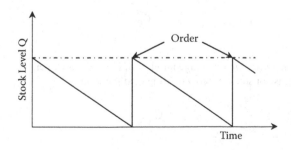

FIGURE 11.1 An inventory problem.

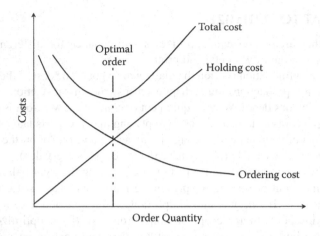

FIGURE 11.2   Economic order quantity.

on hand. These conflicting features are illustrated in Figure 11.2. The stores controller wants to determine which order quantity will minimize the total cost. This total cost can be plotted, as shown in Figure 11.2, and used to solve the problem.

A much more rapid solution, however, is to construct a mathematical model of the decision situation. The following parameters can be defined:

D = total annual demand
Q = order quantity
C_o = ordering cost per order
C_h = stockholding cost per item per year

Total cost per year of ordering and holding stock = Ordering cost per year + Stockholding cost per year

Now,

Ordering cost/year = Number of orders placed per year × Ordering cost per order =

$$\frac{D}{Q} C_o$$

Stockholding cost/year = Average number of items in stock per year (assuming linear decrease of stock) × Stockholding cost per item

$$\frac{Q}{2} C_h$$

Therefore, the total cost per year, which is a function of the order quantity and is denoted $C(Q)$, is

$$C(Q) = \frac{D}{Q}C_o + \frac{Q}{2}C_h \qquad (11.1)$$

Equation 11.1 is a mathematical model of the problem relating order quantity, Q, to total cost, $C(Q)$. The stores controller wants to know the number of items to order to minimize the total cost, that is, the right-hand side of Equation 11.1. The answer is obtained by differentiating the equation with respect to Q, the order quantity, and equating the answer to zero as follows:

$$\frac{dC(Q)}{dQ} = -\frac{D}{Q^2}C_o + \frac{C_h}{2} = 0$$

Therefore,

$$\frac{D}{Q^2}C_o = \frac{C_h}{2}$$

$$Q = \sqrt{\frac{2DC_o}{C_h}} \qquad (11.2)$$

Since the values of D, C_o, and C_h are known, substituting them into Equation 11.2 gives the value of the order quantity Q. You can check that the value of Q obtained from Equation 11.2 is a minimum and not a maximum by taking the derivative of $C(Q)$ and noting the positive result. This confirms that Q is optimal.

Example

Let $D = 1,000$ items, $C_o = \$5$, $C_h = \$0.25$

$$Q = \sqrt{\frac{2 \times 1000 \times 5}{0.25}} = 200 \text{ items}$$

Each time the stock level reaches zero, the stores controller should order 200 items to minimize the total cost per year of holding and ordering stock.

Note that the various assumptions that have been made in the inventory model may not be realistic. For example, no consideration has been given to quantity discounts, the possible lead time between placing an order and its receipt, and the fact that demand may not be linear or known for certain. The purpose of the previous model is simply to illustrate constructing a model and attaining a solution for a particular problem. There is abundant literature about stock control problems without many of these limitations. If you are interested in stock control aspects of maintenance stores, see Nahmias.[1]

It's clear from the previous inventory control example that we need the right kind of data, properly organized. Most organizations have a computerized maintenance management system (CMMS) or an enterprise asset management

(EAM) system. The vast amount of data they store makes optimization analyses possible.

Instead of building mathematical models of maintenance decision problems, software is available to help you make optimal maintenance decisions. This is covered in Chapter 12.

11.5 KEY MAINTENANCE MANAGEMENT DECISION AREAS

There are four key decision areas that maintenance managers must address to optimize their organization's human and physical resources. These areas are depicted in Figure 11.3. The first column deals with component replacement, the second with inspection decisions, including condition-based maintenance and the third with establishing the economic life of capital equipment. The final column addresses decisions concerning resources required for maintenance and their location.

To build strong maintenance optimization, you need an appropriate source, or sources, of data. The foundation for this, as shown in Figure 11.3, is the CMMS/EAM system/enterprise resource planning (ERP) system.

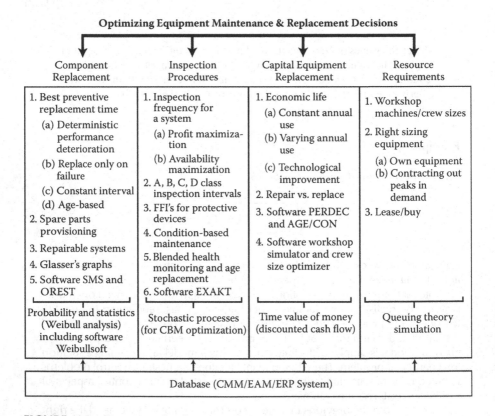

Optimizing Equipment Maintenance & Replacement Decisions

Component Replacement	Inspection Procedures	Capital Equipment Replacement	Resource Requirements

1. Best preventive replacement time
   (a) Deterministic performance deterioration
   (b) Replace only on failure
   (c) Constant interval
   (d) Age-based
2. Spare parts provisioning
3. Repairable systems
4. Glasser's graphs
5. Software SMS and OREST
	1. Inspection frequency for a system
	   (a) Profit maximization
	   (b) Availability maximization
	2. A, B, C, D class inspection intervals
	3. FFI's for protective devices
	4. Condition-based maintenance
	5. Blended health monitoring and age replacement
	6. Software EXAKT
		1. Economic life
		   (a) Constant annual use
		   (b) Varying annual use
		   (c) Technological improvement
		2. Repair vs. replace
		3. Software PERDEC and AGE/CON
		4. Software workshop simulator and crew size optimizer
			1. Workshop machines/crew sizes
			2. Right sizing equipment
			   (a) Own equipment
			   (b) Contracting out peaks in demand
			3. Lease/buy

Probability and statistics (Weibull analysis) including software Weibullsoft	Stochastic processes (for CBM optimization)	Time value of money (discounted cash flow)	Queuing theory simulation

Database (CMM/EAM/ERP System)

FIGURE 11.3 Key areas of maintenance and replacement decisions.

In Chapter 12, we discuss optimization of key maintenance decisions of component replacement, inspection procedures, and capital equipment replacement (Columns 1, 2, and 3 of Figure 11.3). The framework, foundation, or database is discussed in detail in Chapter 5.

Extensive development and discussion of models, including case studies, is provided in Jardine and Tsang.[2]

REFERENCES

1. Nahmias, S., *Production and Operations Analysis,* 3d ed., Irwin/McGraw-Hill, 1997.
2. Jardine, A.K.S., and A.H.C. Tsang, *Maintenance, Replacement and Reliability: Theory and Applications,* CRC Press, 2006.

12 Optimizing Maintenance and Replacement Decisions

Andrew K. S. Jardine

CONTENTS

12.1 CHAPTER OVERVIEW

In this chapter, we explore the various strategies and tools you need to make the best maintenance and replacement decisions. In particular, you need to know the optimal replacement time for critical system components (also known as line-replaceable units, or LRUs), and capital equipment.

At the LRU level, we examine age- and block-replacement strategies. You'll learn how OREST[1] software can help you optimize LRU maintenance decisions to keep costs under control and increase equipment availability.

With capital equipment, it's critical to establish economic viability. We examine how to do this in two operating environments:

- Constant use, year-by-year
- Declining use, year-by-year (older equipment is used less)

In this section you'll discover how to extend the life of capital equipment through a major repair or rebuild, and when that is more economical than replacing it with new equipment. The optimal decision is the one that minimizes the long-run equivalent annual cost (EAC), the life cycle cost of owning, using and disposing of the asset.

Our study of capital equipment replacement includes AGE/CON[1] and PERDEC[1] software, which simplify the job of managing assets.

Finally we look at maintenance resources: what resources there should be, where they should be located, who should own them, and how they should be used. The role that can be played by the theory of queues and simulation will be highlighted.

12.2 INTRODUCTION: ENHANCING RELIABILITY THROUGH PREVENTIVE REPLACEMENT

Generally, preventing a maintenance problem is always preferred to fixing it after the fact. You'll increase your equipment reliability by learning to replace critical components at the optimal time, before a breakdown occurs. When is the best time? That depends on your overall objective. Do you most want to minimize costs or maximize availability? Sometimes the best preventive replacement time accomplishes both objectives, but not necessarily.

Before you start, you need to obtain and analyze data to identify the best preventive replacement time. Later in this section, we present some maintenance policy models of fixed interval and age-based replacements to help you do that.

But, for preventive replacement to work, these two conditions must first be present:

- If cost is most important, the total cost of a replacement must be greater after failure than before. If reducing total downtime is most essential, the total downtime of a failure replacement must be greater than a preventive replacement. In practice, this usually happens.
- The risk of a component failing must increase as it ages. How can you know? Check that the Weibull shape parameter associated with the component's failure times is greater than 1. Often, it is assumed that this condition is met but be sure that is in fact the case. See Chapter 10 for a detailed description of Weibull analysis.

12.2.1   Block Replacement Policy

The block replacement policy is sometimes called the group or constant interval policy, since preventive replacement occurs at fixed times, and failure replacements whenever necessary. The policy is illustrated in Figure 12.1. C_p is the total cost of a preventive replacement, C_f is the total cost of a failure replacement, and t_p is the interval between preventive replacements.

You can see that for the first cycle of t_p, there isn't a failure, while there are two in the second cycle and none in the third and fourth. As the interval between preventive replacements decreases, there will be fewer failures in between. You want to obtain the best balance between the investment in preventive replacements and the economic consequences of failure replacements. This conflict is illustrated in Figure 12.2.

$C(t_p)$ is the total cost per week of preventive replacements occurring at intervals of length t_p, with failure replacements occurring whenever necessary. See Appendix A for the total cost curve equation.

The following problem is solved using OREST software, incorporating both the cost model and Weibull analysis, to establish the best preventive replacement interval.

PROBLEM

There has been a bearing failure in the blower used in diesel engines. The failure has been established according to a Weibull distribution, with a mean life of 10,000 km and a standard deviation of 4,500 km. The bearing failure is expensive, costing ten times as much to replace than if it had been done as a preventive measure. Determine:

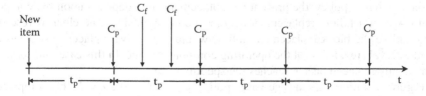

FIGURE 12.1   Constant interval replacement policy.

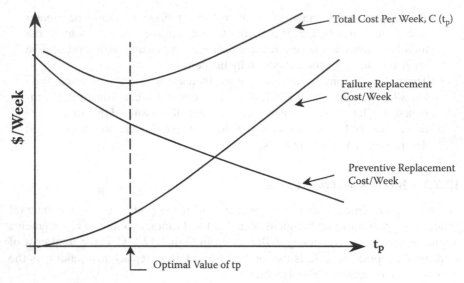

Preventive Replacement Cost Conflicts

FIGURE 12.2   Constant interval policy: optimal replacement time.

- The optimal preventive replacement interval (or block policy) to minimize total cost per kilometer
- The expected cost saving of the optimal policy over a run-to-failure replacement policy
- Given that the cost of a failure is $2000, the cost per km. of the optimal policy

Figure 12.3 shows a screen capture from OREST:

Result

You can see that the optimal preventive replacement time is 3873 kilometers (3872.91 in chart) Also, the Figure provides valuable additional information. For example:

- The cost per kilometer of the best policy is: $0.09/km
- The cost saving compared to a run-to-failure policy is: $ 0.11/km (55%)

12.2.2   AGE-BASED REPLACEMENT POLICY

In the age-based policy, the preventive replacement time depends upon the component's age. If a failure replacement occurs, the component's time clock is reset to zero, unlike the block replacement policy, where preventive replacements occur at fixed intervals regardless of the operating component's age. In this case, the component is only replaced once it reaches the specified age.

Figure 12.4 illustrates the age-based policy. C_p and C_f represent the block replacement policy, and t_p represents the component age when preventive replacement occurs.

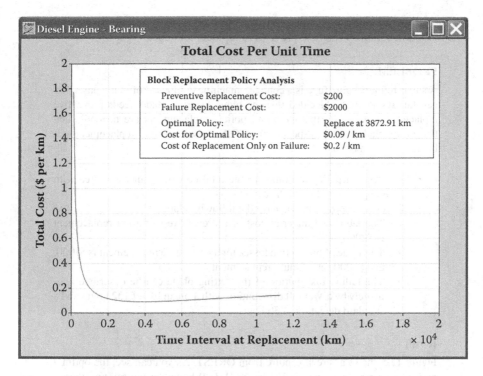

FIGURE 12.3   OREST output: block replacement.

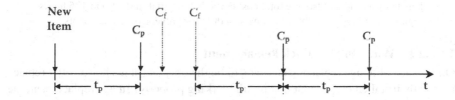

FIGURE 12.4   Age-based replacement policy.

In Figure 12.4, you can see that there aren't any failures in the first cycle, which is t_p. After the first preventive replacement, the component must have failed before reaching its next planned preventive replacement age. After the first failure replacement, the clock is set to zero and the next preventive replacement scheduled. However, before it's reached, the component is again replaced due to failure. After this second failure replacement, the component survives to the planned preventive replacement age of t_p. Similarly, the next replacement cycle shows that the component made it to its planned preventive age.

The conflicting cost consequences are identical to those in Figure 12.2, except that the x-axis measures the actual age (or utilization) of the item, rather than a fixed time interval. See Appendix A for the mathematical model depicting this preventive replacement policy.

The following problem is solved using OREST software, which incorporates the cost model, to establish the best preventive replacement age.

PROBLEM

A sugar refinery centrifuge is a complex machine composed of many parts that can fail suddenly. It's decided that the plough-setting blade needs preventive replacement. Based on the age-based policy, replacements are needed when the setting blade reaches a specified age. Otherwise, a costlier replacement will be needed when the part fails. Consider:

- The optimal policy to minimize the total cost per hour associated with preventive and failure replacements
- To solve the problem, you have the following data:
  - The labor and material cost of a preventive or failure replacement is $2000
  - The value of production losses for a preventive replacement is $1000, and $7000 for a failure replacement
  - The failure distribution of the setting blade can be described adequately by a Weibull distribution with a mean life of 152 hours and a standard deviation of 30 hours.

Result

Figure 12.5 shows a screen capture from OREST. As you can see, the optimal preventive replacement age is 112 hours (111.49 hours on figure), and there's additional key information that you can use. For example, the preventive replacement policy costs 44.8% of run-to-failure ([(59.21 − 32.66)/59.21]), making the benefits very clear. Also, the total cost curve is fairly flat in the 90 to 125 hours region, providing a flexible planning schedule for preventive replacements.

12.2.2.1 When to Use Block Replacement

On the face of it, age replacement seems to be the only sensible choice. Why replace a recently installed component that is still working properly? In age replacement, the component always remains in service until its scheduled preventive replacement age.

To implement an age-based replacement policy, though, you must keep an ongoing record of the component's current age and change the planned replacement time if it fails. Clearly, the cost of this is justified for expensive components, but for an inexpensive one, the easily implemented block replacement policy is often more cost-efficient.

12.2.2.2 Safety Constraints

With block and age replacement, the objective is to establish the best time to preventively replace a component, to minimize the total cost of preventive and failure replacements.

If you want to ensure that the failure probability doesn't exceed a particular value, say 5%, without cost considerations formally being taken into account, you can determine when to schedule a preventive replacement from the cumulative

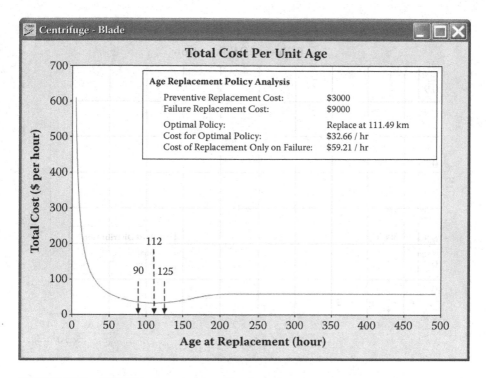

FIGURE 12.5　OREST output: Age-based replacement.

failure distribution. This is illustrated in Figure 12.6. You can see that the preventive replacement should be planned once the item is at 3000 km.

Alternatively, you can preventively replace a critical component so that the risk, or hazard, doesn't exceed a specified value, such as 5×10^{-5} failures per kilometer. In this case, you need to use the hazard plot to identify the preventive replacement age. This is illustrated in Figure 12.7, which shows the component's appropriate preventive replacement age is 4000 km.

12.2.2.3 Minimizing Cost and Maximizing Availability

In block and age replacement, the objective is minimizing cost. Maximizing availability simply requires replacing the total costs of preventive and failure replacement in the models with their total downtime. Minimizing total downtime is then equivalent to maximizing availability. In Appendix A, you'll find total downtime minimization models for both block and age replacement.

12.3　DEALING WITH REPAIRABLE SYSTEMS

Rather than completely replace a failed unit, you may be able to get it operating with minor corrective action. This is what's known as a minimal and general repair. Models for addressing this case are discussed in Jardine and Tsang.[2]

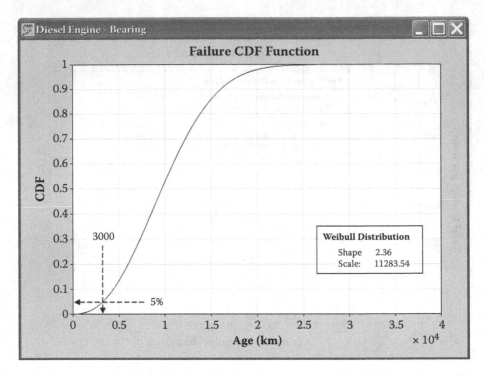

FIGURE 12.6 Optimal preventive replacement age: risk based maintenance.

12.4 ENHANCING RELIABILITY THROUGH INSPECTION

A big part of the maintenance mandate, of course, is to ensure that the system is reliable. One way to achieve this is to identify critical components that are likely to fail and preventively replace them. In Section 12.2, we covered methods to identify which components (line replaceable units) are candidates for preventive replacement, and how to decide when to do it.

An alternative is to consider the system as a whole, and make regular inspections to identify problem situations. Then, carry out minor maintenance, such as changing a component or topping up the gearbox with oil, to prevent system failure. You need to know, though, the best frequency of inspection.

Yet another approach is to monitor the health of the system through predictive maintenance and only act when you get a signal that a defect is about to happen which, if not corrected, will create a system failure. This second approach is covered in detail in Section 12.4.

12.4.1 Establishing the Optimal Inspection Frequency

Figure 12.8 shows a system composed of five components, each having its own failure distribution. (In Reliability Centered Maintenance terminology, these components are different modes of system failure.)

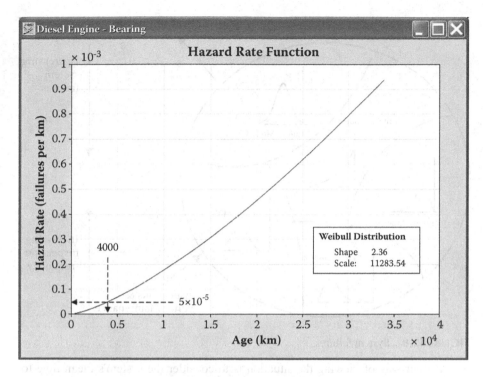

FIGURE 12.7 Optimal preventive replacement age: Hazard limit.

Because the Weibull failure pattern is so flexible, all the components can likely be described as failing, but with different shape parameters. That is, depending on the component, the risk of it failing as it ages can either increase, remain constant, or decrease. The overall effect is that there will be system failures from any number of causes. If you analyze the overall system failure data, again a Weibull would fit. The shape parameter almost certainly would equal 1.0, though, indicating that system failures are occurring strictly randomly. This is what you should expect. The superposition of numerous failure distributions creates an exponential failure pattern, the same as a Weibull with shape parameter of 1.0 (Drenick[3]).

To reduce these system failures, you can inject inspections, with minor maintenance work, into the system, as shown in Figure 12.8. The question is then: How frequently should inspections occur?

While you may not know the individual risk curves of system components, you can determine the overall system failure by examining the maintenance records. The pattern will almost certainly look like Figure 12.9, where the system failure rate is constant, but can be reduced through increasing the inspection frequency.

The risk curve may not be constant if system failures emanate from one main cause. In this case, the system failure distribution will be identical to the component's failure pattern.

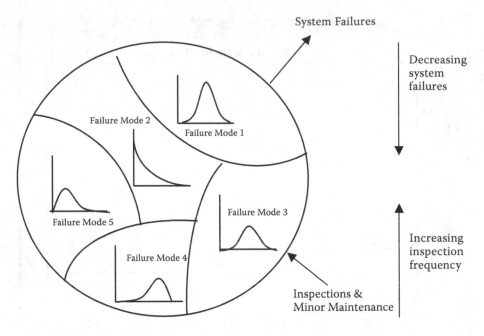

FIGURE 12.8   System failures.

Another way of viewing the situation is to consider the system's mean time to failure (MTTF). As the system failures decline, the MTTF will increase. This is illustrated in the probability density functions in Figure 12.10.

If the optimal inspection frequency minimizes the total downtime of the system, you get the conflicting curves of Figure 12.11. See Appendix A for the underlying mathematical model that establishes the optimal frequency.

Case Study

An urban transit fleet had a policy of inspecting the buses every 5000 kilometers. At each inspection, an A, B, C or D check occurred. An A check is a minor maintenance inspection, while a D check is the most detailed. The policy is illustrated in Table 12.1.

Since the 5000 km. inspection policy was not followed precisely, some inspections took place before 5000 km, and others were delayed. As a result, Figure 12.12 shows that the average distance traveled by a bus between failures decreased as the interval between the checks was increased.

Knowing the average time a bus was out of service due to repair and inspection established the optimal inspection interval at 8,000 km, as Figure 12.13 shows. Note that the total downtime curve is very flat around the optimum, so the current policy of making inspections at the easily implemented 5,000 km interval might be best. Before the analysis, though, it wasn't known whether the current policy was appropriate, or should be modified. As you can see, data driven analysis revealed the answer. The Jardine and Hassounah study provides full details.[4]

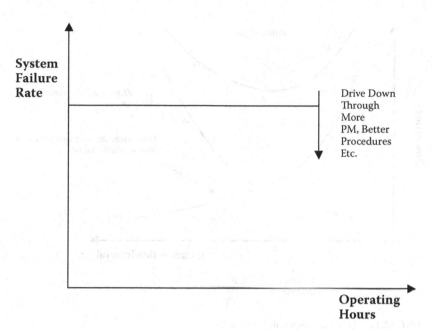

FIGURE 12.9   System failure rate.

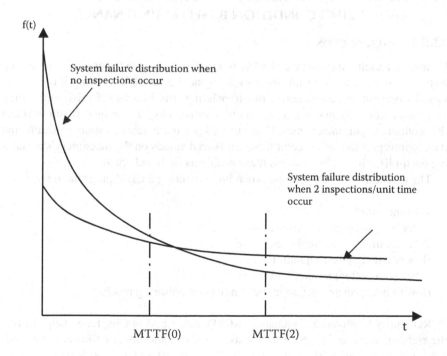

FIGURE 12.10   Inspection frequency versus MTTF.

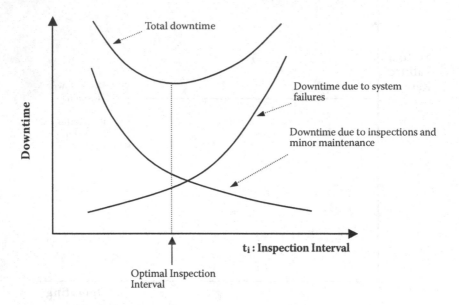

FIGURE 12.11 Optimal inspection frequency.

12.5 ENHANCING RELIABILITY THROUGH INSPECTION: OPTIMIZING CONDITION BASED MAINTENANCE

12.5.1 INTRODUCTION

Condition Based Maintenance (CBM) is an obviously good idea. Of course, you want the maximum useful life from each physical asset before taking it out of service for preventive maintenance. But translating this into an effective monitoring program, and timely maintenance, isn't necessarily easy. It can be difficult to select the monitoring parameters most likely to indicate the machine's state of health, and then to interpret the influence of those measured values on the machinery's remaining useful life (RUL). We address these problems in this chapter.

The essential questions to pose when implementing a CBM program are:

Why monitor?
What equipment components to monitor?
What monitoring technologies to use?
How (what signals) to monitor?
When (how often) to monitor?
How to interpret and act upon the condition monitoring results?

Reliability Centered Maintenance (RCM), described in Chapter 8, helps to find the right answers to the first 3 questions. However, additional optimizing methods are required to handle the remainder. In Chapters 10 and 11, you learned that the way to approach these problems is to build a model describing the factors surrounding maintaining or replacing equipment. In Chapter 10, we dealt with the lifetime of

TABLE 12.1
A, B, C, and D Inspection Policy

Transit Commission's Bus Inspection Policy

Km (1000)	Inspection Type "A"	"B"	"C"	"D"	
5	X				
10		X			
15	X				
20			X		
25	X				
30		X			
35	X				
40			X		
45	X				
50		X			
55	X				
60			X		
65	X				
70		X			
75	X				
80				X	
Total	8	4	3	1	$\Sigma = 16$
Ri	0.5	0.25	0.1875	0.0625	$\Sigma = 1.0$

Where

R_i = No of type I inspections/Total No. of inspections
I = A, B, C or D

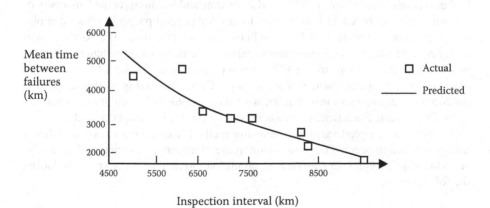

FIGURE 12.12   Mean distance to failure versus inspection interval.

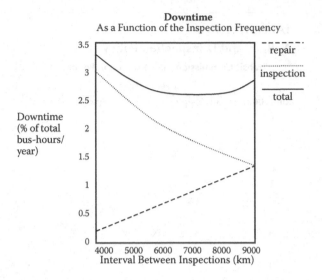

FIGURE 12.13   Optimal inspection interval.

components considered independent random variables, meaning that no additional information, other than equipment age, is used to schedule preventive maintenance.

CBM, however, introduces new information, called condition data, to determine more precisely the most advantageous moment to make a repair or replacement. We extend the models in Chapter 10, Section 10.2, to include the influence of condition data on the remaining useful life (RUL) of machinery and its components. The extended modeling method we introduce in this chapter, which takes measured data into account, is known as Proportional Hazards Modeling (PHM), and the measured condition data are referred to as covariates.

Since D.R. Cox's (1972) pioneering paper[5] on PHM, it has been used primarily to analyze medical survival data. Since 1985, it has been used more extensively, including applications to marine gas turbines, nuclear reactors, aircraft engines, and disk brakes on high-speed trains. In 1995 A.K.S. Jardine and V. Makis, at the University of Toronto, initiated the CBM Laboratory,[6] to develop general purpose software applying proportional hazards models to available maintenance data. The software was designed to be integrated into the plant's maintenance information system to optimize its CBM activities. The result, in 1997, was a program called EXAKT™, now in its 4th version and rapidly earning attention as a CBM optimizing methodology. We produced the examples in this chapter, with their graphs and calculations, using the EXAKT program. For a demonstration version of EXAKT contact OMDEC.[7]

Industry has adopted various monitoring methods that produce a signal when a failure is about to occur. The most common are vibration monitoring and oil analysis. Moubray[8] provides an overview of condition monitoring techniques, including the following:

Vibration analysis
Ultrasonic analysis

Ferrography
Magnetic chip detection
Atomic emission spectroscopy
Infrared spectroscopy
Gas chromatography
Moisture monitoring
Liquid dye penetrants
Magnetic particle inspection
Power signature analysis

As Pottinger and Sutton[9] said, "much condition-monitoring information tells us something is not quite right, but it does not necessarily inform us what margins remain for exploitation before real risk, and particularly real commercial risk, occurs." In this chapter, you will learn how to accurately estimate equipment health, using condition monitoring. The goal is to make the optimal maintenance decision, blending economic considerations with estimated risk.

Early work on estimating equipment risk dealt with jet engines on aircraft.[10] Oil analysis was conducted weekly and, if unacceptable metal levels were found in the samples, the engine was removed before its scheduled removal time of 15,000 flying hours. A PHM was constructed by statistically analyzing the condition data, along with the corresponding age of the engines that functioned for the duration, and those that were removed due to failure (in this case, operating outside tolerance specifications). Three key factors, from a possible 20, emerged for estimating the risk that the engine would fail:

Age of engine
Parts per million iron (Fe) in the oil sample at the time of inspection
Parts per million chrome (Cr) in the oil sample at the time of inspection

The PHM also identifies the weighting for each risk factor. The complete equation used to estimate risk of the jet engine failing was:

$$\text{Risk at time of inspection} = \frac{4.47}{24,100}\left(\frac{t}{24,100}\right)^{3.47} e^{0.41Fe+0.98Cr}$$

where the contribution of the engine age towards the overall failure risk is

$$\frac{4.47}{24,100}\left(\frac{t}{24,100}\right)^{3.47}$$

(this is termed a Weibull baseline failure rate) and the contribution to overall risk from the risk factors from the oil analysis is $e^{0.41Fe+0.98Cr}$.

The constants in the age contribution portion of the risk model 4.47, 24,100, and 3.47 are obtained from the data and will change depending on the equipment. They may even be different for the same equipment if operating in a different environment. The key iron and chrome risk factors are also equipment and operating environment specific. In this case, it was the absolute values of iron and chrome that were used. In other cases, rates of change or cumulative values may be more meaningful for risk estimation. By carefully analyzing condition monitoring data, along with information about the age and reason for equipment replacement, you can construct an excellent risk model.

Optimizing maintenance decisions usually requires that more than just risk of failure is taken into account. You may want to maximize the operating profit and/or equipment availability or minimize total cost. In this section, we assume your objective is to minimize the total long term cost of preventive and failure maintenance. Besides determining the risk curve, then, you must get cost estimates for both prevention and failure replacement, and failure consequences.

Being able to detect failure modes, which gradually lessen functional performance, can dramatically impact overall costs. This, therefore, is the first level of defense in the RCM (Reliability Centered Maintenance) task planning logic in Figure 12.14.

The logic diagram in Figure 12.14 shows that condition based maintenance is preferred if the impending failure can be "easily" detected in ample time. This proactive intervention is illustrated in the P-F interval[11] in Figure 12.15. Investigating the relationship between past condition surveillance and past failure data helps develop

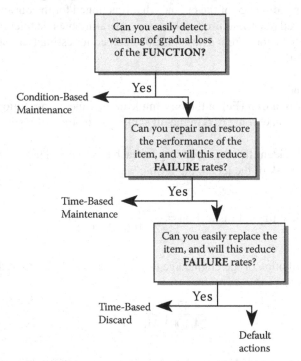

FIGURE 12.14   RCM logic diagram can use Figure 12.1 in Maintenance Excellence, 2001.

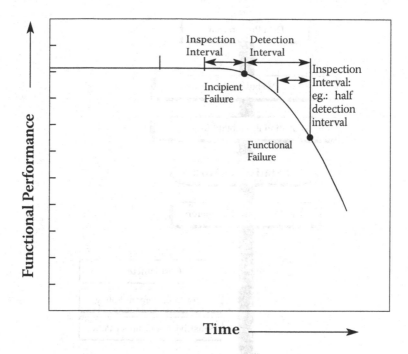

FIGURE 12.15   The P-F interval. This is 12.2 in original book, page 329.

future maintenance management policy and specific maintenance decisions. As well, modern, flexible maintenance information management systems compile and report performance, cost, repair, and condition data in numerous ways.

Maintenance engineers, planners, and managers perform Condition-Based Maintenance by collecting data that reflect the state of equipment or component health. These condition indicators (or covariates) can take various forms. They may be continuous, such as operational temperature, or feed rate of raw materials. They may be discrete, such as vibration or oil analysis measurements.[13] They may be mathematical combinations or transformations of the measured data, such as measurement change rates, rolling averages, and ratios. Since it's hard to know exactly why failures occur, the condition indicator choices are endless. Without a systematic means of judging and rejecting superfluous data, you will find CBM far less useful than it should be. Proportional hazards modeling is an effective approach to information overload because, based on substantial historic condition and corresponding failure age data, it can decipher the equipment's current condition, and make an optimal recommendation.

In this section, we describe, with examples, the key steps of the proportional hazards modeling process as provided in Figure 12.16. An integral part of the process is statistical testing of various hypotheses. This helps avoid the trap of blindly following a method without adequately verifying whether the model suits the situation and data.

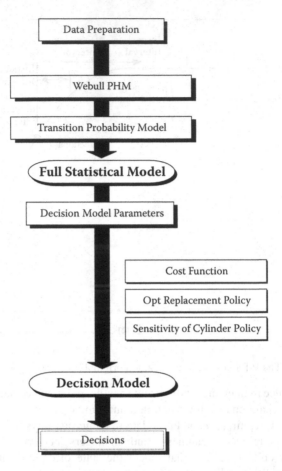

FIGURE 12.16   PHM flow diagram. This is Figure 12.3 on p. 331 of *Maintenance Excellence*.

12.5.1.1   Step 1: Data Preparation

No matter what tools or computer programs are available, you should always examine the data in several ways.[14] For example, many data sets can have the same mean and standard deviation and still be very different. That can be of critical significance. As well, instrument calibration and transcription mistakes have likely caused some data errors. You may have to search archives for significant data that's missing. To develop accurate decision models, you must be fully immersed in the operating and maintenance context. You must know the failure and repair work order process. Properly collected and validated subjective data that reflect all that is currently known about a problem is invaluable, but it's not enough. You need as well to collect sound objective data about the problem or process, for a complete analysis and to confirm subjective opinion.[15]

Generally, the maintenance engineer or technologist starts out with an underlying model based on the type of data, where the observations came from, and previous experience. After obtaining or converting the data into well-structured database tables, you must verify the model in stages before plunging ahead. This is where the tools of descriptive statistics—graphics (frequency histograms, cumulative frequency curves and numbers (mean, median, variance, skewness,)—are very useful. Some of these EXAKT™ methods are described in the following sections.

Although use of powerful computerized maintenance management systems (CMMS) and enterprise asset management (EAM) systems is growing, too little attention has been paid to data collection. Existing maintenance information management database systems are under-used. A clear relationship between accurately recorded component age data and effective maintenance decisions has been lacking. Trades people need to be educated about the value of such data, so that they meticulously record it when they replace failed components. Consistent with the principles of Total Productive Maintenance described in Chapter 7, maintenance and operational staff are the true custodians of the data and models it creates.

12.5.1.1.1 Events and Inspections Data

Unlike simple Weibull analysis described in Chapter 11 and applied in Section 12.2, PHM requires two types of equally important information: *Event* data, and *inspection* data. Three types of events, at a minimum, define a component's lifetime or history:

The Beginning (B) of the component life (the time of installation)
The Ending by Failure (EF)
The Ending by Suspension (ES), ie, a preventive replacement

Additional events should be included in the model if they directly influence the measured data. One such event is an oil change. Into the model, you should input that, at each oil change, some covariates—such as the wear metals—are expected to be reset. Periodic tightening or re-calibrating the machinery may have similar effects on measured values and should be accounted for in the model.

Example 1

In a food processing company, shear pump bearings are monitored for vibration. In this example, 21 vibration covariates from the shear pump's inboard bearing—represented in Figure 12.17—are reduced to only the 3 significant ones shown in Table 12.2.

The bearing measurements were taken in three directions: axial, horizontal and vertical. In each direction, the fast fourier transform of the velocity vector was taken in five frequency bands. The overall velocity and acceleration were also measured. This provided a total of 21 vibration measurements from a single bearing. Example 1 analyzes these 21 signals using EXAKT™ software, concluding that only three of the vibration monitoring measurements are key risk factors that need to be considered. There are two different velocity bands in the axial direction, and one velocity band in the vertical direction.

By combining the proportional hazard model with economic factors, you can devise a replacement decision policy, such as the one represented by Figure 12.18. The cost of a failure repair, compared to a preventive repair, was input into the decision model, and the ratio of preventive cost to failure cost was 9:1. On the

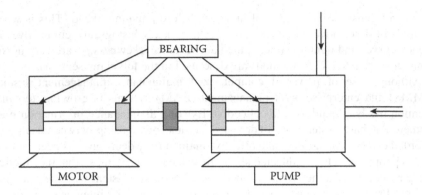

FIGURE 12.17   Shear pump.

TABLE 12.2
Significant Covariates

Indent	Date	Working Age	VEL_1A	VEL_2A	VEL_1V
B1	29-Sep-94	1	0.09	0.017	0.066
B1	08-Nov-94	41	0.203	0.018	0.113
B1	24-Nove-94	57	0.142	0.021	0.09
B1	25-Nov-94	58	0.37	0.054	0.074
B1	26-Nov-94	59	2.519	0.395	0.081
B2	11-Jan-95	46	0.635	0.668	0.05
B2	12_jan-95	47	1.536	0.055	0.0078
B3	19-Apr-95	97	0.211	0.057	0.144
B3	04-May-95	112	0.088	0.022	0.079
B3	06-May-95	114	0.129	0.014	0.087
B3	29-May-95	137	0.225	0.021	0.023
B3	05-Jun-95	144	0.05	0.017	0.04
B3	20-Jul-95	189	1.088	0.211	0.318
B4	21-Jul-95	1	0.862	0.073	0.102
B4	22-Jul-95	2	0.148	0.153	0.038
B4	23-Jul-95	3	0.12	0.015	0.035
B4	24-Jul-95	4	0.065	0.021	0.018
B4	21-Aug-95	33	0.939	0.1	0.3

graph, the composite covariate Z—the weighted sum of the three significant influencing factors, are points plotted against working age. If the current inspection point falls in the dark (red) area, the unit should be repaired immediately, since there is a good chance that it will fail before the next inspection.

If the current point is in the medium gray (green) area, the optimal decision to minimize the long run cost is to continue operating the equipment and inspect it at the next scheduled inspection. If the point is in the light (yellow) zone, the optimal decision is to keep operating it but preventively replace the component before the

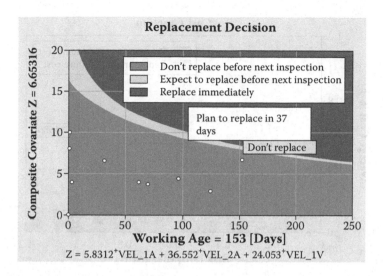

FIGURE 12.18 Optimal condition based replacement policy.

next scheduled inspection. The chart also indicates how much longer (remaining useful life) the equipment should run before being repaired or replaced.

Using the Figure 12.18 decision chart, total maintenance cost was reduced from $59.46/day to $26.83/day—an impressive 55% saving. Also, by following the recommendation on the chart, the mean time between bearing replacement is expected to increase by 10.2%.

12.5.1.2 Step 2: Building the PHM

Examining proportional hazards, you can see that they're an extension of the Weibull hazard function described in Chapter 10 and applied in Section 12.2.

$$h(t) = \frac{\beta}{\eta}\left(\frac{t}{\eta}\right)^{\beta-1} e^{\gamma_1 Z_1(t) + \gamma_2 Z_2(t) + \ldots + \gamma_n Z_n(t)} \qquad (12.1)$$

The new part factors in (as an exponential expression) the covariates $Z_i(t)$. These are the measured signals at a given time t of, for example, the parts per million of iron or other wear metals in the oil sample. The *covariate parameters* γ_i specify the relative impact that each covariate has on the hazard function. A very low value for γ_I indicates that the covariate isn't worth measuring. You'll find that software programs provide valuable criteria to omit unimportant covariates.

To fit the proportional hazards model to the data, you must estimate not only the parameters β and η, as we did in the simple Weibull examples in Chapter 10, but also the covariate parameters γ_i.

Remember that "condition monitoring" isn't actually monitoring the equipment's condition per se, but variables that you *think* are related to it. Those variables or "covariates" influence the failure probability shown by the hazard function, $h(t)$ in Equation 12.1. From the model you construct, you want to learn each covariate's degree of influence (namely the size of the covariate parameters γ_i), based on past data.

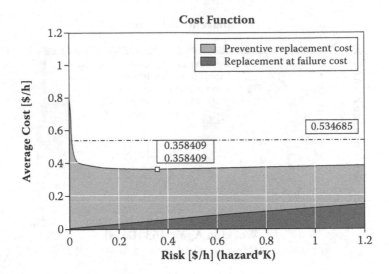

FIGURE 12.19   Cost function.

12.5.2 THE OPTIMAL DECISION

12.5.2.1 The Cost Function

In Chapter 10, you learned how models optimize an objective such as the overall long run cost to maintain a system. An analogous process is used in Proportional Hazards Modeling to optimize the cost function. Once again, we compare the costs for a preventive versus failure replacement and ask the software to calculate the cost function graph, Figure 12.19, which shows the minimum cost associated with an optimal replacement policy.

The cost function is the sum of the costs due to preventive replacements (upper part) and failure replacements (lower). Obviously, when the risk level goes up, the cost of failure replacements also increases. When the risk level rises to infinity, the cost function increases to the cost of the failure replacement policy indicated by the value at the right hand edge of the graph. You want to select the lowest possible risk level (horizontal axis), without increasing the total maintenance cost per hour. Naturally, zero risk would entail infinite cost. An infinite risk is tantamount to a "run-to-failure" policy whose cost is indicated by the dashed line.

12.5.2.2 The Optimal Replacement Decision Graph

The Replacement Decision graph, Figure 12.20, reflects the entire modeling exercise to date. It combines the proportional hazards model results, the transition probability model, and the cost function into the best decision policy for the component or system in question.

The ordinate is the composite covariate, Z—a balanced sum of covariates that statistically influence failure probability. Each covariate's contribution is weighted by its influence on the failure risk in the next inspection interval.

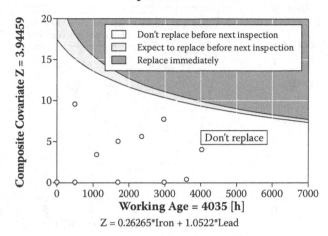

FIGURE 12.20   The optimal replacement decision graph.

A major advantage of this system is that a single graph combines the information you need to make a replacement decision. The alternative is to examine the trend graphs of perhaps dozens of parameters and "guess" whether to replace the component immediately or a little later. The Optimal Replacement Decision Graph recommendation is the most effective guide to minimize maintenance cost *in the long run*.

12.5.3 SENSITIVITY ANALYSIS

How do you know that the Optimum Replacement Decision Graph constitutes the best policy, considering your plant's ever-changing operations? Are the assumptions you used still valid and, if not, what will be the effect of those changes? Is your decision still optimal? These questions are addressed by sensitivity analysis.

The assumption you made in building the cost function model centered around the relative costs of a planned replacement versus those of a sudden failure. That cost ratio may have changed. If your accounting methods don't provide precise repair costs, you had to estimate them when building the cost portion of the decision model. In either case, these uncertain costs can create doubts about whether the Optimal Replacement Decision Graph policy is well founded.

The sensitivity analysis allays unwarranted fears, and indicates how to obtain more accurate cost data.

Figure 12.21, the Hazard Sensitivity of Optimal Policy graph, shows the relationship between the optimal hazard or risk level and the cost ratio. If the cost ratio is low, less than 3, the optimal hazard level would increase exponentially. We need, then, to track costs very closely to substantiate the benefits calculated by the model. On the other hand, if the cost ratio is in the 4 to 6 range, the curve is fairly flat, and the optimal replacement decision graph is accurate.

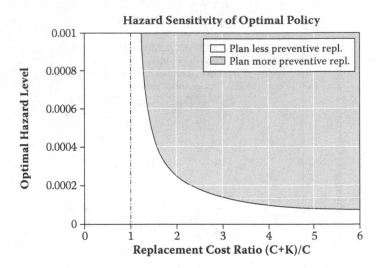

FIGURE 12.21   Sensitivity of optimal policy to cost ratio.

The Cost Sensitivity of Optimal Policy graph, Figure 12.21, has two lines.

Solid line: If the actual (C + K)/C (the cost of a failure replacement divided by the cost of a preventive replacement) differs from that specified when you built the model, it means that the current policy (as dictated by the Optimal Replacement Graph) is no longer optimal. The solid line tells how much % more you're paying above the optimal cost/unit time originally calculated.

Dashed line: Again, assume the actual cost ratio has strayed, and you want to rebuild the model using the new (C + K)/C. The dashed line tells how much your new optimal cost would change if you follow the new policy. The sensitivity graphs assume that only Cf (failure repair cost) changes and Cr (planned repair cost) remains the same.

12.5.4   CONCLUSION

In this section, we have explored a new approach to presenting, processing, and interpreting condition data. You have seen the benefits of applying proportional hazards models to condition monitoring and equipment performance data in several different industries, including petrochemical, mining, food processing, and mass transit. Table 12.3 summarizes the advantages of CBM optimization by proportional hazards modeling in 4 companies. Good data and the increasing use of software will fuel even greater maintenance progress in the coming years.

TABLE 12.3
Summary of Recorded Benefits

Industry	Data Reduction of Key Condition Indicators	Cost Savings Over Run-to-Failure Policy or Simple Age Based Maintenance Policy	Average Extension In Replacement Life
Mining	21 oil analysis measurements, 3 found to be significant.	25%	13%
Mass transit	A single color observation used.	55%	More frequent maintenance inspections but less extensive repairs required.
Food processing	21 vibration signals, 3 found to be significant.	5%	
Petro chemical	12 vibration signals, 2 found to be significant.	42%	

12.6 ENHANCING RELIABILITY THROUGH ASSET REPLACEMENT

Eventually, it becomes economically justifiable to replace an aging asset with a new one. Since it's usually years between replacements, rather than weeks or months, as it often is for component preventive replacement, you must consider the fact that money changes in value over time. This is known as discounted cash flow analysis. Figure 12.22 shows the different cash flows when replacing an asset on a 1, 2, and 3 year cycle.

To decide which of the three alternatives would be best, you must compare them fairly. You can do this by converting all cash flows associated with each cycle to today's prices, or their current value. This is the process known as discounting cash flows. Also, to be fair, you must compare all possible cycles in the alternatives over an infinite period of time. While this may seem unrealistic, it isn't. It actually keeps the analysis straightforward, and is used in the following section.

12.6.1 ECONOMIC LIFE OF CAPITAL EQUIPMENT

There are two key conflicts in establishing the economic life of capital equipment:

- the increasing operations and maintenance costs of the aging asset
- the declining ownership cost in keeping the asset in service, since the initial capital cost is being written off over a longer time period.

These conflicts are illustrated in Figure 12.23, where fixed costs (such as operator and insurance charges) are also depicted by the horizontal line:

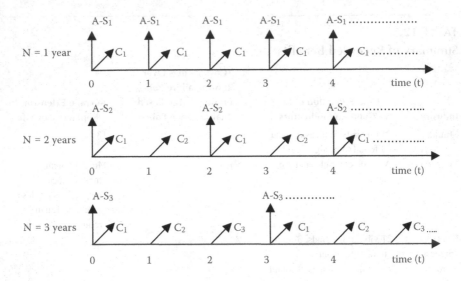

FIGURE 12.22 Asset replacement cycles.

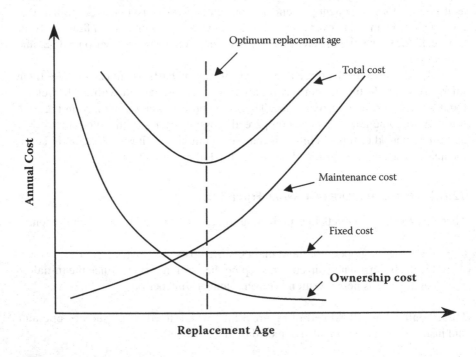

FIGURE 12.23 Economic life problem.

The asset's economic life is the time when the total cost is minimized. You will find the total cost curve equation, along with its derivation, in Appendix A. Rather than directly rely on the economic life model, you can use software that incorporates it. The following problem is solved using PERDEC software, which contains the economic life model provided in Appendix A.

PROBLEM

Canmade Inc. wants to determine the optimal replacement age for its materials handling turret side loaders, to minimize total discounted cost. Historical data analysis has produced the information in Table 12.4 (all costs are in present-day prices). As well, the cost of a new turret side loader is $150,000, and the interest rate, for discounting purposes, is 12%.

Solution

Using PERDEC produces Figure 12.24, which shows that the economic life of a side loader is 3 years, with an annual cost of $79,973. This amount would be sufficient to buy, operate and sell side loaders on a 3 year cycle. This is the optimal decision. Note that the amount in the annual budget to fund replacements would be calculated based on the number and age of side loaders and based on a three-year replacement cycle.

TABLE 12.4
Maintenance Cost and Resale Value

Year	Average Operating and Maintenance Cost ($/year)	Resale Value at End of Year ($)
1	16,000	100,000
2	28,000	60,000
3	46,000	50,000
4	70,000	20,000

Description

Number of Years 4   Acquisition Cost 150,000   Best Year   3

Parameters

Age of Vehicle[s]	Q&M Cost	Resale Value	Resale Value[%]	EAC
1 Year Old	16,000	100,000	66,6667	85,920
2 Years Old	28,000	60,000	40,0000	84,712
3 Years Old	46,000	50,000	33,3333	79,973
4 Years Old	70,000	20,000	13,3333	87,176

FIGURE 12.24 PERDEC optimal replacement age.

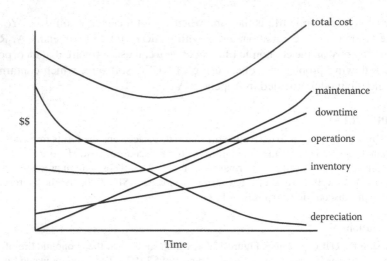

FIGURE 12.25 APWA economic life model: establishing the economic life of equipment where utilization varies during its life.

While the above example has only considered operating and maintenance (O & M) costs and resale value, you must be sure to include all relevant costs. Figure 12.25, taken from an American Public Works Association publication[16] shows that, for example, the cost of holding spares (inventory) and downtime is included in the analysis.

The calculation of this section has used the "classic" economic life model that assumes equipment is being replaced with similar equipment. It also assumes that the equipment is steadily used year by year. You may find this useful for some analyses, such as forklift truck replacement, if the equipment's design isn't impacted significantly and its use is constant.

Some equipment isn't used steadily, year to year. You might use new equipment frequently and older equipment only to meet peak demands. In this case, you have to modify the classic economic model, and examine the total cost of the group of similar equipment, rather than individual units. Examples where this applies include:

* Machine tools, where new tools are highly used, to meet basic workload, and older tools used to meet peak demands, say during annual plant shut downs.
* Materials handling equipment, such as older fork lift trucks in a bottling plant, where they are kept to meet seasonal peak demands.
* Trucking fleets undertaking both long distance and local deliveries. New trucks are used on long haul routes initially then, as they age, they are relegated to local deliveries.

The following is an example of establishing the economic life of a small fleet of delivery vehicles, using AGE/CON software:

PROBLEM

A company has a fleet of 8 vehicles to deliver their products to customers. The company uses its newest vehicles during normal demand periods, and the older ones to meet peak demands.

In total, the fleet travels 100,000 miles per year, and these miles are distributed, on average, among the 8 vehicles as follows:

Vehicle number 1 travels 23,300 miles / year
Vehicle number 2 travels 19,234 miles / year
Vehicle number 3 travels 15,876 miles / year
Vehicle number 4 travels 15,134 miles / year
Vehicle number 5 travels 12,689 miles / year
Vehicle number 6 travels 8,756 miles / year
Vehicle number 7 travels 3,422 miles / year
Vehicle number 8 travels 1,589 miles / year

Determine the optimal replacement age for this class of delivery vehicle.

Solution

You must first establish how often the vehicle is used as it ages. (For the underlying mathematical model of when best to replace aging equipment, see Appendix A. It features a case study that establishes the economic replacement policy for a large fleet of urban buses).

The utilization data will look like Figure 12.26. The trend of Figure 12.26 can be described by the equation of a straight line:

$$Y = a - bX$$

where "Y" is the "miles/year" figure and "X is the "Vehicle number"—Vehicle number 1 is the one most utilized. Vehicle number 8 is the least utilized.

Using the actual figures given above, you can establish from AGE/CON (or by plotting the data on graph paper or using a trend fitting software package) that the equation in this case would read:

$$Y = 26,152 - 3,034X$$

Next you must establish the trend for Operating and Maintenance (O&M) costs.

For *Vehicle 1*, (your newest vehicle, and the one used the most) you need the following information:

Miles traveled last year (already given)	23,300
O & M cost last year, say	$3,150
Cumulative miles on the odometer to the mid-point of last year, say	32,000

As you can see, the O & M cost/mile is $ 0.14 for Vehicle 1. Do the same for all 8 vehicles.

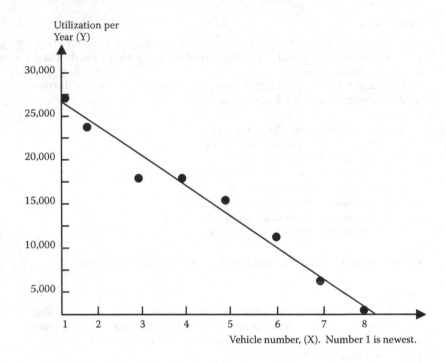

FIGURE 12.26 Vehicle utilization trend.

Vehicle 8 (the oldest, and the one least used) may have data that looks like this:

Miles traveled last year (already given)	1,589
O & M cost last year, say	$765
Cumulative miles on the odometer to the mid-point of last year, say	120,000

In this case, the O & M cost/mile is $0.48 for Vehicle 8. A plot of the trend in O & M cost would look like Figure 12.27. Each vehicle's O & M cost is represented by a "dot" on the graph. Vehicles 1 and 8 are identified in the diagram. The straight line is the trend that has been fitted to the "dots".

The equation you get this time is:

$$Z = 0.0164 + 0.00000394T$$

where

Z = $/mile,
T = Cumulative miles travelled

The two trend lines are both used as input to AGE/CON

Note

In both cases, a straight (linear) relationship existed for $Y(X)$ and $Z(T)$ so that the fitted lines read $Y = a - bX$ and $Z = c + dT$. Often a polynomial equation will give

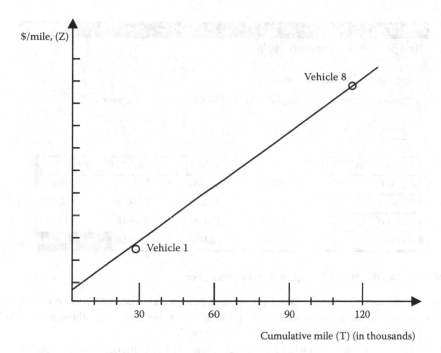

FIGURE 12.27 Trend in O & M cost.

a better "fit" to a particular series of data. These polynomial equations can be generated by using a standard statistical package such as Minitab or SPSS.

To solve the problem, you require additional information:

Assume that a new delivery vehicle costs $40,000. The resale values for this particular type of vehicle are:

1-year old vehicle	$28,000
2-year old vehicle	$20,000
3-year old vehicle	$13,000
4-year old vehicle	$6,000

The interest rate for discounting purposes is 13%. Figure 12.28 from AGE/CON shows that the optimal replacement age of a delivery vehicle is 4 years with an associated equivalent annual cost (EAC) of $14,235.

To implement this recommendation, you'd likely need to replace a quarter of the fleet each year, so that the same number of vehicles would be replaced each year. All vehicles, then, would be replaced at the end of their fourth year of life.

12.6.2 BEFORE AND AFTER TAX CALCULATIONS

In most cases, you conduct economic life calculations on a before-tax basis, and this is always the case in the public sector, where tax considerations are not applicable. In

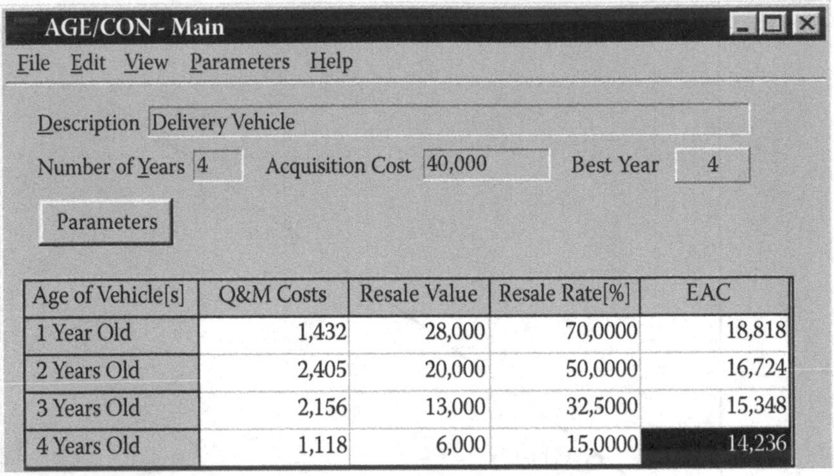

AGE/CON - Main   _ □ ✕

File   Edit   View   Parameters   Help

Description │Delivery Vehicle

Number of Years │4   Acquisition Cost │40,000   Best Year   │4

Parameters

Age of Vehicle[s]	Q&M Costs	Resale Value	Resale Rate[%]	EAC
1 Year Old	1,432	28,000	70,0000	18,818
2 Years Old	2,405	20,000	50,0000	16,724
3 Years Old	2,156	13,000	32,5000	15,348
4 Years Old	1,118	6,000	15,0000	14,236

FIGURE 12.28   AGE/CON optimal replacement age.

the private sector, your financial group can help decide the best course of action. In many cases, the after-tax calculation doesn't alter the decision, although the EAC is reduced when the result is in after-tax dollars.

Figure 12.29 illustrates what happened when a Feller Buncher was replaced in the forestry industry. The data used is provided in Tables 12.5 and 12.6.

You can see from Figure 12.29 that the general trend of the EAC curve remains the same, but it veers downwards when the tax implications are included in the economic life model. In before-tax dollars, the economic life of the Feller Buncher is five years, while in after-tax dollars, the minimum is eight years. Note that, in both cases, the total cost curve is flat around the minimum. In this example, a replacement age between five and eight years would be good for either a before-tax or after-tax calculation.

In Appendix A, you'll find two forms of the EAC model, one for before-tax analysis, the other for after-tax, taking into account corporation tax and capital cost allowance. Both models are in AGE/CON, which provided the graphs in Figure 12.29.

If you are making calculations after-tax, take care that all relevant taxes and current rules are incorporated into the model. One example is the Buttimore and Lim report[17] that deals with the cost of replacing shovels in the mining industry on an after-tax basis. It includes not only corporation tax and capital cost allowances, but federal and provincial taxes applicable to the mining industry at the time.

12.6.3   THE REPAIR VERSUS REPLACEMENT DECISION

You may be facing a sudden major maintenance expenditure for equipment, perhaps due to an accident. Or, you may be able to extend the life of an asset through a major overhaul. In either case, you have to decide whether to make the maintenance expenditure or dispose of the asset and replace it with a new one.

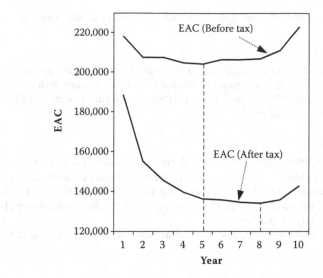

FIGURE 12.29   Economic life: before and after tax calculation (Feller Buncher data).

TABLE 12.5
Feller Buncher: Base Data

Acquisition cost of Feller Buncher	$ 526,000
Discount rate	10%
Capital Cost Allowance (CCA)	30%
Corporation Tax (CT)	40%

TABLE 12.6
Feller Buncher: Annual Data

Year	O&M Cost	Resale Value
1	8,332	368,200
2	60,097	275,139
3	107,259	212,423
4	116,189	169,939
5	113,958	104,189
6	182,516	95,085
7	173,631	85,981
8	183,883	83,958
9	224,899	73,842
10	330,375	40,462

Here's an approach that can help you make the most cost effective decision:

PROBLEM

A major piece of mobile equipment, a front-end loader used in open pit mining, is 8 years old. It can remain operational for another 3 years, with a rebuild costing $390,000. The alternative is to purchase a new unit, costing $1,083,233. Which is the better alternative?

Solution

You need additional data to make an informed decision that will minimize the long run equivalent annual cost (EAC). Review historical maintenance records for the equipment; to help forecast what the future O & M costs will be after rebuild, and obtain O & M cost estimates and trade-in values from the supplier of the potential new purchase.

Figure 12.30 depicts the cash flows from acquiring new equipment at time T where:

- R is the cost of the rebuild
- $C_{p,i}$ is the estimated O & M cost of using the present equipment after rebuild in year i, $i = 1, 2, ..., T$
- A is the cost of acquiring and installing new equipment
- T is the time at which the change-over occurs from the present equipment to a new equipment. $T = 0, 1, 2, 3$
- $S_{p,T}$ is the trade-in value of the present equipment at the change-over time, T
- $C_{t,j}$ is the estimated O & M cost associated with using the new equipment in year j, $j = 1,2, ..., n$
- S_n is the trade in value of the new equipment at age n years
- n is the economic replacement age of the new equipment

The necessary data for the current equipment is provided in Table 12.7, and The necessary data for the new equipment is provided in Table 12.8. The interest rate for discounting is 11% and the purchase price for the new equipment is: $A = \$1,083,233$. Evaluating the data gives the new equipment an economic life of 11 years, with an associated EAC of $494,073.

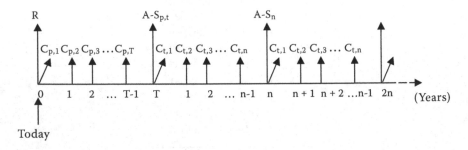

FIGURE 12.30 Cash flows associated with acquiring new equipment at time T.

TABLE 12.7
Cost Data: Current Equipment

$C_{p,1} = \$138,592$	$S_{p,0} = \$300,000$
$C_{p,2} = \$238.033$	$S_{p,1} = \$400,000$
$C_{p,3} = \$282,033$	$S_{p,2} = \$350,000$
	$S_{p,3} = \$325,000$

TABLE 12.8
Cost Data: New Equipment

$C_{t,1} = \$38,188$	$S_1 = \$742,500$
$C_{t,2} = \$218,583$	$S_2 = \$624,000$
$C_{t,3} = \$443,593$	$S_3 = \$588,000$
$C_{t,4} = \$238,830$	$S_4 = \$450,000$
Etc.	Etc.

TABLE 12.9
Optimal Change-Over Time

	Change-Over Time to New Loader, T			
	T = 0	T = 1	T = 2	T = 3
Overall EAC ($)	449,074	456,744	444,334	435,237
				↑ Minimum

Solution

To decide whether or not to rebuild, calculate the EAC associated with not rebuilding (i.e., $T = 0$), replacing immediately, rebuilding then replacing after 1 year (i.e., $T = 1$), and rebuilding then replacing after 2 years (i.e., $T = 2$), rebuilding and replacing after 3 years (i.e., $T = 3$), and so on. The result of these various options is shown in Table 12.9. The solution is to rebuild and then plan to replace the equipment in 3 years, at a minimum equivalent annual cost of $435,237.

Note

In a full-blown study, the rebuilt equipment might be kept for a longer time. See Appendix A for the model used to conduct the above analysis. The same model is used in the following section.

12.6.4   TECHNOLOGICAL IMPROVEMENT

If a new, more technically advanced model of equipment you are using becomes available, you will have to weigh the costs and benefits of upgrading. See Appendix A for a basic model to evaluate whether or not to switch. The Buttimore and Lim case study dealing with shovel replacement in an open pit mining operation[16] shows that a better technical design improved productivity. (This is an extension of the model in Appendix A).

12.6.5   LIFE CYCLE COSTING

Life cycle costing (LCC) analysis considers all costs associated with an asset's life cycle, which may include :

- Research and development
- Manufacturing and installation
- Operation and maintenance
- System retirement and phase out

Essentially, when making decisions about capital equipment, be it for replacement or new acquisition, reflect on all associated costs. Figure 12.31 is a good illustration of what can be involved.

The iceberg (Blanchard[17]) in Figure 12.31 shows very well that, while costs like upfront acquisition are obvious, the total cost can be many times greater. In the airline industry, the life cycle cost of an aircraft can be five times its initial purchase. A Compaq Computer Corp. ad states that 85% of computer costs are usually hidden—going to administration (14%), operations (15%) and the bulk—about 56%—to asset management and service and support costs.

In the economic life examples covered in this section, we take an LCC approach, including costs for:

- Purchase price
- Operations and maintenance cost
- Disposal value

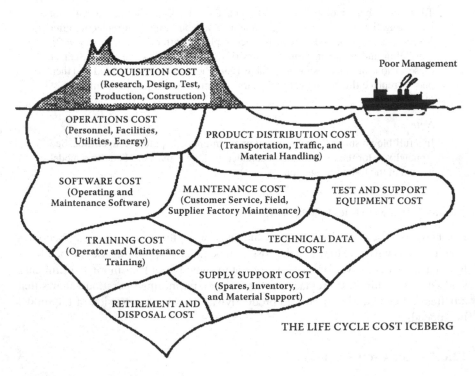

FIGURE 12.31   The life-cycle cost iceberg.

Of course, if necessary, you may have to include other costs that contribute to the LCC, such as spare parts inventory and software maintenance.

12.7 RESOURCE REQUIREMENTS

When it comes to maintenance resource requirements, you must decide what resources there should be, where they should be located, who should own them and how they should be used.

If sufficient resources aren't available, your maintenance customers will be dissatisfied. Having too many resources, though, isn't economical. Your challenge is to balance spending on maintenance resources such as equipment, spares and staff with an appropriate return for that investment.

12.7.1 ROLE OF QUEUING THEORY TO ESTABLISH RESOURCE REQUIREMENTS

The branch of mathematics known as queuing theory, or waiting line theory, is valuable in situations where bottlenecks can occur. You can explore the consequences of alternative resource levels to identify the best option. Figure 12.32 illustrates the benefit of using queuing theory to establish the optimal number of lathes for a Workshop.

In this example the objective is to ensure that the total cost of owning and operating the lathes and tying up jobs in the Workshop is minimized. For a model of this decision process, see Appendix A.

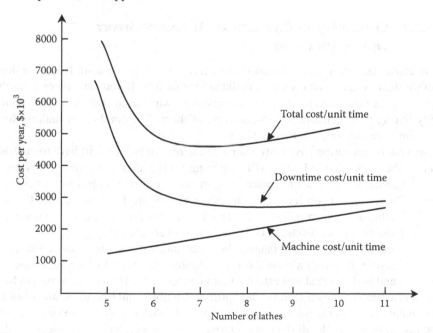

FIGURE 12.32 Optimal number of machines in a workshop.

12.7.2 Optimizing Maintenance Schedules

In deciding maintenance resource requirements, you must also consider how to use resources efficiently. An important consideration is scheduling jobs through a Workshop. If there is restricted Workshop capacity, and jobs cannot be contracted out, you must decide which job should be done first when a workshop machine becomes available.

Sriskandarajah et al.[18] presents a unique and highly challenging maintenance overhaul scheduling problem at the Hong Kong Mass Transit Railway Corporation (MTRC). In this case, preventive maintenance keeps trains in a specified condition, taking into account both the maintenance cost and equipment failure consequences. Since maintenance tasks done either "too early" or "too late" can be costly, how maintenance activities were scheduled was important.

Smart scheduling reduces the overall maintenance budget. Establishing a schedule, it's essential to acknowledge constraints, such as the number of equipment/ machines that can be maintained simultaneously, as well as economic, reliability and technological concerns. Most scheduling problems are a combination of factors. Because of this complexity, a heuristic technique known as Genetic algorithms (GAs) was used in this case to arrive at the global optimum.

The performance of the algorithm compared well with manual schedules established by the Hong Kong Mass Transit Railway Corporation. For example, total costs were reduced by about 25 percent. The study supports the view that genetic algorithms can provide good solutions to difficult maintenance scheduling problems.

12.7.3 Optimal Use of Contractors (Alternative Service Delivery Providers)

The above decision process assumed that all the maintenance work had to be done within the railway company's own maintenance facility. If you can contract out the maintenance tasks, you must decide, for instance, whether to contract out all, some, only during peak demand periods, or none of them. The conflicts of making such decisions are illustrated in Figure 12.33.

If your organization hasn't any maintenance resources, you will have to contract out all the work. The only cost will be paying for the alternate delivery service. If your organization has a maintenance division, though, there will be two cost components. One is a fixed cost for the facility, shown in Figure 12.33 to increase linearly as the facility size expands. The second cost is variable, increasing as more work is handled internally, but leveling off when there is over capacity.

The optimal decision is a balance between internal resources and contracting out. Of course, this isn't always the best solution. You have to assess the demand pattern, and both internal and external maintenance costs. If it is cheaper not to have an internal maintenance facility, the optimal solution would be to contract out all the maintenance work. Alternatively, the optimal solution could be to gear up your own organization to do all the work internally. See Appendix A for a model of this decision process.

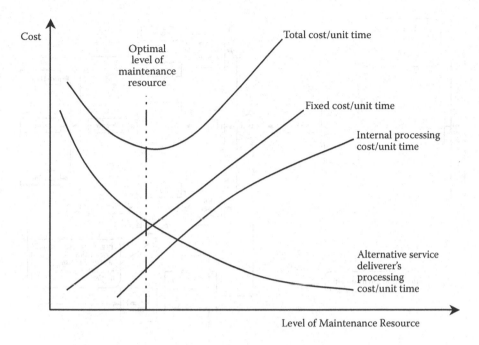

FIGURE 12.33 Optimal contracting-out decision.

12.7.4 ROLE OF SIMULATION IN MAINTENANCE OPTIMIZATION

Common maintenance management concerns about resources include:

- How large should the maintenance crews be?
- What mix of machines should there be in a workshop?
- What rules should be used to schedule work through the workshop?
- What skill sets should we have in the maintenance teams?

Some of these questions can be answered by using a mathematical model. For complex cases, though, a simulation model of the decision situation if often built. You can use the simulation to evaluate a variety of alternatives, then choose the best. There are many simulation software packages available that require minimal programming.

Many resource decisions are complex. Take, for example, a situation where equipment in a petrochemical plant requires attention. If the maintenance crew is limited by size, or attending to other tasks, the new job may be delayed while another emergency crew is called in. Of course, if there had been a large crew in place initially, this wouldn't have occurred. Since there can be many competing demands on the maintenance crew, it's difficult to establish the best crew size.

This is where simulation is valuable. Figure 12.34 illustrates a case where simulation was used to establish the optimal maintenance crew size and shift pattern in a petrochemical plant. For illustration purposes, statistics are provided for the failure

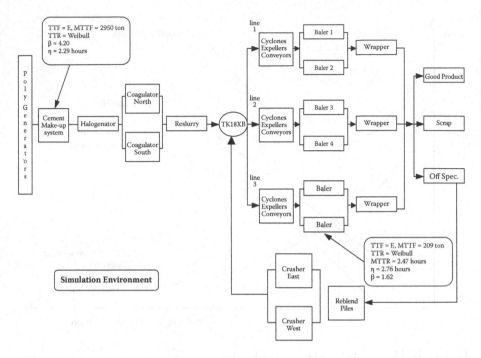

FIGURE 12.34 Maintenance crew size simulation.

pattern of select machines, along with repair time information. The study was undertaken to increase the plant's output. The initial plan was to add a fourth finishing line, to reduce the bottleneck at the large mixing tank (TK18XB), just prior to the finishing lines. The throughput data, though, showed that once a machine failed, there was often a long wait until the maintenance crew arrived. Rather than construct another finishing line, it was decided to increase maintenance resources. By using simulation, the throughput from the plant was significantly increased with little additional cost. Crew size increase was a much cheaper alternative and achieved the required increased throughput.

12.7.5 Software That Optimizes Maintenance and Replacement Decisions

In this chapter, we have referred to several software packages: OREST for component replacement decisions, EXAKT for CBM optimization, AGE/CON for mobile equipment replacement and PERDEC for fixed equipment replacement. There are other good software tools on the market, and more becoming available all the time, all of which you can find on the Web.

ACKNOWLEDGMENT

Section 12.5 on CBM is heavily based on the chapter in CBM written by Murray Wiseman in the first edition of this book.

ENDNOTES

1. OREST—Details at http://www.banak-inc.com.
2. A.K.S. Jardine and A.H.C. Tsang, *Maintenance, Replacement, and Reliability: Theory and Applications*, CRC Press, Taylor & Frances, 2006.
3. Drenick, R.F., "The Failure Law of Complex Equipment," *Journal Society Industrial Applied Math*, Vol. 8, pp. 680–690, 1960.
4. Jardine, A.K.S. and Hassounah, M.I., "An Optimal Vehicle-Fleet Inspection Schedule," *Journal of Operations Research Society*, Vol. 41, No. 9, pp. 791–799, 1990.
5. Cox, D.R., Regression models and life Tables (with discussion). J. Roy. Stat. Soc. B, 34, 187–220, 1972.
6. www.mie.utoronto.ca/cbm.
7. www.omdec.com.
8. RCM II 2.2 second edition, John Moubray, ISBN 0 7506 3358 1 Butterworth-Heinemann, 1999.
9. K. Pottinger and A. Sutton, Maintenance Management: An Art or Science, Maintenance Management International, 1983.
10. M Anderson, A.K.S. Jardine and R.T. Higgins, The use of concomitant variables in reliability estimation, *Modeling and Simulation*, 1982, 13: 73–81.
11. Nowlan F.S. and Heap, H., Reliability-centered Maintenance. Springfield Virginia, National Technical Information Service, U.S Department of Commerce, 1978.
12. "Proportional Hazards Modelling, a new weapon in the CBM arsenal," Murray Wiseman and A.K.S. Jardine, Proceedings Condition Monitoring '99 Conference, University of Wales, Swansea 12–15 April 1999.
13. Statistical Methods in Reliability Theory and Practice, Brian D. Bunday, Ellis Horwood Limited, 1991.
14. On the application of mathematical models in maintenance, Philip A. Scarf, European Journal of Operational Research, 1997.
15. Green, H. and Knorr, R.E., "Managing Public Equipment," The APWA Research Foundation, pp. 105–129, 1989.
16. Buttimore, B. and Lim, A., "Noranda Equipment Replacement System," *Applied Systems and Cybernetics*, Pergamon Press, Vol. 2, pp. 1069–1073, 1981.
17. Blanchard, B.S., *Logistics Engineering and Management*, Fifth Edition, Prentice Hall, 1998.
18. Sriskandarajah, C., Jardine, A.K.S. and Chan, C.K., Maintenance Scheduling of Rolling Stock Using a Genetic Algorithm, *Journal of the Operational Research Society*, 1998; 49:1130–1145.

13 A Maintenance Assessment Case Study

Don Barry

CONTENTS

The maintenance excellence assessment described in the previous chapters and shown in detail in the Appendix of this book has been applied to hundreds of maintenance operations seriously looking for areas to improve and looking to prioritize their identified action items.

Maintenance Excellence Pyramid

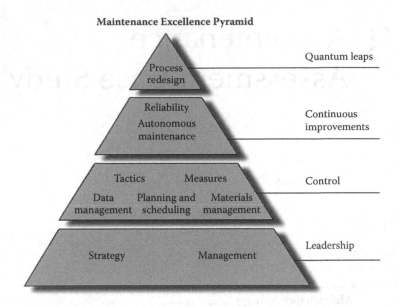

FIGURE 13.1　Maintenance excellence pyramid.

In this chapter, we will review a modified real example of a maintenance self-assessment report completed by a third-party consulting company and examine how this kind of report can help maintenance and operations experts identify actions to improve their operation. The case study has been adopted from a real report, where company names have been changed, and the recommendations described later in this chapter come from maintenance and operations experts who understand elements of leading practices in the 10 categories of the maintenance excellence pyramid (shown in Figure 13.1).

The ABC Consulting Company was engaged to perform a site assessment of UTO's maintenance management practices at their Base Valley operations and to compare current UTO practices with best practices. UTO Base Valley operations include three production plants (i.e., Falcon, West End, and Argile), mining site (Lake Operations), central services and utilities, materials and logistics, and railroad.

13.1　SITE OVERVIEW

The Base Valley operations produce a number of bulk chemical products. These are commodity products with little or no differentiation from competing products other than price. Owing to its remote location, shipping costs are a significant component of the price to their customers. This places considerable pressure on keeping manufacturing costs down.

The remote location and small local community affect the site's ability to hire locally, although once hired, employees tend to be long term. The remote location also affects the local supplier support, since the nearest major center is 100 miles away.

The plant's location in California imposes close scrutiny from the state on environmental issues, particularly air pollution. This affects the operation of the coal-fired steam boilers in the utilities operations.

Bill Jackson and Associates has financed the purchase of a number of companies through bank loans. Bank assessment would focus on expected corporate cash flow first and then asset value. Banks would place restrictive covenants to ensure there is no significant change to flow or asset value without their consent. The expected outcome is a corporate focus on cash flow conservation, with tight capital spending and an overall short-term focus on spending.

13.2 METHODOLOGY

The methodology used involved the following:

- Participation by each site in a self-assessment using a standard questionnaire
- On-site walk-throughs, mapping, and interviews by ABC Consulting to determine what practices and processes were being used at each site
- Comparison of current UTO practices with leading practices
- Recommendations
- Presentation of the findings to the site
- Report on findings

The areas are assessed at each site corresponding with the following 10 elements of maintenance best practices:

- Maintenance strategy
- Maintenance tactics
- Reliability analysis/engineering
- Performance measures/benchmarking
- Information technology
- Planning and scheduling
- Materials management
- Maintenance processes reengineering
- Organization and human resources
- Autonomous maintenance

Process mappings were carried out for the whole valley and correspond to actual practices at UTO Base Valley operations. The mappings reflect what and how the processes are executed. Process maps do not necessarily exist for all strategic elements examined. Most of the maps pertain to planning and scheduling (work management) and materials management.

In this process UTO would expect to receive a report, which outlines the following:

- UTO practices based on ABC Consulting observations and mapping
- UTO self-assessment results

- A description of the best practices
- A set of recommendations for implementation

What follows is a list of findings derived from the maintenance excellence questionnaire and the follow-up interviews of the participants.

13.3 FINDINGS

13.3.1 MAINTENANCE STRATEGY

Current situation:

- Their developed vision is communicated.
- A maintenance council has been formed, and their mission statement has been defined and communicated. The goal of the maintenance council is to establish a world-class maintenance and reliability organization.
- UTO uses a combination of centralized services with area based workforce crews.
- UTO is ISO-9002 certified.
- Some sites (utilities, lake) share common vision and values between operations and maintenance.
- The maintenance strategy is not clearly defined.
- Long-term objectives relate only to cost and head count, not to long-term improvement initiatives.
- There is a lack of clear and consistent vision for asset management throughout the valley.
- Key value drivers are defined but not effectively communicated or well understood at all levels.
- Maintenance and capital expenditures appear inadequate with unclear definitions and objectives.
- Production complains that maintenance does not view them as a customer and does not provide good customer service.

13.3.2 ORGANIZATION

Current situation:

- There is good area organization design with some central support.
- There is some centralized computerized maintenance management system (CMMS) support and maintenance policy (council).
- The ratio of trades to supervisors (crew size) is very low at 6 to 8:1.
- There is a high number of planners, 1:6 to 10 and a lack of maintenance engineers.
- The number of levels within maintenance is high in some areas at 4.
- The backlog is not measured by man-hours or crew-weeks.

- Some technical support and predictive maintenance (PdM) is lacking.
- Some maintenance supervisors would like to have a pool of mechanics they could pull from for shutdowns, vacations, and so forth. There is no backlog planning.

13.3.3 HUMAN RESOURCES

Current situation:

- The workforce is not unionized, except for railroad employees.
- Training program for trades is in the early stage of development.
- Leadership training for supervisors is well under way; classes are held once a month for all supervisory staff.
- The formal policy on training is 60 hours/year but is not strictly applied; however, most trades indicated that when training was requested, it was usually granted.
- A performance bonus is in place for exempts.
- Industrial relations policy appears positive; some concerns raised by supervisors to open-door policy.
- Safety is a high priority.
- Maintenance staffing level is adequate, capable, and experienced.
- Overall overtime represents approximately 10% of total man-hours and is well distributed among trades and areas.
- Contractors are used for special projects and are held accountable for their work on the same basis as UTO employees.
- Horizontal communications is generally effective but sometimes variable.
- The morale of the maintenance workforce is currently low, mainly due to the rumors of outsourcing of maintenance processes.
- No skill- or performance-based incentives for trades.
- Technical documentation is missing or out of date; there is a lack of discipline in documentation.
- Shifts vary from 10 hours for 4 days per week to 8 hours for 5 days per week depending on the department and the service. The workforce is therefore reduced on Mondays and Fridays, and maintenance and operations shift hours do not match.
- Many trades indicated they would like to have the apprenticeship program reinstated.
- Contractors are used to replace UTO employees who leave (e.g., retirements, long-term disabilities).
- Vertical communications is often complex as a result of the large number of hierarchical levels.

13.3.4 MAINTENANCE TACTICS

Current situation:

- The preventive maintenance (PM) program is widespread and comprises up to 45% of work.
- Condition-based maintenance is well started.
- CSI 2120 portable devices are used to take readings for vibration analysis, and Commodity Systems Inc.'s Master Trend software is used for data analysis. Most people indicated they were very satisfied with the vibration analysis services they receive. Some people indicated they would like to have more feedback from the Vibration Analysis crew.
- Maintenance performs thermography on electrical equipment.
- Basic oil analysis is performed in-house at Argile and is otherwise sent out to contractors.
- Many employees were trained to perform laser alignment. Not all sites have laser alignment equipment.
- Monthly equipment availability for Boron is approximately 96–99%.
- Emergencies represent 10–30% of work.
- Schedule compliance is approximately 80%.
- There is no formal reliability-based program for determining the correct PM routines to perform. PMs are defined based largely on vendor recommendations and intuition and to a lesser extent equipment history in Revere.
- Little effectiveness analysis of tactics used or systematic approach or formal and systematic use of equipment histories.
- Lubrication team at West End (WE) is used primarily as helpers for maintenance, so lubrication is not always done; failures due to lack of lubrication were reported.
- Central Services PM their own equipment (e.g., vehicles, jib cranes, fire extinguishers, grinders, electric boxes).
- PMs represent on average approximately 30% of total man-hours. The Railroad team members indicated that they didn't have time to perform PM tasks in the summer.
- At Argile, maintenance crews spend a lot of time performing demand maintenance on the fluid bed dryers at the Bi-Carb.

13.3.5 PERFORMANCE MEASURES AND BENCHMARKING

Current situation:

- Maintenance performance is measured for inputs (i.e., costs), with some process measures (i.e., labor distribution, schedule compliance) and output measures (i.e., availability, production).
- Internal comparisons of best practices by plant were started. Five years prior, the maintenance best practice team was created with members from various

Jackson companies. The purpose of this team was to identify maintenance strengths and weaknesses and, by sharing this knowledge, to improve the overall effectiveness of maintenance within the Jackson group.

- Labor and material costs are accumulated and reported against key systems and equipment but are difficult to access in Revere.
- Key measures and value drivers are trended but are not effectively communicated. Tracked key value drivers include safety (number of incidents), %Availability, %Emergency, %PM, %Compliance, %Schedule Efficiency, %OT, %Machines in alarm, %Expenses ($Actual/$Budget).
- Value drivers are not well understood by the workforce.
- External best practice benchmarking is not formally done.
- Goals are often not set or well communicated.

13.3.6 INFORMATION TECHNOLOGY

Current situation:

- Revere (CMMS system) is fully functional although not always easy to use. It is an older system and is not as user-friendly as the newer systems.
- Most equipment is recorded in Revere.
- A project management system (MS project) is used to plan and schedule shutdowns.
- Many supervisors indicated that they would like to look at the equipment history but that it is very lengthy and complicated to get into.
- Data analysis of inventory using the current system is limited.
- CMMS and financial systems are not fully integrated.
- Labor times have to be entered twice: once in Revere to report time against memos (work orders) and a second time in the payroll system.

13.3.7 PLANNING AND SCHEDULING

Current situation:

- Most of the work is covered by a work request, which is then made into a memo (WO) within Revere.
- Approval process is rapid and effective. Some approvals are done electronically.
- Weekly planning meetings are held between maintenance and production.
- Major shutdowns are planned well in advance through monthly meetings, which become weekly meetings as the shutdown approaches. Utilities manager; maintenance and production superintendents; and engineering, mechanical, and electrical and instrumentation supervisors are usually present at these planning meetings.
- Work estimates are done quickly in most cases (without really assessing the needs) and sometimes not done at all.
- WO priorities are well defined but often abused.

- Work backlog is not formally tracked or tracked in meaningful units. (Most people could tell us how many pages of backlog they had but did not know how much time it represented.)
- Work management is overdone. (Trades often have to report to more than one person.)
- Different sites plan their work and often have conflict regarding use of common resources, especially cranes.

13.3.8 MATERIALS MANAGEMENT

Current situation:

- Stores management has recently been contracted out to a third-party company, which is currently in a transition period.
- Stock-keeping units (SKU): approximately 30,000 line items in stock for a total value of $7.7 million. The objective is to eliminate $700,000 of obsolete stock to get a stock value of $7 million and eventually $5 million by decreasing initial spare parts and overstocked parts stocks.
- Stock rotation is approximately 1.6.
- Service level is evaluated to 97% by stores.
- Accuracy of inventory is high.
- ABC analysis is performed regularly; items are categorized as A, B, C, or D. A items are counted monthly; B items are counted quarterly; C items are counted every six months; and D items are counted yearly.
- Procurement credit cards are made available to some individuals to accelerate the direct purchase process.
- Most departments depend on parts stocked in their area to have parts readily available, especially when there is some distance from stores (e.g., West End); this material is not tracked once it has left stores and increases material holding costs.
- Order quantities are not based on economic order quantity (EOQ).
- Complaints were formulated about part descriptions. Parts are often named using inconsistent naming techniques, making them difficult to locate in the system. The resulting descriptions are not complete enough to uniquely identify the part.
- Standardization of parts and equipment is not formally done.
- Material purchases done by Central Services have to be approved by the client first. Some supervisors complain that the work hasn't been done yet, only to realize they forgot to approve the purchase requisition (PR).
- A total of 5% of material is verified when received at stores; 95% is verified by the customer, who has to return material to stores when there is a problem.
- A central tool crib is available, but discipline is lacking when it comes to returning the tools. They are often stored in individual lockers.

13.3.9 AUTONOMOUS MAINTENANCE

Current situation:

- Partnering is exhibited in some areas only (e.g., back shifts).
- "Helping-hand" concept is used only in certain operations or only under certain conditions (e.g., Lake).
- Operators are willing to perform simple maintenance tasks, and maintenance doesn't mind transferring these tasks.
- Although maintenance is not represented on every shift, maintenance trades respond to call outs after hours quickly and with minimum effort from operations.
- Self-directed work teams are not used.
- Little opportunity for autonomy.
- Management communication often one way only.
- No formal multiskilling or cross-skilling done.
- Some maintenance trades said they would like to receive training on operations to be able to evaluate equipment criticality.
- When maintenance work is required after hours, operations have to decide what support is needed. There is no on-shift maintainer to take care of this task.
- Argile operators said that they often operate different production lines, which makes it more difficult for them to develop a sense of ownership.
- Argile operators also reported that some operators change the parameters of operation of their equipment, which increases the production on their shift but which has a negative effect on the production capacity of the next shift. Because of the strong production focus of the company, these operators are praised instead of being reprimanded. Some operators feel it would be better to produce at a constant rate without changing the parameters. There is no consistency in operating strategy or philosophy.

13.3.10 RELIABILITY MANAGEMENT

Current situation:

- Availability is tracked (i.e., total duration of failures).
- Equipment histories are kept by saving completed WOs but are not used systematically for reliability analysis.
- Under the quality program, root-cause analysis is performed for major failures.
- Some equipment redesign is done to improve reliability in specific areas (e.g., Pump Crew, Utilities, Lake).
- PM tasks are based largely on manufacturer's recommendations and experience.
- Change order process: to get as many people as possible involved in the decision process when a change is considered, engineering, maintenance, production, safety, and environment representatives have to sign the change

order process form, which describes the changes considered. Even with this procedure, there were many complaints regarding new or revised installations and equipment.

- Routes for lubrication and vibration analysis are established by the crews responsible for these tasks and are, for the most part, kept on the computer.
- Reliability (mean time between failures [MTBF]) and maintainability (mean time to repair [MTTR]) are not tracked.
- Operations and maintenance both indicated that they seldom have sufficient input into engineering projects. They feel that communication among maintenance, operations, and engineering could be improved.
- There are only a few maintenance engineers assigned to maintenance departments.
- The maintenance department received some criticism for not solving problems at their root causes. The tendency is to fix the problems that arise quickly, often cutting corners to save production time.

13.3.11 MAINTENANCE PROCESS REENGINEERING

Current situation:

- Most maintenance and materials processes have been documented, but they are not always followed.
- No formal process is in place to review and revise current processes to eliminate non–value-added activities.
- Activity costs are not measured.

13.4 ADDRESSING THE FINDINGS

Now this company ultimately did review these findings, along with leading practices, and developed a set of opportunities and recommended actions from their combined findings. They then worked to prioritize their findings into to categories such as "benefit" and "cost to implement" and from there developed a roadmap for managed business transformation going forward. This approach often helps executives appreciate that the whole set of business issues are more likely to come out and be addressed. As such, the management and staff will feel that their issues were, at the least, considered, and that this contributes to managing change acceptance when the final roadmap is communicated.

Without revealing their exact business strategy or going into the cultural dynamics that will affect every business, we have given these findings to a group of maintenance experts and asked them what they would recommend given the previous report. For the sake of this case demonstration, three or four recommendations were identified for each section of the assessment.

The first thing they observed is that the findings were not all tidy and wrapped in a way that made the recommendations clear and obvious. In other words, if you were looking for planning and scheduling issues and opportunities, you really needed to look through the whole report to determine whether you captured all the issues identified from the interviews. This is common since sometimes ideas arise from stakeholders that affect other areas of the pyramid.

A second observation is outlined in the following section.

13.4.1 STRATEGY

The first request of the team of maintenance experts was this: What should the primary goal of this maintenance assessment be for this business?

With only 10 to 15 minutes to answer, the thoughts and hopes that were returned included the following:

- Holistic analysis of strengths, weaknesses, opportunities, and threats (SWOT)
- An understanding of how to bridge financial and manufacturing constraints
- An understanding of the company's viability
- An understanding of how this company can affect production optimization and equipment influence on availability, costs, and safety or environment
- An understanding of what the production strategy should be going forward and how it affects productivity or employee satisfaction

These are lofty goals. The maintenance assessment will help you get part of the way there. However, you will, no doubt, also require an understanding of the company's change history and the commitment of the management team to understand how some of the recommendations that will arise can be accepted and implemented. The team did establish a chart of the SWOTs (Figure 13.2) to call out what they thought would summarize the most apparent health of UTO.

To align with the observed business imperatives, the maintenance strategy should support the business requirement to improve operations and maintenance delivery with a focus on cash flow and then asset value. The team developed the following strategic recommendations:

Strengths	Weaknesses
• Competent maintenance staff • Safety is a priority • Training is a priority	• CMMS usability • Operations and maintenance not working together on production issues • No tech support
Opportunities	Threats
• Gains to be made by sharing a common vision across the company sites • Develop maintenance strategies that align with the local operating context • Integrate an enhanced CMMS solution with their financial system	• Lack of available resources (supplies and skilled personnel) • Employee morale • Metrics to track activity costs needed to support cash flow concerns

FIGURE 13.2 Example of SWOT analysis that could apply to case study.

- Develop a clear vision for asset management and a defined PM program.
- Identify how asset management (and critical assets) contribute to cash flow.
- Identify pain points that affect production, sustainment, and short- versus long-term goals.
- Clearly communicate the asset management strategy to management and staff.
- Develop craft skills for future needs, including cross-training so that flexibility can be leveraged.

An analysis should also look at their "go-to-market" strategy to understand how the cost of distribution and manufacturing really contributes to their competitive pricing model. What are the value drivers to their success in the market?

13.4.2 ORGANIZATION

Their maintenance management had a few opportunities identified by the team. There is a defined desire to review the maintenance team mix, including the following:

- Consider the ratio of supervisors versus crafts (may be too many supervisors versus workers).
- Consider adding schedulers and reducing the number of planner roles.

To develop a stronger resource pool in their organization:

- Consider cross-training the crafts to create the ability to leverage existing staff.
- Create a craft pool to help manage workload balance and backlog.
- Review the need and benefit to an apprenticeship program.
- Entertain the concept of an "on-call" maintenance supervisor for after-hours maintenance coordination.

Create an environment of communication and visibility of key business processes by doing the following:

- Leverage the CMMS across the company to maintain maintenance standards.
- Publish issues and planned change actions to keep all stakeholders informed and to improve employee morale.
- Promote and celebrate the perception that they had strong maintenance staff and are well trained.

13.4.3 MAINTENANCE TACTICS

Generally maintenance tactics did not draw a great deal of recommendations from the group members. They acknowledged there was a displayed use of preventive and predictive maintenance and that mobile workforce tools were in use to capture some of the equipment readings. They suggested that a root-cause failure analysis could be done to focus on the high-end emergencies (30% of total maintenance). In addition, short-term activities suggested were tightening PM execution across the company and understanding why some PMs were missed.

13.4.4 KEY PERFORMANCE INDICATORS

Plan to educate the staff and to communicate the key measures and value drivers that are the current focus. Tracked key value drivers include Safety (number of incidents), %Availability, %Emergency, %PM, %Compliance, %Schedule Efficiency, %OT, %Machines in alarm, %Expenses ($Actual/$Budget).

13.4.5 INFORMATION TECHNOLOGY

With an understanding of leading practices and efficiencies to be gained from managing data: the team recommended replacing or upgrading the existing CMMS system and having it integrate with their financial system. Improvements in maintenance history data and resource management will help them assess what needs to be looked at to initiate a work order and to manage the resource backlog.

13.4.6 PLANNING AND SCHEDULING

Schedule compliance is 80%, so on the surface UTO appear to have a healthy planning and scheduling maintenance operation. However, they seem to lack long-term planning, which could come about with successful short-term execution that can be managed in the new CMMS solution. There is a need to understand why work orders that have well-defined priorities are often overruled by operations. This could be a smoldering fuse to a real problem or, in any event, a process that needs to be corrected.

Short-term quick hits could include the following:

- Reassign a planner to a scheduler role to review backlog and to coordinate crane availability.
- Establish roles and responsibility rules and lines of authority between planners and schedulers, maintenance, and operations.
- Set up daily operations meetings for daily schedule alignment.
- Planners should attend shutdown meetings.
- Engineers should be assigned to support planners.
- Begin measuring backlog by craft hours.

13.4.7 MATERIALS MANAGEMENT

- Establish visibility of parts availability across the company.
- Reduce material carrying costs.
- Ensure content management completes on SKU identification.
- Develop a "service-level agreement" with the parts outsource company.
- Notify supervisor on purchase requisitions not approved.

13.4.8 Autonomous Maintenance

- Establish a maintenance control reporting system.
- Engage operations, maintenance, and engineering in PM validation reviews.
- Support total productive maintenance (TPM) with a maintenance strategy education program.

13.4.9 Reliability-Centered Maintenance

- Operate a context-specific maintenance task assignment.
- Identify key critical assets and do reliability-centered maintenance (RCM) analysis.

13.4.10 Business Process Reengineering

- Consider routes for maintenance personnel once RCM analyzed tasks and frequencies have been assigned.
- Train crafts on more than one discipline for maintenance delivery flexibility.

Section IV

Achieving Maintenance Excellence

14 Real Estate, Facilities, and Construction

Andrew Carey and Joe Potter

CONTENTS

14.1   INTRODUCTION

Property is often one of the largest items on both an organization's profit and loss account and its balance sheet. By moving from typical to best practice, businesses can improve value for money and reduce their real estate and facilities costs by up to 20% while still improving services and consistency of performance. Property activities also frequently encompass complex construction projects that, if managed more effectively, can be delivered more quickly and at lower cost. Construction activities also have a critical effect on the life-cycle asset and operations costs of a building. In addition, real estate and facilities (REF) management change can be used to leverage business transformation, thus enhancing agility, improving customer service, and contributing to staff retention.

Chapter Summary: This chapter focuses on the management of all aspects of real estate and facilities management. It includes sections on process and organization design, procurements, information and communication technology systems and optimization.

14.2 OVERVIEW

14.2.1 DEFINITIONS

14.2.1.1 Real Estate

Real estate and property are referred to interchangeably in this text. Real estate is defined as the buildings, land, and associated ancillary assets (e.g., roads, parking lots) that an entity may own or lease to run its enterprise.

14.2.1.2 Facilities Management

Facilities management (FM) includes all support services for the built environment and the management of their impact upon the organization, people, and the workplace. Facilities management therefore applies equally to all types of facility, including but not limited to office, industrial, retail, and residential accommodation.

Facilities management services are often categorized as either "hard" or "soft," with the majority of suppliers having grown from either a technical hard or a services-oriented soft base.

The major hard categories include the following:

- Building fabric maintenance
- Mechanical and electrical maintenance
- Minor projects

Soft facilities management services include the following:

- Cleaning
- Pest control
- Catering
- Manned security
- Office services
- Waste disposal

It is sometimes appropriate to procure soft and hard facilities management from separate providers. Nevertheless, the overall approach to service delivery, including such considerations as performance and environmental management, can be common to all services and contracts.

14.2.2   CONSTRUCTION

Construction involves the building of new property assets or the extension or refurbishment of existing property assets. It is sometimes referred to as *capital projects*; however, this term infers the traditional finance approach for such projects and can, at times, be misleading. For the purposes of this chapter, construction is treated as a subset of real estate (for new build) or facilities management (extensions and refurbishments).

14.2.3   BUSINESS DRIVERS

There is a strong synergy between the disciplines of real estate, facilities management, and construction. Businesses that understand the relationships between these three activities use holistic REF strategies to support their business drivers.

Whether a business is a retailer, an office-based company, or a manufacturer, the relationship between its operational and REF strategies can be critical to its success. For example, the world of investment banking may appear far removed from facilities management, but if the cooling and ventilation fails at the data center serving the trading floor, the link between the two becomes immediately apparent.

Major businesses are increasingly locating their operations in the most economically advantageous parts of the world. It is common to hear of firms relocating their manufacturing capability to Eastern Europe or China, while Bangalore has become synonymous with outsourced customer service centers.

Despite this evident mobility in both core business areas and other support services, REF services are often procured and managed on a predominantly local basis. Individual facilities service categories are commonly sourced separately, often on a site level, with the inherent lack of leverage and increase in administrative workload that such decentralization entails. Service levels vary widely and can be difficult to compare. Data on estates, and the associated services and spending, is often poor and held only locally, if at all. Benchmarking between sites and suppliers is therefore unrealistic, and demand management discipline is very difficult to implement.

This inefficient state of affairs has not been helped by the historic inability of much of the supply market to deliver consistent multisite or multinational service. Moreover, client managers have often struggled to see benefits from developing a consistent company-wide approach to REF services procurement and management.

Clearly, many REF services have a strong local delivery component. However, the management of such services can increasingly be procured on a more aggregated level, leading to significant economies of scale. Notably, REF service providers in both the United States and Western Europe are increasingly extending their global reach. There are now, therefore, opportunities for significantly more efficient estate operating models for organizations that are both aware of them and willing to embrace the necessary change.

14.2.4 REAL ESTATE AND FAILURE

REF management typically accounts for at least 15% of a company's cost base. IBM's experience is that organizations can expect to save between 5% and 20% of this spending through a review of REF supply and demand management. This contributes directly to bottom-line results.

Whether the drivers are regional, national, or international, experience has demonstrated that the following are the most common reasons for change:

- *Reducing operating costs*, as part of an organization-wide cost-saving initiative
- *Improving service quality*, often as a reaction to changes to the business that mean its operations no longer meet the business need
- *Increasing service consistency* since the support services should operate in a similar manner to the other business functions
- *Increasing focus on the core business*, thus avoiding management distraction on support and noncore activities
- *Mergers and acquisitions*, with the consequent drive for rapid and effective rationalization
- *Sustainability*, with the need to reduce carbon footprint

The typical benefits of any change program are summarized in Figure 14.1.

Savings come from both supply- and demand-side economies. Leveraging spending by aggregating contracts allows suppliers to pass on economies of scale and simplifies communication, typically providing a much better quality of asset and spending data. This data in turn facilitates further savings through better demand management but frequently requires a more robust information and communication technology (ICT) solution to do so. Simplified lines of communication, together with the frequent outsourcing of service delivery elements, allows staff to be more productively deployed elsewhere.

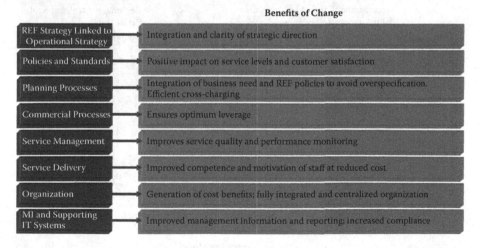

Benefits of Change

REF Strategy Linked to Operational Strategy	Integration and clarity of strategic direction
Policies and Standards	Positive impact on service levels and customer satisfaction
Planning Processes	Integration of business need and REF policies to avoid overspecification. Efficient cross-charging
Commercial Processes	Ensures optimum leverage
Service Management	Improves service quality and performance monitoring
Service Delivery	Improved competence and motivation of staff at reduced cost
Organization	Generation of cost benefits; fully integrated and centralized organization
MI and Supporting IT Systems	Improved management information and reporting; increased compliance

FIGURE 14.1 Typical benefits of a REF change program.

FIGURE 14.2 The three REF components.

14.2.5 Fit with Business Organization Model

The first stage of any change program is a review of the existing situation: REF needs to be reviewed in two different ways. First, the relationships between the three main components of estates, as shown in Figure 14.2, should be considered in the context of overall REF strategy and financing options. Second, each component (e.g., should be considered individually). By way of explanation, the replacement of a lightbulb is not a matter for great strategic debate, but the financing options for the contracts responsible for replacing such lightbulbs probably are.

When considering an organization's REF strategy, business models have a significant impact, as shown in Figure 14.3.

As businesses move toward a more globalized operating model, both their internal and external REF supply chain may have to develop in multiple strategic directions:

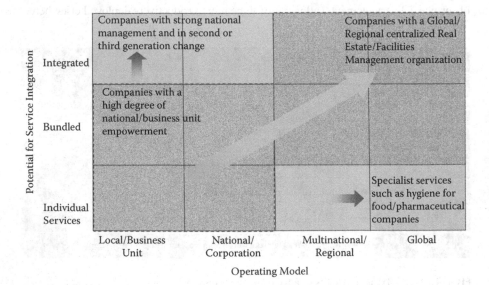

FIGURE 14.3 Service integration versus business reach.

- Acting as an agent across multiple service lines, with varying degrees of self-performance
- Focusing on specialist services areas

It is rare that companies can move from a locally focused individual service model to a fully integrated model without moving through a number of generations of change. Currently, very few companies can genuinely claim to have a truly global and integrated model of REF service delivery.

14.3 PROCESS AND ORGANIZATION

14.3.1 OVERVIEW

Before considering the appropriate organizational structure, some basic operating model decisions should be established. Clearly, the organization design must explicitly respond to the key business drivers (e.g., increasing shareholder value). Figure 14.4 shows further areas typically to be considered at this design stage.

Usually the decisions adopted in each of these areas will indicate a focus on supporting the core business as the customer rather than serving external customers. For example, the "product to offer" is typically the provision of REF services to support the business, not the exploitation of physical assets as a source of profit in their own right (although this is a consideration for some corporations, such as retailers with multi-use develpoments). The underlying economic model for the organization also needs careful thought. If it is to be a profit center in its own right, consideration should be given to whether this could create conflict of other parts of the organization.

Having established the basic operating model, it is then necessary to consider a number of overarching principles need to be considered in the high-level organization design. Typically these consist of the following:

	Typical/ Best Practice	Alternative Options
Business Drivers	Minimize costs/ increase shareholder value	Maximize asset value/ asset disposals
Customers to Serve	Internal only	Internal and third party
Products to Offer	Management of physical assets and services	Service delivery asset management
Activities to Perform	Outsource	All in-house
Economic Model	Cost recovery center	Profit center

FIGURE 14.4 Typical operating model decision matrix.

- Alignment with the organizational and management structure of the core business
- Inclusion of policies and standards to guide decision making
- A strong planning process to translate strategy to action
- Effective demand challenge
- Clearly agreed on interfaces and performance management
- Appropriate supporting ICT systems
- Recruitment or retention of capable and motivated staff

Later sections of this chapter cover a number of these topics, but two key areas are covered in more detail in Sections 14.3.2 and 14.3.3.

14.3.2 ALIGNMENT WITH CORE BUSINESS ORGANIZATIONAL AND MANAGEMENT STRUCTURE

14.3.2.1 Delivery Framework

Figure 14.5 highlights a delivery framework that can be used in the development of more detailed processes (i.e., process maps from levels 2 to 4). The model distinguishes between centralized "long-term planning" activities and the local management of delivery ("maintaining").

14.3.2.2 Review Business Need

A fundamental tenet of any REF delivery model is for the estate to satisfy operational requirements. There is a risk that, while it is reasonable to apply genuine constraints from a REF perspective, an estate can become too inflexible. Clearly, a prerequisite

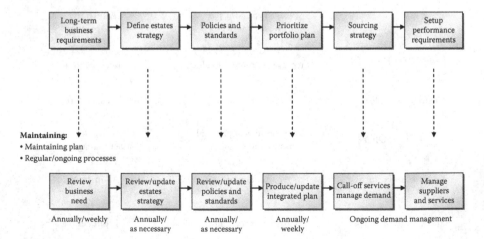

Long-term Planning:
- Establishing portfolio plan
- 1+ years planning/projects

| Long-term business requirements | Define estates strategy | Policies and standards | Prioritize portfolio plan | Sourcing strategy | Setup performance requirements |

Maintaining:
- Maintaining plan
- Regular/ongoing processes

| Review business need | Review/update estates strategy | Review/update policies and standards | Produce/update integrated plan | Call-off services manage demand | Manage suppliers and services |
| Annually/weekly | Annually/ as necessary | Annually/ as necessary | Annually/ weekly | Ongoing demand management | |

FIGURE 14.5   REF delivery framework.

for meeting the operational requirement is to understand it. This requires constant and effective communication between REF managers and their clients.

Strictly speaking, operational units should identify their own requirements. In practice, however, this can be difficult to translate into REF requirements. One solution is to embed "informed client" functions within operational units to help to manage the interface between operations and estates. These functions vary dramatically in size and scope to match the needs of the business, and the concept, though powerful, invariably needs to be tailored to specific client needs. Another option is to embed an "account management" function within the REF organization itself that has specific focus on particular operational units.

14.3.2.3 Review/Update REF Strategy

A REF strategy forms a key part of the link between the REF organization and the rest of the business. It should extract the key themes from the business strategy, develop a vision, set priorities for the estate, and define the delivery model to achieve the vision according to explicit priorities. It should also link to ICT infrastructure, risk management, and financing strategies.

14.3.2.4 Review/Update Policy and Standards

The REF function must direct both high-level policy (e.g., tenure and group-wide issues, such as sustainable development and health and safety) and provide standards to be used as technical design requirements (e.g., specifications for sprinkler systems). Policies and standards should be derived from both business needs and external policy.

14.3.2.5 Produce/Update Integrated Plan

The REF organization must ensure that business requirements are captured and translated into coordinated and fully budgeted project plans.

14.3.2.6 Call-Off Services/Manage Demand

Call-off services from existing suppliers need to be managed while ensuring that demand for those services is controlled in line with agreed processes, policies, and standards.

14.3.2.7 Manage Suppliers and Services

This activity entails maintaining and developing working relationships with suppliers to deliver planned work and providing an appropriate level of service to fulfill policy standards and to meet budgets (e.g., the development and implementation of procurement strategies) while monitoring and challenging performance where necessary.

14.3.3 EFFECTIVE DEMAND CHALLENGE

At its highest level, the provision of REF services is no different from the delivery of any other service or product; that is, it is the satisfaction of demand through supply.

Figure 14.6 shows the basis for any relationship between end users (i.e., staff members in the business unit who generate demand for accommodation and services) and

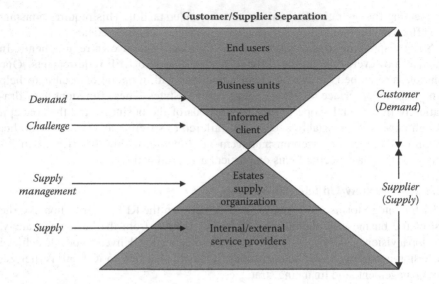

FIGURE 14.6 Supply and demand model.

the suppliers of those services (i.e., the internal or external contractors working for the internal REF management organization).

This conceptual model clearly separates supply (i.e., the REF organization and its third-party suppliers) from demand (i.e., the operational business units who are the ultimate customers of REF services). The upper portion of the diagram focuses on customer requirements, whereas the lower part (i.e., the estate management organization and its supply chain) is structured toward managing supply and the organization's suppliers.

The interface between the supply side and the demand side has to be designed in line with the business operating model. This may mean that the "informed client" role actually resides within the supply organization as a form of "account" or "service manager," or it may mean that it lies within the client business as previously indicated. However, irrespective of its location, the informed client must "own" and perform its role of communicating the REF requirements of the business to the REF organization.

Demand challenge concerns ensuring that the requested services genuinely are required and that the client's needs cannot be satisfied in an alternative, more efficient way. A simple example is the cleaning of offices: the offices need to be maintained with an appropriate level of hygiene, but this does not necessarily mean that they have to be cleaned in the same way every night to maintain that level. Once the REF organization understands the genuine (often implicit) needs of its clients, it can set about either providing internally or procuring services to satisfy them, along with those of its other clients, in the most cost-effective way.

Experience has shown that the greatest overall savings in cost often come from exercising effective demand control and that savings in other areas, including supplier unit costs, may be of a far lower relative value.

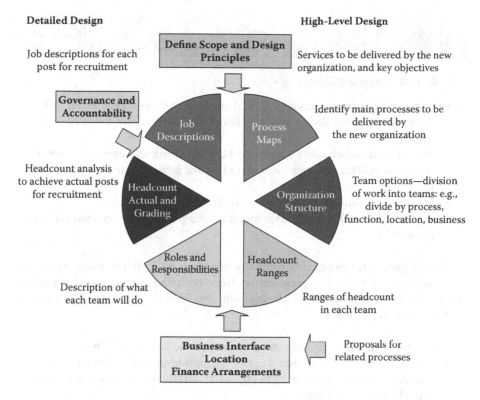

FIGURE 14.7 Organization design delivery model.

14.3.4 Organization Design Methodology

Figure 14.7 shows a typical high-level approach to organization design. Once the scope and design principles (including the desired operating model previously discussed) have been established, the design team steps around the wheel clockwise, performing the activities indicated. It may well be necessary to go around the wheel more than once, since there is an iterative element to this process. This is particularly the case when this exercise is being carried out in parallel with procurement or ICT work streams.

The skill base of the existing organization and its ability to recruit are aspects that should be considered with particular care when designing an organization.

The high-level delivery model depicted in Figure 14.7 can be broken down into individual stages as set out and described in Sections 14.3.5 and 14.3.6. Each of these stages is explained in turn.

However, before embarking on this process, it is worth noting the following:

- There is no single "right answer" for REF organization design.
- The rigorous use of business cases concentrates the minds of all involved on the benefits of the change.

FIGURE 14.8   Organization design stages.

- Organization change is not easy: realistic expectations should be set from the outset, and the design process should be methodically followed stage by stage.
- All affected stakeholders should be kept informed and engaged: change leaders should take the time to explain how the new organization will work for each of them.
- Structured communications support stakeholder engagement.
- A strong focus should be maintained on the practical rather than on the theoretical.

During project initiation, it is generally beneficial to establish a steering group as soon as possible, with representation from both the existing REF organizations and the business units interfacing with the REF organization. The benefits of this approach are that it

- Allows immediate engagement with many of the key stakeholders
- Facilitates discussion of each stage of the high-level design development
- Can achieve early buy-in from many of the staff members who may be running the new organization

14.3.5   INITIAL STEPS (PHASES 1 TO 3)

Figure 14.9 illustrates these first steps of the process.

14.3.5.1   Design Principles

Typical design principles used in designing business processes and organization structures for a new REF structure are presented in Figure 14.10.

14.3.5.2   Existing Processes

A process is a group of logically related activities that produce defined results. The detailed consideration of existing processes helps to determine the functions that the new organization also needs to perform.

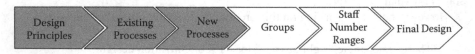

FIGURE 14.9   Organization design: initial steps.

Typical Summary Design Principles	
Category	Description
Client	• Service the client businesses. • Serve selected external customers where appropriate. • Be responsive to individual business unit needs. • Operate with a formalized performance agreement between businesses and the estate management organization(s). • Interface with other policy setting organizations involved in estate matters.
Operating model	• Fit with the corporate operating model. • Be scaleable and flexible in order to accommodate volume, service, and service level changes in a timescale to meet business needs.
Service delivery	• Operate with a generally outsourced service delivery model unless there is a compelling reason not to (either operational risk or cost). • Retain a sufficient level of in-house expertise in estate management, facilities management, and construction to enable, as a minimum, an informed and skilled internal supply and external relationship management function to be performed. • Retain a sufficient level of in-house expertise in program and project management, planning, and budgeting to enable business needs to be understood and translated into deliverable property programs.
Procurement strategy	• Maximize the leverage of the spending across the relevant estates market from which services and products are procured. • Maximize the efficiency of Business' capital and maintenance speed across estates. • Use best in class procurement processes, technology, and supplier management techniques.
Compliance (including risk management)	• Ensure compliance with corporate and, where relevant, wider government policies on accommodation issues. • Ensure compliance with corporate processes including project and program management. • Effective risk management of accommodation services.

FIGURE 14.10 Typical design principles.

In this phase, the existing processes need to be quickly but comprehensively reviewed to reach clear conclusions on the following:

- Where process gaps currently exist
- Which processes exist today but are unlikely to be sufficient or suitable to guide and meet the needs of the new organizational design
- What processes do work and could or should be used in the new organization or else cannot be changed for other reasons (e.g., business or regulatory)

The concept of *maturity profiling* can also be used as a tool to compare current client practice with recognized "best practice" asset management processes. A sample maturity profile is shown in Figure 14.11, relating to FM.

An organization, or its constituent parts, is mapped by type of process (e.g., planning processes, service delivery), and its "maturity" is determined on the continuum of "innocence" to "excellence." A comparison with where the business feels it should be in each of these process areas builds the momentum for change as well as focusing attention on where existing processes need to be improved or expanded in the new organization.

14.3.5.3 New Processes

The identification and design of new processes should be guided by the following:

- The design principles for the new organization
- Lessons learned from the review of existing processes (business/regulatory).
- Gaps identified by maturity profiling
- Consultation with the selected business and REF stakeholders through structured workshops
- The need to fit with the corporate operating model (which may also be undergoing change)
- A realistic appraisal of people's skills

New processes can be developed by first producing high-level process maps aligned to relevant categories or functions. These are then tested and further developed with customers through the steering group or else on a more individual basis with designated process owners and champions. In each case, process maps should be supported by a clear and simple explanation of the process purpose, reporting requirements, and issues to be reviewed.

14.3.6 FINAL STEPS (PHASES 4 TO 6)

Figure 14.12 illustrates the final steps of the process. The final steps in the process and organization design develop the detailed organization and its supporting structures.

14.3.6.1 Groups

At this stage, a high-level organization structure should be developed, including options for future groups based around the detailed processes and creating the shortest possible linkages between activities. "Lean" techniques can be used to support this activity.

The generation of the high-level organization should comprise a broad framework, to be further refined later, for the following:

- Teams and functions should be based on the division of work and business units' structure: this should include a high level definition of team roles and responsibilities.

Facility Management Maturity Matrix

The maturity profile for an organization shows pockets of "best practice" and significant opportunities for improvement

	Managing Processes (Overview)				
	Innocent	Awareness	Understanding	Competence	Excellence
FM strategy linked to operational strategy	Link not considered	Ad hoc discussions	Occasional process	Clear process—infrequent application	FM strategy aligned to support overall operational objectives
Policies and standards	Ad hoc policy development; low levels of enforcement	Some policies in place; H&S and manufacturer led	Key policies further developed; moderate levels of enforcement	Many policies in place; high enforcement levels	Value engineered policies in place to deliver service levels—enforced
Planning processes	Local budgets—often stopped or reduced	Local planning for FM; not coordinated with P&S or business need	Planning linked to business need; some demand challenge	Strong link with business need and P&S; demand challenge in place	Periodic planning round integrated with business need and FM policies; efficient intracharging
Commercial processes	Local appointments	Some bundling and grouping	Centralized purchasing in key areas; contract development in house	Central purchasing; contract structures further developed	Appropriate contract structures in place; suppliers actively managed
Service management	In house—detached from rest of organization	In house—better integrated—few formalized contract management processes	Active management of service provision	Formalized management structures and reporting	Contracted services delivered to contract; in house services effectively managed; Key Performance Indicator regime
Service delivery	Unproductive and high cost + extensive use of specialists	Some adjustments to cost of operation—output quality varied	Right sized, market rate workforce; output quality varied	Market standard organization; output quality addressed	Well trained, efficient and motivated labor force and supervision; market cost
Organization	Local organizations with much variation	Standardized local operations	"Designed in" regionalization and centralization	Organization built around end to end processes	Centralization and outsourcing as appropriate
Management information and supporting information technology systems—FM related	Inconsistent and ad hoc use of MI across organization	MI supported by ad hoc IT plus use of some standard systems	Common MI across some functions; not fully aligned with IT	Common MI supported by networked IT	Integrated centralized and networked systems designed to support MI strategy

FIGURE 14.11 Typical maturity profile.

FIGURE 14.12   Organization design: final steps.

- Location should be based upon the REF portfolio and customer spread as well as recognized best practice to determine national and regional functional requirements.
- Governance and accountability should take place together with decision-making structures.
- Business interfaces should consider the numbers and types of interface with other parts of the organization as well as external entities.

It may be beneficial at this juncture to create and present or workshop day-to-day activity scenarios to test the basic understanding of how the new organization, its processes, and the fundamental interrelationships between functions and teams will work.

14.3.6.2   Staff Number Ranges

Following the high-level design, staff number ranges for the whole organization should be developed and broken down to a team level.

Future staff number ranges can be generated by considering the benchmark information and the change in current workload that would result from the implementation of the new processes. The information on groups and staff numbers can then be used to populate the final design. The result is usually a range that will need to be refined with key stakeholders in the period leading up to final design. A grading profile for the organization should also be drafted and approved by key stakeholders as a target profile for the final design.

There will often be a requirement for some revision and further delicate stakeholder management to produce a detailed structure. It is therefore important to obtain, before proceeding any further, stakeholders' agreement that the general method used in the high-level design has been both transparent and robust.

14.3.6.3   Final Design

The main objective of the final design phase is to generate the actual headcount and grading for each team and each individual position within the new organization. This will inform the selection or recruitment process to fill those posts. At this stage, the roles and responsibilities of all posts within teams and functions need to be detailed and job descriptions produced, preferably in coordination with the organization's human resources department.

Depending on the organization, consideration should be given to a two-stage selection and recruitment approach to test robustness and to achieve more stakeholder buy-in. This typically involves recruiting the new senior management team first and then giving them a say in the size and shape of the new teams they will be required to manage. This will in turn require further iterations depending on the specific skills and capabilities of the individuals selected for each position.

Final design can thus become an extended process and requires considerable work to gain full agreement. It can involve revaluation and reassessment of previous work and analysis. However, it is important for there still to be a process of continuous improvement. Detailed processes (at process mapping levels 3 to 4) can be left to the new staff and managers to shape and fine-tune to some degree.

Finally, decisions on the following areas must be confirmed:

- The location of each team, to inform job descriptions and advertisements
- Governance and accountability arrangements
- Performance measures
- Financial arrangements for budget transfers and allocations
- Training arrangements

14.4 PROCUREMENT

14.4.1 OVERVIEW

Typical drivers for organizations to undertake a program of procurement activity (i.e., outside the normal churn of periodic contract renewal) include cost-reduction initiatives, process improvement, the establishment of shared services centers, and consolidation following merger and acquisition activity.

Procurement activity typically follows six key steps:

- Strategy
- Sourcing
- Specification
- Tender process
- Mobilization
- Contract award

However, REF procurement can have extremely complex service levels, performance and pricing mechanisms, financing, funding, and risk transfer arrangements. As such, the processes and approaches used differ significantly from the commodity-like buying of many other services, and the potential for complexity related to each step should not be underestimated.

These generic steps as they apply to procuring REF services are explained in detail in the following section.

14.4.2 HIGH-LEVEL PROCUREMENT APPROACH

14.4.2.1 Procurement Strategy

The implementation of a REF procurement strategy by a business can significantly contribute to maximizing the value for money achieved from its REF assets and associated activities. The use of category management as an approach can provide structure and focus to this critical stage.

Figure 14.13 outlines a typical category management approach. To improve the chances of a successful production of a category plan, the following questions have to be satisfactorily answered:

- Are key stakeholders involved throughout the development of the strategy?
- Is there a clear understanding of the breakdown of historic and future third-party spends, by service, geography, and line of business?
- Is the relative importance and "total cost" to the business of each area of spending understood?
- Are the cost drivers and dynamics within each supply market (i.e., both technical and geographic) clear?
- Are the strategies and capabilities, by service and by geography, of suppliers understood?
- What are the implications of the balance of power between the business and each supply market?
- Have radical new options been identified by the incorporation of supply market insights within the business strategies and using the full set of procurement levers?
- Have opportunities been logically prioritized, trading off business risk against business impact?
- Is there a multiyear plan of prioritized opportunities?
- Is it clear what should be managed at global, regional, national, and local levels?
- Will there remain within the business sufficient competent resources to manage, effectively challenge, and eventually reprocure the services within scope?
- Are there clear owners and processes for managing key supplier relationships?

14.4.2.2 Sourcing Strategy

If the category plan recommends some form or reprocurement of services, then a sourcing strategy will be required.

The development of the sourcing strategy is intimately related to the services to be purchased. Within REF services, this may vary quite considerably between the various components of estates management, FM, and construction services. In all cases, the following typically need to be considered:

- The degree of risk transfer desired
- The form of contract (e.g., framework, annual fixed payment)
- The minimum contract length and any intention to extend if services prove acceptable
- The specific dynamics and traits of the supply market in the geographies within scope
- Anticipated trends in these markets over the anticipated duration of the contracts

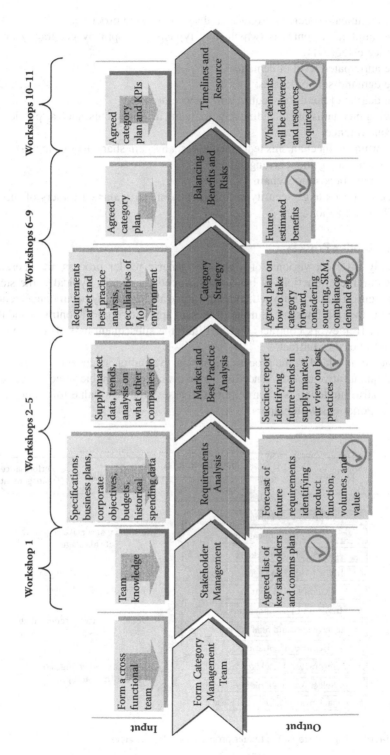

FIGURE 14.13 Typical steps in delivering a category plan.

- The technical aspects of products and services being procured
- The number of contracts (which may typically be split by geography or service category)
- The anticipated level of demand
- The demand-side processes
- The degree of flexibility required in service provision
- Pricing mechanisms (including any financing arrangements, such as public-private finance)
- Performance mechanisms (e.g., incentives, any gain share arrangements)
- Contract management arrangements
- The incumbent supply chain
- Specific needs (e.g., security requirements) of particular customers of the REF organization

14.4.2.3 Tender Process

Many subtly different variants of a broadly generic tender process are operated by the procurement functions of businesses around the world. Typically, the steps shown in Figure 14.14 are followed, although some activities may occur concurrently to reduce the overall timescale. Similar processes apply for one-off contracts and the formation of framework contracts that facilitate the subsequent call-off of services during the life of the contract.

Complexity is added to the process during the evaluation stages with both qualitative and quantitative measures needing to be evaluated to enable selection of suppliers. Scenario modeling is often required to assess the best value for money and facilitate negotiations with bidders.

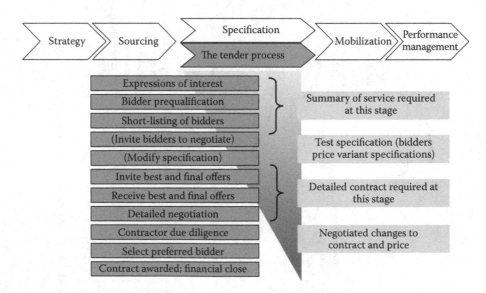

FIGURE 14.14 Typical steps in a tender process for REF services.

14.4.2.4 Specification

In simple terms, the specification needs to translate business requirements into user requirements that a supplier can understand and deliver for an agreed price. Traditional "input-based" specifications in all three REF areas of estates management, facilities management, and construction are now tending to be replaced by "output-based" specifications. These output-based specifications allow the supplier greater flexibility to deliver their solutions, albeit with a generally higher level of risk transfer. Provided that responsibilities can adequately be defined, more sophisticated service providers generally welcome such output-based specifications since these typically allow the provider to use their technology and process skills to offer a significant efficiency savings, much of which can be passed on to the client. This helps the service provider achieve competitive advantage.

Specifications, with their associated service levels, are typically a major component of a suite of documents issued to a limited number of preselected suppliers to produce a firm offer to deliver the required services. This stage is often known as a request for quotation (RFQ) or request for proposal (RFP).

Figure 14.15 shows the typical structure of an RFP document for an FM contract. Where external consultants are involved in the production of an RFP, they typically focus on developing the specification, its service levels and the performance measurement system and payment mechanisms.

The client business REF function generally uses its standard terms and conditions (modified where necessary to reflect alternative procurement approaches) and produces the asset data—which can include a list of properties in scope, site and floor plans, photographs, matrices of services to be provided at each location, existing contractual information, and detailed or aggregated spend data.

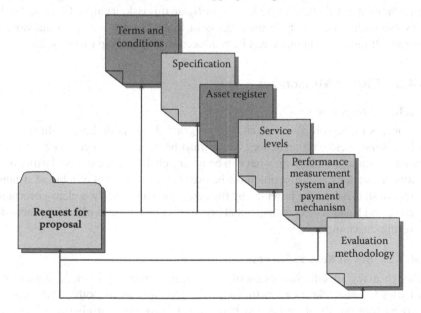

FIGURE 14.15 Typical structure of a facilities management request for proposal.

While some outline indicators may be provided to assist tenderers, the detailed evaluation methodology is not usually issued to suppliers: however, it must be developed in parallel with the rest of the documentation to ensure that all requested information is both relevant and taken into consideration when selecting a preferred bidder.

14.4.2.5   Mobilization

A good mobilization phase can greatly contribute to a successful contractual relationship. This phase includes such activities as due diligence, site setup, and document review. It can also include the transfer of staff from the client organization or from incumbent suppliers to new providers. With service contracts (e.g., facilities management), this phase ensures that service continuity risk is adequately managed. With delivery contracts this phase can ensure that all site procedures (e.g., health and safety) are in place and fully operational. Often a contract guide can be produced to further help this process.

14.4.2.6   Contract Award

Ultimately, the contract should be able to inform both parties of their, and others', responsibilities with regard to their joint relationship. While the REF industry has tried to reduce some of the inherent complexity by the production of standard documentation, the practical reality is that major contracts often require significant periods of time to reach a financial close.

One consideration often missed in contracts is that the individuals involved in day-to-day contract management from both a client and supplier side have rarely been involved in the initial negotiations. There is therefore a risk that the original intent of particular contract clauses can be lost. To mitigate this risk, it can be beneficial either to involve such parties early in the process or to ensure that a detailed handover is affected. Often a contact guide can be produced to further help this process.

14.4.3   CHANGE MANAGEMENT

14.4.3.1   Overview

The success of any procurement change program depends on having three factors in place. First, a clear rationale for change must be developed to generate the necessary momentum to embark on what can be a very challenging exercise lasting many months. Second, a clear direction must be provided, in the form of a broad strategic outline. Both these can be provided by the category plan. Finally, a clear governance structure and basic project management processes need to be in place for practically delivering the change.

14.4.3.2   Seven Keys to Success

Although many specific factors should be considered in any implementation plan, IBM uses its own "Seven Keys to Success" methodology as both a guidance and reporting tool on all of its change management programs: it also provides early warning of likely problems. In brief, this process entails ensuring that seven key criteria are satisfied throughout the program life cycle:

1. Stakeholders are committed: The program must be supported at the highest level to have the necessary credibility to drive through change. Appropriate governance structures must also be in place.
2. Business benefits are being realized: Both tangible and "softer" program benefits should be monitored against targets.
3. Work and schedule is managed: A realistic work schedule should be established at the outset with all stakeholders and regularly monitored throughout the implementation.
4. Team is high performing: The program team members require not only the right mix of skills and experience but also the time to focus away from their day jobs. The team needs to be motivated and confident.
5. Scope is realistic and managed: The scope needs to be agreed at the outset. Nevertheless, such programs inevitably require modifications during implementation; these should be managed through a rigorous change control process.
6. Risks are being mitigated: Program risks should be identified, regularly reviewed, and actively managed.
7. Delivery organizations' benefits are realized: A program can be threatened if external delivery organizations (e.g., consultants and the FM suppliers themselves) are not making their own anticipated margins. Facilities management is a service industry and should be procured as such.

These considerations are common to all change programs, indeed to the majority of projects of any kind. However, REF change programs, particularly when these are cross-border, present their own specific challenges.

14.4.3.3 Governance

REF, and in particular facilities management, delivery can affect people in a more personal way than most other services, since it directly influences individuals' working environment—for example, their desk, chair, ambient lighting, and temperature. Any change in responsibilities, particularly where outsourcing is involved, will be perceived as a threat to established roles and power bases. Change projects in this field are especially vulnerable to resistance and require particularly rigorous governance processes. Senior management should itself first be convinced of the need for any change: it should then ensure that both adequate program resources are available and that more junior stakeholders in all geographies actively contribute to its success.

Appropriate governance structures will depend on the internal organization of a given business but typically include a core project team that reports to an executive sponsor while liaising with in-country teams. IBM advocates formalizing the structure of such in-country teams, with a steering group representing all major stakeholders meeting on a regular basis and guiding the local development and implementation project. Figure 14.16 shows a typical governance structure for a cross-border REF procurement project.

14.4.3.4 Communication

The existence of a REF program, as well as its terms of reference, should be communicated as soon as possible. Ground rules should be established, for example, that no site or business line will have to pay more for its REF services than before or that service levels will not deteriorate.

The coopting of key personnel at both core and country level to shape the strategy will help to obtain their personal buy-in and turn them into advocates for the program. Regular communication, appropriately targeted at each stakeholder group, is also important to maintain morale and momentum. The use of output specifications is more efficient than a traditional input-based approach but requires careful explanation to end users.

Successes should be publicized and celebrated, and the need to achieve short-term "quick wins" that could compromise the longer-term outcome should be resisted.

14.5 INFORMATION AND COMMUNICATION TECHNOLOGY SYSTEMS

14.5.1 Overview

The introduction to this chapter highlighted that organizations are increasingly centralizing the management of their property assets and are thus treating them as strategic, rather than as local, assets. Furthermore, they are also increasingly outsourcing the delivery of their REF services.

The former trend means that it is no longer feasible to operate on the traditional basis of all asset knowledge being held locally. Consequently, corporate-level systems must be employed to ensure consistent and effective management of property assets. At the same time, the trend toward outsourced service delivery means that an organization no longer holds detailed information about its own portfolio's performance. This is increasingly now in the hands of outsourced providers, using their own systems to manage their element of service delivery to the estate. This fragmentation of data systems and ownership is increasingly driving organizations to consider the benefits of using common systems for all users. This section addresses the issues arising from these trends.

14.5.2 System Types

14.5.2.1 Introduction

In ICT terms, there are two types of systems that support REF functions: (1) *transactional*; and (2) *property performance management*.

14.5.2.2 Transactional Systems

Transactional systems are generally used by organizations that directly manage REF activity themselves. They are built around off-the-shelf software packages

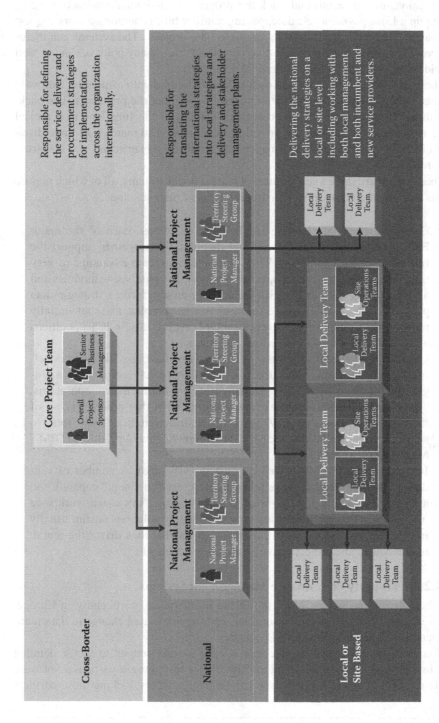

FIGURE 14.16 Typical governance structure for a cross-border REF procurement project.

organizations use to record and track the progress of individual transactions (e.g., setting up a lease payment schedule, paying a utility bill, generating a work request to repair a broken window, updating progress on a project). The systems can therefore provide a "snapshot" of progress across all services at any point in time as well as hold historical records for each individual activity.

These systems rely on the business rules configured into whichever software package is being used to ensure data integrity throughout the processes. The packages themselves are covering an increasing number of the functions in a typical property department and thus are providing functionality to manage all three disciplines of REF services: estates management, facilities management, and construction projects.

There are in turn three main variants of transactional systems, all of which require details on individual transactions to drive the rest of their solution:

- Enterprise resource planning (ERP)-based solutions. Each of the major ERP vendors (i.e., SAP, Oracle) has modules that, together, support the full range of REF functions. They offer the significant advantage of very tight integration with their respective financial and purchasing modules and allow property data to be fully included in an enterprise's strategic data pack. Though extremely powerful, they can be complex, and thus initially more costly, to implement. However, their life-cycle cost in ICT operations and maintenance terms can redress the balance.
- Best of breed (BoB) solutions. These systems have usually been developed from an initial specialist package that perhaps supported a single function, such as lease management or maintenance operations. The systems marketplace has now matured such that practically all of the major BoB packages have grown their functionality "footprints" to support multiple REF functions. An example of such a product is IBM's Maximo software.
- Integrated workplace management systems (IWMSs). A number of vendors in this marketplace have taken their "cross-functional" capability to a higher level, maturing into what Gartner* terms "integrated workplace management systems." They are characterized by solutions (again, usually modular) based around a central asset register that then drives the rental, maintenance, and project activities across the portfolio.

14.5.2.3 Property Performance Management Systems

An alternative trend is now emerging in the marketplace, particularly in Europe. This recognizes the growth in outsourcing and the associated change in data management that such a business model generates.

Traditionally, there have only been two ways for an occupier to access detailed data in a heavily outsourced situation. The first is to impose a particular software solution (or set of solutions) on its service delivery partners, locking those partners

---

* Gartner Magic Quadrant for IWMS in North America, December 20, 2005.

into a defined software solution and the ways of working that go with it. This allows strategic property data to be kept in-house, but at the expense of having to maintain often complex transactional systems and potentially stifling supplier innovation. Alternatively, many organizations allow their delivery partners the freedom to use their own preferred systems but then suffer restrictions in their ability to extract the data that they need, often having to mine data through myriad inconsistent spreadsheets to extract data for key metrics such as total cost of ownership.

Consequently, property "performance management" solutions have started to emerge that, while extracting the key data from service delivery organizations, do not attempt to replicate the detailed transactional processes and business rules functionality that drive transactional systems.

14.5.2.4 System Type Summary

Both types of system have their place: transactional systems are most appropriate where property is managed directly or systems are imposed on the supplier, whereas performance management solutions should be considered where much of the delivery is outsourced. However, whatever the solution, the diagram of a typical REF series of functions, as represented by Figure 14.17, is a useful starting point in developing any ICT strategy.

This diagram notably introduces the topic of environmental management. This has been historically included within either estates management or facilities management, or sometimes treated separately altogether and managed together with the core functions of, for example, a manufacturing plant due to its business-critical importance. However, it is useful to address it as a separate discipline from the ICT context.

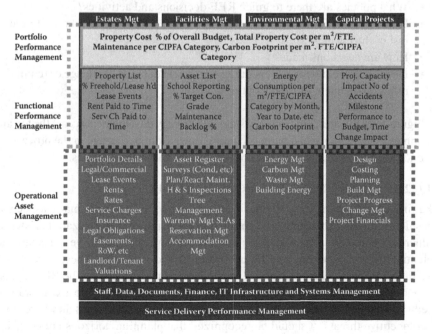

FIGURE 14.17 REF functions.

14.6 REAL ESTATE AND FACILITIES OPTIMIZATION

14.6.1 Overview

With the correct REF organization in place, aligned procurement strategies and detailed management information systems corporations can then truly address the optimization of their property portfolio.

An initial step in this process is the production of an REF strategy and an REF portfolio plan, and the techniques involved in these two steps are explored in this section.

14.6.2 Asset Strategy

Asset strategy is about translating the overall corporate strategy into a real estate/physical infrastructure strategy that will support the aims of the business. It can be described as a process to ensure the development of physical asset strategies, policies, and portfolio management plans that support the core business and optimize performance while providing an appropriate level of flexibility. This covers the strategy itself and also the policies and principles that govern all of the client's real estate and physical infrastructure activity.

Many key questions are to be considered in assessing an asset strategy:

- Is there an existing asset strategy in place?
- How does it support the core business?
- How is flexibility addressed?
- What policies are there to guide REF decisions and activities?
- How is the REF function organized, both internally and in relation to the rest of the business?
- How are decisions made?
- Does the current structure encourage sufficient demand challenge or require robust assessment of supply options?

Figure 14.18 illustrates the four key stages of strategy development across the top and the inputs, interventions, and outputs coming down from each stage in the process.

Each of these stages is reviewed in detail in the following sections.

14.6.2.1 Stage 1: Review Business Environment

The first stage to consider is the business environment. There are many aspects to consider when thinking about how physical assets support an organization. External influences (political, economic, social, and technological) need to be considered as potential drivers to existing and future strategies and processes. These drivers will also define business objectives and identify constraints imposed by the REF assets.

One key driver is how much flexibility the business needs. Most corporations need to remain agile enough to take advantage of new opportunities and to respond to competitive threats. It should be recognized that planning horizons are considerably shorter than they used to be. Flexibility is therefore an ongoing requirement.

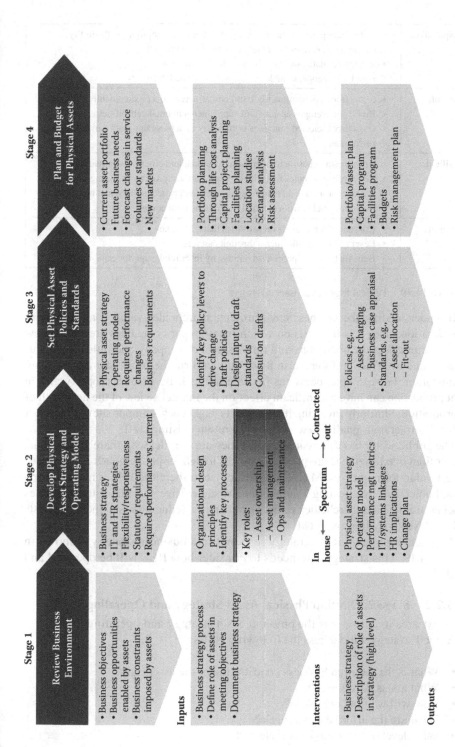

Stage 1

Review Business Environment

Inputs
- Business objectives
- Business opportunities enabled by assets
- Business constraints imposed by assets

Interventions
- Business strategy process
- Define role of assets in meeting objectives
- Document business strategy

Outputs
- Business strategy
- Description of role of assets in strategy (high level)

Stage 2

Develop Physical Asset Strategy and Operating Model

- Business strategy
- IT and HR strategies
- Flexibility/responsiveness
- Statutory requirements
- Required performance vs. current

- Organizational design principles
- Identify key processes
- Key roles:
  - Asset ownership
  - Asset management
  - Ops and maintenance

In house ← Spectrum → Contracted out

- Physical asset strategy
- Operating model
- Performance mgt metrics
- IT/systems linkages
- HR implications
- Change plan

Stage 3

Set Physical Asset Policies and Standards

- Physical asset strategy
- Operating model
- Required performance changes
- Business requirements

- Identify key policy levers to drive change
- Draft policies
- Design input to draft standards
- Consult on drafts

- Policies, e.g.,
  - Asset charging
  - Business case appraisal
- Standards, e.g.,
  - Asset allocation
  - Fit-out

Stage 4

Plan and Budget for Physical Assets

- Current asset portfolio
- Future business needs
- Forecast changes in service volumes or standards
- New markets

- Portfolio planning
- Through life cost analysis
- Capital project planning
- Facilities planning
- Location studies
- Scenario analysis
- Risk assessment

- Portfolio/asset plan
- Capital program
- Facilities program
- Budgets
- Risk management plan

FIGURE 14.18 Stages of strategy development.

Responsive:	• Anticipates potential future changes and provides appropriate flexibility • Customizes services to fit customer needs, adding value • Aggregates data, turning it into useful information • Enables its people to make rapid, well-informed decisions
Variable:	• Shifts from a predominantly fixed cost structure to a predominantly variable one • Builds for average capacity and supplements internal capabilities externally • Outsources selected functions completely to achieve optimum variability and performance
Resilient:	• Knows own exposure to operational, market, and other risks in real time • Effectively distributes risk with strategic partners/proactively manages remainder • Builds robust, "self-healing" capabilities (assets, technology, and processes) • Recovers quickly from external disruptions to operations
Focused:	• Concentrates resources on activities that add value • Leverages scale efficiencies through partners • Benefits from competitive advantage by insourcing superior capabilities

FIGURE 14.19 Typical high-level asset strategy.

In understanding the business environment, shareholder value is an important consideration, since it is the basis on which many major companies are run. A new strategy can affect the operations and financing of a company and influence shareholder value by affecting future cash flows and, in turn, the future share price of an organization. For example, decisions on whether to lease or own property and in what proportion can have a significant effect on the market's view of the net worth of a corporation. Similarly investing in new assets (e.g., a new campus site) can make it easier to attract high-quality new staff and to retain existing staff.

The public sector has different goals, yet they are parallel with shareholder value. Rather than maximizing share price, the aim is to deliver public goods and services at best value for money (VfM). A new strategy can therefore help achieve departmental objectives by affecting a number of the same drivers available to companies seeking to influence shareholder value. For example, reducing asset operating costs and use levels may improve VfM.

A typical high-level REF strategy response to a business strategy is shown in Figure 14.19. This strategy also needs to consider how REF will be operated and funded.

14.6.2.2 Stage 2: Develop Physical Asset Strategy and Operating Model

The next area to consider is the physical asset strategy and operating model.

A good strategy will address the following issues:

• What is the required business contribution of the asset class?
• What assets do we need?
• How will we fund our assets?
• What are the affordability criteria?
• What level of risk transfer is preferred?

Voice of the customer analysis (*interviews, questionnaires*)	Maturity profiles	Away days
Competitor analysis "Porters 5 forces" analysis	PEST analysis (political, economic, sociocultural, and technological)	Workshops
Shareholder value analysis (private sector)	SWOT analysis (strengths, weaknesses, opportunities, threats)	Supplier discussions

FIGURE 14.20 Typical tools and approaches for strategy development.

- What are the required high-level specifications or capabilities?
- What asset volumes or capacity do we need?
- What broad locations do we need the assets in?
- What performance and reliability standards do we need?
- What asset-related services and service standards are required?
- What flexibility are we likely to need to change any of the above?
- How does this compare with existing asset base and performance?

Tools and techniques are available to develop a new strategy. Figure 14.20 includes a nonexhaustive list of tools and approaches that can help in the understanding of a business environment and thus the development of a new strategy.

Careful consideration needs to be given to the degree of consultation required with the business; if, as usual, they need to "own" the finished strategy, then activities like workshops and structured interview programs are invaluable in helping to build this. Development of the asset management organization's operating model is an important part of the strategy, since it will impact many aspects of asset performance. This has been covered in the previous section on organization design.

14.6.2.3 Stage 3: Set Physical Asset Policies and Standards

The next area of consideration is the policies and standards that should apply to the physical assets. Following is a list of areas that guide decision making and should be covered when setting policies and standards for physical assets.

- Level of flexibility required
- Investment and funding policies
- Appraisal policies
- Asset management and maintenance policies
- Internal charging policies
- Service levels required
- Risk management policies
- Business specific standards

Flexibility needs to be embedded in asset management policies to ensure that flexibility issues are addressed when making key decisions on sourcing and supply

	Demand Policies	Supply Policies
Acquisition	1. How the business units articulate demand for new accommodation	2. How Group Property (GP) considers flexibility and property risks 3. How GP appraises property supply options 4. How GP fits out client accommodation
Management	5. How business units articulate demand for FM services and set service levels 6. How GP procures services 7. How business units occupy the accommodation efficiently 8. How internal charging operates	9. How GP plans property supply 10. How business units budget for property 11. How GP manages relationships with business units 12. How GP manages external suppliers 13. How GP manages business critical buildings 14. How GP manages multi-occupied buildings
Change	15. How business units articulate changes in their requirements	16. How GP manages moves/churn 17. How GP manages capital expenditure projects
Disposals	18. How business units identify surplus accommodation	19. How GP manages surplus properties

FIGURE 14.21 Typical policy framework.

options. It is necessary to strike the right balance between the level of flexibility the client requires and the cost of providing that flexibility, recognizing that there may be trade-offs.

The diagram shown in Figure 14.21 illustrates an example of a high-level property policy framework.

14.6.2.4 Stage 4: Plan and Budget for Physical Assets

The final area to consider is the planning and budgeting for physical assets. Real estate and physical infrastructure plans need to be prepared in the context of an agreed strategy and policy framework.

14.6.3 Planning Framework

The planning process is iterative and will usually consist of three core elements, as shown in Figure 14.22.

A portfolio plan, which establishes individual properties or other physical facilities, is required to meet business need. It should also describe how flexibility should be provided to meet changes in business needs and to avoid unnecessary legacy costs should existing or planned facilities become surplus to requirements. The plan sets out change to be made to the existing real estate or physical infrastructure portfolio to meet current and future business need.

An FM plan will usually flow from the portfolio plan. It describes the services and service levels required to maintain and operate the portfolio of properties and

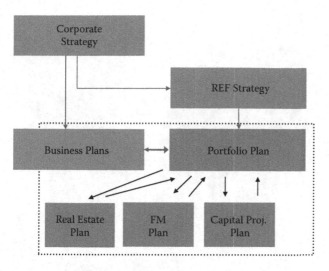

FIGURE 14.22 Typical planning framework.

other physical facilities in line with business need. The FM plan describes how these facilities management services should be provided and includes a facilities management budget.

The capital plan (CP) also flows from the portfolio plan. It sets out a prioritized and phased program of the capital projects required to meet agreed business needs.

The real estate plan also flows from the portfolio plan. It sets out the prioritized and placed program of aquisitioned and disposals of property and assets within the client's control.

None of these plans should be looked at in isolation; all can be impacted by radical changes in corporate strategy or business plans.

The key is to maintain flexibility and ensure that all documents remain "living" and flexible to change.

15 Information Technology Service Management Life Cycle

Brian Helstrom and Ron Green

CONTENTS

At their core, all organizations have the same objective: to achieve their business mission in the most effective manner possible. And, while most institutions rely on technology tools to facilitate the achievement of this mission and its related business goals, the tools they rely on often prove to be more of a hindrance than a help.

As a result, many organizations have turned to process models as a way to take a more structured approach to information technology (IT) management. Two such models, total life-cycle asset management (TLAM) and the Information Technology Infrastructure Library (ITIL), created by Britain's Office of Government Commerce (OGC), provide the guidance that IT organizations crave.

In concert, TLAM and ITIL can help organizations reduce operational costs, improve transaction efficiencies, enhance customer experience, and better meet the business mission. Despite their complementary capabilities, few businesses have merged these two methodologies. However, organization leaders are beginning to recognize that, with thoughtful deployment, TLAM and ITIL can be woven together to make IT assets even more effective in serving and achieving the business mission and goals.

IT service management consists of many different parts that all add up to provide the solutions from IT to support the organization. Over the last 20 years, the IT infrastructure has become as important to the operation of an organization as the other classes of assets addressed in this book. IT assets are managed like other assets; however, they often use different terminology. IT people like to have their own language and have adapted terminology that fits the nature of their delivery of services. However, when we take the time to analyze the IT languages and terms they describe strategies and services that are not all that different from the maintenance strategies and services we have already been talking about throughout this book. This chapter will discuss some of the unique language used to describe IT service management while showing the parallelisms with other forms of maintenance and why it is part of this text.

15.1   INTRODUCTION

IT service management has become an icon of sorts as the global community encircles itself with ITIL and COBIT, just to name a few. The idyllic solution of optimal management of the services of IT has been the topic of many books, seminars, and actions. So, why consider IT service management as a chapter in a book on maintenance excellence? IT, like any other part of an asset-based organization, requires asset management that delivers the goal of successful production. What sets it apart, of course, is that through IT the assets life cycle is not much different from that of those big pieces of equipment that exist in the manufacturing field whose obsolescence is generally much higher and that are extremely expensive to maintain. IT assets, however, can have a relatively shorter lifespan in the organization but over time can be just as expensive. An organization's IT assets are focused on providing value to the company more through the services they enable in the equipment itself. That is, the assets are more like services augmented by the people who support the technology in providing those services. This is the foundation of IT service management: delivery and support of IT functions to the business through technology.

15.2   MAINTENANCE IN A SERVICES BUSINESS

Unlike a plant that produces products (e.g., oil, gas, cars, electricity), IT is definitely a services-based business since its only purpose is to provide technology to the business it services. So as we go about maintaining IT assets, we are in turn managing the services that support the business. IT has no function if the business has no use for information technology, so this idealizes what a services business is: its sole purpose is to provide service that enables the business.

Understanding that a service business is there only to support its customers, we can better appreciate the focus of maintenance in this type of structure. Maintenance must be done to ensure that we, as a service business, can provide the needed services our customers require. This is independent of whether the customer is internal to the organization. In the discussions here, the customer is the user of the IT service and may very easily be part of the same organization or company. Likewise, IT service business could be a department that manages IT for the same organization or it could be a separate company.

In understanding this premise we can quickly understand what the level of importance must be in the services business. Maintenance in a services business must be focused around providing a level of service that is appropriate to the customer's demands. This may entail servicing assets and equipment that provide services as well as servicing the services themselves. The assets that provide services, especially in IT, usually are layered across many component pieces such as in the network, which is made up of routers, hubs, switches, multiplexers, authentication devices, firewalls, bridges, and miles of cable. Within IT these services layer upon other services to form a complete offering to the customer. The service level to support the customer must be measured across all these components, which by definition is not additive but multiplicative. That is, 80% of component A when serialized with 80% of component B does not equal 80% for A + B together but instead equals 64% for A + B. This measurement is what causes the greatest problem in supporting services to a customer. We cannot just measure the maintenance of the single component, since it is not the single component that offers the service. Therefore it is crucial to measure the complete delivery of service as it is expected by the business.

To support this you must ensure both that your organization has the necessary tools on hand—tools that can support key business aims—and that these assets function efficiently to full capacity. However, securing smooth, always working IT asset functionality can prove difficult.

Add to this conundrum the global business environment, which, by its very nature, requires distributed IT capabilities that can function across the enterprise and across the geographic expanses of that enterprise. As a result, keeping track of IT assets enterprise-wide—where these assets are, how they are used, who is using them—can become an issue. Often, in just keeping your organization's technology assets up to date and functional, focus can drift away from the core mission in an effort to ensure IT assets work the way they need to.

Whether organizations are based in the corporate world or within a government agency, a number of challenges must be dealt with when it comes to business technology assets. First, most enterprise executives must confront a growing gap between what is actually in the IT asset portfolio and what is needed to achieve the business mission and support related strategies. But bridging this gap requires careful thought and planning, since pressures to reduce costs seem to grow increasingly stringent.

Additionally, after September 11, not only does it make good business sense for organizations to better account for and manage their assets and information, but it is also now often a legal requirement. An increasing number of national and international regulations and security directives must be complied with—for example, Basel II and the Sarbanes-Oxley and the U.S. Patriot Acts. Couple all of this with inadequate or skewed investment in staff and IT infrastructure, and the result is a misalignment between an organization's asset base and productive use of those assets.

15.3 UNDERSTANDING ITIL AS A BASELINE

ITIL is the most widely accepted approach to IT service management in the world. The library is a compilation of books developed for Britain's OGC that consisted of a cohesive set of best practices for managing IT in any organization. ITIL was the result of years of effort by a number of public and private sector firms and consultancies to develop a comprehensive collection of processes, frameworks, techniques, and tools from which to measure and manage IT.

The foundational baseline of ITIL is the delivery of service management and is centered on IT service delivery and IT service support. Each of these is again subdivided into process frameworks. IT service support consists of the service desk, incident management, problem management, change management, release management, and configuration management. The other side of IT service management is IT service delivery, which consists of availability management, capacity management, IT financial management, continuity management, and service-level management.

This chapter will not describe in detail the ITIL process and its framework but instead will attempt to explain how these various concepts can be redefined in the terminology of the plant maintenance realm. Readers who want more information on ITIL can find this through the OGC Web site (http://www.itil.co.uk/) or through the IT Service Management Forum (ITSMF) (http://www.itsmf.com/index.asp).

Yes, it is quite clear that the terminology between plant maintenance and IT is different. But what is most interesting to note is that the resulting high-level actions and activities are quite similar and align in quite similar ways. For example, a very simple approach to terminology differences between IT and plant maintenance can be seen in Figure 15.1:

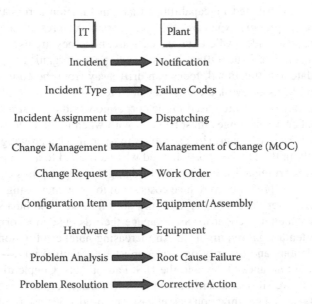

FIGURE 15.1 ITIL terms versus industrial terms.

- *Incident management* in IT is similar to a repair notification in the plant environments, where issues need to be addressed as they occur and productive operations restored as quickly as possible.
- *Problem management* in IT deals with problem analysis and root-cause determination through to problem resolution to correct deficiencies in the IT operations. This does not vary much from the plant and its definition for root-cause failure and corrective action, which addresses deficiencies in the operational characteristics of its equipment.
- *Release management* in IT deals with large or collective change to the IT environment in the same way a plant would approach a *turnaround* or *major overhaul*. These often consist of wholesale changes in the environments with large project efforts introduced to make the changes or a collection of changes that work in tandem to make for a large effort that is incorporated in one time slice.

These simple, high-level comparisons illustrate that ITIL is not very different from the traditional plant management methods for optimizing support services to the organization. ITIL still aligns with the full scope of asset life cycle and fits the capability maturity model for excelling within the scope of impact to the business. It is therefore evident that ITIL is an extension of these principles of uptime to the organization but with a tighter focus around a more service-based approach to the delivery of IT services. After all, IT is really a service-based component of the business. IT provides business services to the rest of the business so that it can be productive or delivers other services to the business in driving the core business functional operations. Maintenance really does the same thing, ensuring productive use of the equipment for core business functional operations. Both are services to the business, and both represent a necessary component of enabling production (business success).

15.4 APPLYING ITIL FIRST TO IT ASSET MANAGEMENT

Gartner estimates that enterprises that begin an asset management program experience up to a 30% reduction in cost per asset in the first year (e.g., people, process, and technology costs) and continued savings of 5% to 10% annually over the next five years. These savings can be found by recovering assets rather than having to buy new ones and eliminating unused assets that have costs involved (e.g., maintenance costs for unused equipment).

ITIL offers a framework of processes that can be used to develop and support an enterprise IT asset management (ITAM) program. Gartner Inc. analyst Patricia Adams stated, "IT Asset Management is 80% process and 20% tools." This reinforces why developing processes around the framework provided by using ITIL is the first step in implementing any ITAM solution or tool. You cannot have one without first implementing the other. ITIL provides the processes and identifies the people roles that should be defined so that you can develop a clearly defined and successful ITAM solution.

In an effort to better understand some of the features that will lead to a complete solution for the organization, it is necessary to consider the capability maturity model, which measures and identifies the level of capability an organization has reached with regard to its IT asset management practice.

The TLAM model is a way to describe IT asset management best practices within the framework of ITIL. The ITIL implicitly contains the maturity model concept in its presentation of best practices. Maturity of asset management practices is an underlying theme for many of the ITIL concepts. The ITIL functions defined in the following sections illustrate the interface between the asset management strategy and the business processes that support it. Each of these functions describes criteria for a successfully managed process and allows you to identify the level of maturity within a maturity model for which these functions apply.

15.4.1 CONFIGURATION MANAGEMENT

Broadly, configuration management is the identification, control, maintenance, and verification of configuration items (CIs) in a configuration management database (CMDB). Examples of CIs include individual assets, business processes, aggregations, and virtually allocated resources.

Within the maturity model an "aware" state requires the following developments within the organization:

- Instantiation of a CMDB
- Definition of CIs for shared infrastructure
- Conducting a discovery of shared infrastructure and entering results into CMDB
- Established control and verification of processes for the CMDB

In contrast, the "capable" state requires the following:

- Creation of interfaces among existing CMDBs
- Feeding a central CMDB with validated information
- Propagating identification, control, verification, and maintenance processes throughout the organization

15.4.2 INCIDENT MANAGEMENT

Incident management is the effort to minimize the impact events have on services and the mission. Incidents can be equipment failures in managed assets, service outages, acquisitions, or the sudden discovery of undocumented assets. Incident management deals with effects and symptoms. Incident management in IT is similar to the break-fix dispatch repair process in industrial maintenance practices. Both have a close link with a related function, problem management, that results in the analysis and resolution of causes. Often, problem definition will arise from the root-cause analysis of an incident.

Within the maturity model an "aware" state requires the following be in place:

- Proper maintenance service functions for the asset management system as well as assets within the mission dependency chain
- Incident reporting triggers for repair and maintenance processes
- Incident records that remain available to verify maintenance efforts

However, the "capable" state requires that incident management is closely tied with change and configuration management to track every asset's suppliers, maintenance history, methods, licensing, and usage.

15.4.3 Change Management

Change management is the regulation and oversight of requests for change and change processes. This applies to any changes to assets, processes, or organization. The "aware" state is implied in having a change process, but, more importantly, the "capable" state requires that the implemented processes are such that any change automatically results in inventory updates and triggers configuration management.

15.4.4 Financial Management

Financial management is the process of managing and reporting the costs, funding sources, availability of funds, budgets, and return on investment for assets. The maturity model, which identifies the "aware" state, requires that the financial reporting processes and supporting infrastructure exist such that groups and individuals charged with finance and accounting responsibilities can access all relevant data and that the asset costs, maintenance costs, and replacement costs are tracked with CIs in the CMDB. In contrast, the "capable" state requires that the financial management policies are explicitly part of all asset management processes and that the incidents and changes trigger financial effects and reporting without additional intervention.

15.5 MOVING INTO ADDITIONAL ASSET CLASSES

Asset and service management is a set of processes and practices used to manage the performance, in an optimal fashion, of all critical assets in accordance with the requirements and expectations of the organization. This can be accomplished through an effort in asset classification that delivers a common repository for assets and services within a common business model. That is, we enable a view of all assets from the perspective of the physical relationship, the service it performs, and the visible part it plays in the business.

Classifying assets in this way allows us to understand the criticality of a unique piece of equipment in relation to the scheme of the business rather than to the value of the individual asset. The ongoing innovation of assets themselves along with external forces (regulatory governance) is causing the business and organizations to implement tighter controls around assets. To manage the risks implicated by these changes, three areas of focus affect how we classify assets:

- The *interdependency* that exists between assets is on the rise; therefore, it is no longer practical to manage assets independently.
- The *boundaries* between asset classes are fading; therefore, it is necessary to treat equipment, buildings, and IT assets as an ecosystem where operational assets not only have embedded IT assets, but they are dependent on IT assets to function well and likewise IT assets are dependent on operational assets such as heating, ventilation, air conditioners, and power conditioners to function properly.
- Influencing pressures, such as globalization and regulation, have created the need for more transparency of assets throughout the asset's life cycle.

15.6 THE EVOLUTION OF TOOLS

Businesses need to more efficiently use and reuse existing technology. But, simultaneously, they must find a way to deliver services even more effectively. Total life-cycle asset management is one avenue companies can use to streamline efficiencies and cut costs. And TLAM practices provide a framework to assist with meeting these goals by helping IT staff better optimize and align IT investments that also support the enterprise's overall mission and strategies.

The TLAM methodology takes a holistic approach to asset management. It calls for reviewing virtually everything IT related. Beginning with IT strategy and planning, evaluation and design, acquisition and building, TLAM also looks at operation, maintenance, modification, and disposal across the enterprise.

Using TLAM, your firm can categorize asset classes based on similar management attributes. For example, assets might be categorized together if they are financial in nature, if they fit into a particular space within the IT structure, or if a particular user group tends to rely on them.

TLAM helps you look across your IT portfolio, enabling you to assess each asset throughout its life cycle. TLAM can be used to evaluate asset management strategies and best practices at each stage of life, with a focus on better managing total life-cycle cost.

For example, when it comes to looking at real properties, an organization might conclude after completing a total life-cycle analysis that constructing a data center is more attractive than leasing one. Or a chief information officer (CIO) working for a state agency might conclude that, based on maintenance cost projections, procuring a new system is preferable to repairing and maintaining existing servers.

Using TLAM guidance, structure is wrapped around IT asset management. A series of key IT management phases, which span the technology life cycle, are implemented systematically (Figure 15.2):

FIGURE 15.2   Asset management total life-cycle.

- The *strategy* phase focuses on gaining an understanding of an asset's role and the value it brings to the organization.
- The *planning* phase looks at how to best integrate assets into business plans.
- The *evaluate/design* phase occurs when assessing product performance or design.
- The *create/procure* process compares asset purchase requirements with asset capacity requirements.
- The *operate* phase includes all processes required to keep an asset operational.
- The *maintain* phase encompasses any process required to sustain an asset.
- The *modify* phase supports the effective reuse of existing assets.
- The *disposition* phase ensures that asset disposal complies with all security, legal, and contractual requirements.

ITIL also looks at managing the full asset life cycle and has been evolving to raise the level of applicability in support of a changing world and leveraging the best practices of system and service management. Its evolution has taken it from system-based through processes to the new scheme that is services based on managing and supporting technology assets. ITIL V3 is very much aligned with the concept of maintenance excellence: strategy, tactical, continuous improvement. ITIL V3 takes a full-scale approach to the definition of service management through the delivery of IT service by focusing on the best practices for service life cycle and the service capabilities within its principles. Under this new release of ITIL, it now moves from its infancy of system management through the definition of process-based service management to where it focuses on best practices throughout the service life cycle. This service and solution life-cycle management is composed of five core volumes:

- Service strategy
- Service design
- Service transition
- Service operation
- Continual service improvement

Each of these volumes has a focus at different levels on the overall target of improved service (maintenance) excellence. Using ITIL in its new form will align well with integration within the business to COBIT and TLAM to deliver the best tools that will fulfill the needs of the organization to manage the IT assets and services of the organization to its optimal value. It reinforces the alignment with the goals of maintenance excellence in supporting IT assets: strategic, tactical, and continuous improvement across the organization within the information technology realm of the organization.

15.7 SERVICE-LEVEL MANAGEMENT: THE OPPORTUNITY TO RECOVER COSTS

Service-level management consists of the processes of planning, coordinating, drafting, agreeing, monitoring, and reporting on *service-level agreements (SLAs)*. It also consists of periodic reviews of *service achievements* to assure that the identified services are maintained or improved in a cost-justifiable manner. The SLAs provide the foundation for managing the relationship between the service provider and the business (customer).

Through using these reviews the SLA can be used to identify where improvements have resulted in improved costs controls and overall business efficiencies (savings) that can be used to recover the costs of implementing these technology solutions. Furthermore, the SLA can be used as a basis for *charging* and assists in demonstrating what value is being received for the money.

One of the most effective ways of managing services is to manage the assets that derive or enable those services. Leveraging proven practices and maximizing capabilities of assets will allow for the further enhancement and controls that are needed to manage the IT services being provided and thereby meeting the drive of service-level management and supporting the overall mission of the organization.

Where TLAM and ITIL capabilities overlap, they can dramatically improve an enterprise's ability to meet its overall mission. TLAM provides a methodology for monitoring and streamlining IT assets and capabilities throughout each asset's lifetime, while ITIL offers insights on how to best implement and manage these assets in support and delivery of IT services. Organizations can weave together TLAM and ITIL strengths to generate an overarching view of their ITAM capabilities.

ITAM helps your organization better combine its financial and mission objectives. The benefits include the following:

- More efficient data sharing across the enterprise
- Enhanced asset visibility and security
- Improved planning around technology refresh and upgrade needs
- More efficient technology repair and deployment
- Lowering of total life-cycle costs
- A better view on audit compliance
- More uptime to support the mission
- Doing more with less

TLAM and ITIL have many areas of overlap. By merging the two frameworks, your organization gets the best from both. Use the TLAM model to address internal IT asset issues. And look to ITIL guidance for a view on how your IT organization is working and to identify where and how IT services can be better managed.

For example, your IT systems department might be most concerned about the purchase, use, and maintenance of a particular PC, while customers or internal end users, such as those in the tax department answering customer questions, care most about the service that the PC can provide. By merging TLAM and ITIL guidance and capabilities, you can better unite user and business interests—ensuring that hardware, software, and service requirements are more effectively met.

The benefits of merging ITIL and TLAM are the most useful—and powerful—when it comes to enterprise change management. Using the ITAM methodology, you can better plan and manage change, using the best industry standardized practices from beginning to end.

For example, for an organization to anticipate or document movement of its IT assets, there must first be an understanding of and ability to track the available technology tools. However, most organizations today rely on spreadsheet data or have to make numerous, ad hoc phone calls to find this information. Because of the inherent inefficiencies in this approach, competing initiatives to capture the same information often result. Using TLAM and ITIL in tandem can help coordinate asset and change management efforts and drive efficiencies both in near-term projects and long-term processes and technologies.

TLAM's asset repository, a component of the TLAM create/procure process, overlaps with ITIL's CMDB, which administers IT infrastructure change. TLAM's asset repository data—that is, information about an organization's IT tools, where they are, what they are used for—form the cornerstone of ITIL's change management database. Without these records, the configuration management database is incomplete and ultimately ineffective in supporting essential ITIL processes. Additionally, asset repository data can support other essential ITIL processes, including configuration management and release management.

15.8 SUMMARY

The IT service management life cycle consists of two very unique pieces: the assets, which are not all that different from other enterprise assets; and the processes, which as we see are also quite similar to the processes we would use for other maintenance solutions. Maintenance manages keeping assets serviceable, and in this regard TLAM is a fundamental part of that. Allowing maintenance to start to focus on service to the organization and how the physical assets of that organization support delivering service to the organization is the next level of understanding in driving to maintenance excellence. IT is, of course, already a services-based part of the business, and whether it is serving a single customer or multiple customers, the rules of operation remain the same. Providing good asset management and good service support results in optimization of the assets that deliver IT services.

REFERENCE

Adams, P. Three tools of an asset. Gartner, Inc., Management Program. 2001.

16 Information Technology Asset Management

Ron Green and Brian Helstrom

CONTENTS

The focus of this chapter is to look at what information technology (IT) assets are and why we would want to manage them. We will spend some time here to gain an understanding of why we include IT within the realm of maintenance excellence and why IT asset management (ITAM) has become a part of the enterprise asset that is used to deliver production to the organization.

The ability to manage and optimize IT assets is critical to delivering cost-effective support to the business. A multidisciplinary approach to managing these assets and processes provides IT organizations with the capability to minimize costs while maximizing return on assets and achieve targeted service levels of support to the business.

All IT assets have a full life cycle similar in form to the other assets within an organization. They move through the same cycle of events from planning, to procuring and implementing, to maintenance and upkeep through to disposition and replacement. The difference is that IT assets generally expend their assets life cycle over a much shorter time period and are undergoing more radical changes to support the newer technology and the changes in the way we use technology. But, like other plant assets, IT assets require continuous service and maintenance to operate effectively and at peak performance.

16.1 INTRODUCTION

While we are on the topic of enterprise assets it is important to consider one of the expansive areas in every business today: IT assets. The challenge here is not so much in capturing these assets as it is in managing them and understanding the level at which to manage them. Within the IT division of your company there are many views concerning what an IT asset is and what it really consists of. But in the next few pages we are going to attempt to elaborate on exactly what an IT asset is and why we need to consider these in our asset management strategy.

Let's review what comprises a typical enterprise asset. An enterprise asset is something in the organization that we need to manage because it is critical to the viability of our business and without which our company will suffer in either performance or capability to deliver to our customers. Therefore the enterprise asset has a value both in terms of financial consideration and in terms of where it lies within the company's ability to deliver its output to the marketplace.

In today's world this clearly describes how IT assets belong within the overall enterprise view. Most organizations today cannot survive or deliver to their customers without the help of technology. Information technology is embedded, overseeing, and supporting the ability of an organization to deliver to its customers.

To achieve a sustainable ITAM program, an organization must leverage and apply four guiding principles:

- Reduce the total number of ITAM tracking methods.
- Simplify and standardize processes across the organization.
- Consolidate control and accountability for IT assets.
- Establish a "single source of truth" for ITAM data.

The fundamental component for a successful transition to ITAM and long-term health of an enterprise asset management (EAM) program will be the governance component. The asset management maturity model tasks and capabilities illustrated in this chapter and the recommendations for the way forward are heavily dependent on a strong governance structure within the enterprise.

16.2 WHAT IS ITAM?

ITAM is exactly what it implies: the management of the assets that make up the workings of the business's underlying technology delivery capabilities. These include the software, the hardware, the network, and the data—each with its own unique level of importance to the business. But ITAM is more than just managing all the computer equipment for the information technology group. It is also about managing the integrated technology that supports the business, from the networks and equipment that manage the machinery in the plant to the portable devices used to update inventory and logistics in the warehouses. IT is no longer a back office or management tool; it has become fundamental in the operations of an organization. As computer and information technology becomes more integrated into the stream of every part of the business, it now affects the tradespeople on the shop floor as well as the production workers' ability to do their job in managing business; ITAM is becoming mainstream.

16.3 WHY INCLUDE ITAM IN MAINTENANCE EXCELLENCE?

For assets to have longevity to the organization we need to assure that they are operating at their peak performance. To do this it is important to maintain the equipment and perform maintenance tasks on them. The thing that makes maintenance on an IT asset different is we do not often use wrenches to do the maintenance. But this does not change the fact that maintenance must be performed. Therefore, understanding that maintenance is required on IT assets and that IT assets are core to the excellence of business performance, then why would an organization not want to include IT asset management as part of the drive toward achieving maintenance excellence? If we are planning on performing a standard of excellence in maintenance, then it should also apply to the IT assets. Maintenance excellence in IT is just as important as maintenance excellence in the manufacturing environment. In fact, a number of the processes we use to manage maintenance of equipment are also the same as in managing IT. We just use a different label for it.

IT assets are now widespread throughout most organizations and are already expanding within the very workings of the shop floor of the production plant. Through SCADA and other statistical control systems that use microprocessors and are running on common networks, the plant floor has become more capable of sharing its statistics of operation with the management stream for MIS reporting and analysis. These systems are now integrating with other major components of the once isolated interactions of management information systems (MISs) to provide detailed information to the plant and production engineers and to offer more control of the delivery processes of an organization. As the shop floor starts to look more like IT and IT starts to embrace the operations of the plant floor, we begin to respect the need to include such technology assets as an integral part of the productivity of plant operations. This, by its very nature, implies the need to include IT assets within the overall maintenance excellence program.

IT assets today do not act in singularity but consist of complex integrated components of hardware, software, networks, and data. As such, this complexity becomes very limiting when any one of these component parts starts to perform poorly or fail. Proper maintenance can detect and prevent such failure and keep your systems from running poorly. Like the systems in a car, the systems in IT must all work together to keep the company moving. Many people have experienced frustration when a car component starts to perform poorly and fails as a result of delaying necessary maintenance. Many have heard the mantra, "Pay me now or pay me later," from the mechanics when it comes to maintaining a car. As a good maintenance strategy on a car helps to prevent failures and keep your cost lower, the same too is true regarding all asset maintenance, including IT asset maintenance.

IT asset maintenance is fundamentally becoming part of the overall operational plant. Many organizations are realizing that there are synergistic advantages in combining operational assets management with IT assets management, especially the underlying IT infrastructure that supports the management and control of many of the operational assets. To improve manageability, many operational assets are taking on IT attributes such as microprocessors, operating systems, and Internet Protocol (IP) addresses. Plant assets such as generators, power meters, and instrumentation are increasingly being networked and managed with IT software. Statistical processing controllers have been integrated into the shop floor for a great many years and are now being networked to provide continuous feedback to the business. Since these plant assets behave like IT assets, there is an opportunity to leverage IT business processes such as software distribution and patch management onto the shop floor to provide better overall asset management. This reasoning can be used to combine IT asset management with operations maintenance excellence.

16.4 WHAT IS THE VALUE OF ITAM?

IT assets have become an integral part of the delivery of production to any plant for the managing of orders and shipments to the functional delivery of running the machines through automation. IT assets have spread throughout the organization, and managing these assets and their subcomponent parts in an effective manner can make the difference between a successful organization and one that will become obsolete or fail to survive.

IT assets are now, more than ever, an integrated component of the organization, and, with the need to find efficiencies in delivering productivity in a global economy, IT technology and its implementation to the productivity chain have become inseparable.

ITAM brings value to the organization by providing a means to manage these embedded and integrated technology systems within the organization and facilitates having a single set of strategies within a business for managing the assets for IT.

16.5 PROCESS IS THE KEY TO SUSTAINED PERFORMANCE

A lot of time has been spent in forming and managing IT assets. In fact, a whole library of process frameworks has been built for the sole purpose of managing IT assets. This framework of processes has been incorporated into the Information Technology Infrastructure Library (ITIL) and has become the de facto standard for the processes involved in the management of IT assets. The entire framework is not just for asset management but entails much more, since it delves into the delivery of service management to the community that uses the IT assets within an organization (more on this in Chapter 15). ITIL has, in its own growth, adapted some of the very concepts that have been covered in this book in striving to achieve the ideas of maintenance excellence. The use of strategic, tactical, and continuous improvement to the managing of the IT assets and the delivery of services to the organizations are fundamentally embedded in the principles and practices of ITIL.

16.6   UNDERSTANDING THE ROI OPPORTUNITY

The return on investment (ROI) of ITAM is not all that different from the ROI for other enterprise assets. The criticality of the IT assets and the need to track and manage the maintenance of those assets can identify what the gain of proper management is of those assets. In evaluating the ROI it is important to assess the value placed on needing these IT assets. If these assets are not working, what is the impact to the business? What are your current maintenance costs for managing these assets, and what improvement would you make in better managing these assets? Understanding the ROI on any ITAM implementation means understanding the value the IT assets have within the delivery and management of the business as a whole.

Another way of stating the ROI equation of IT is to understand the cost of the IT infrastructure not existing. Should an organization be unable to create product, pay employees, provide regulatory reports, or purchase/manage inventory, the impact would be quite far-reaching. For this reason, the ROI equation must be considered in a broader context. Owing to this extended impact, the solution must be evaluated against business continuity. In addition, from the outset, redundancy, backup, and manual processes should be part of all planning activities.

Poorly managed IT can result in much money wasted on licensing of software, improper care, and maintenance of computer equipment, which could lead to early replacement or lost assets. Proper management of IT assets and resources can save a large company millions of dollars.

16.7   WHAT TYPES OF TOOLS SHOULD YOU CONSIDER?

Many of the tools that can be used for managing enterprise assets are the same tools we can use for managing IT assets. Since the issues of IT assets are that of managing service to the various components and supporting the productive activities for business, it is just as critical to manage these assets. The differences in the tools needed for assets management extend beyond just managing the physical assets, they also encompass the complete configured system of components, which includes software, hardware, networks, services, and data.

In addition, a number of specialized tools can be leveraged to assist in managing IT assets. Tools that can be loaded onto a system that will go out and discover all the assets on a network can be quite helpful in getting started. Generally, these same tools can be used for managing the configuration of the software and devices on the network. IT has been doing this type of management to their environment for years, and this can be easily expanded to manage these new application devices that are appearing in the plant operations too.

Even at the time of this writing, we are seeing further convergence of tools that will discover and monitor the IT infrastructure. It is anticipated that there will be further convergence in the tool sets and that software asset management capabilities will become more refined over time, adding more comprehensive capabilities.

ITIL is a tool with many ideas and processes that can be implemented to manage the services and components associated with the IT assets throughout the organization. Within ITIL are contained concepts around housing configuration management information in a database. Understanding what configuration to use extends beyond just the physical components but also goes into adding information about how the parameters and setting within that asset have been set up to operate to maximum effectiveness for the organization. This is because technology assets are not always the same even though they may look and act the same: a computer on one desk may be very different in its settings from the same computer on another desk.

Standardization of how IT assets are deployed, managed, and configured is an important tool that a company may wish to deploy across its organization. This would include tools that might restrict the asset from being used for purposes other than that for which it was intended. For example, restricting a computer from browsing outside the bounds of the organization (i.e., browsing the Internet) might be needed to assure the computer is not being used inappropriately, which could affect its ability to do what it was intended to do.

16.8   HOW DO YOU BEGIN AN ITAM PROGRAM?

Beginning an ITAM program can seem overwhelming. One fundamental driver has to be continuous improvement. Understanding that you can't solve all past ills in a single step will go a long way toward setting expectations for all involved. Establishing the roles and empowering a person or position to create and enforce policy is key. These policies should work within the existing job frameworks as much as possible. Creating redundant, cumbersome processes will result in poor results.

Starting an ITAM program begins with identifying two very important pieces: the depth of the detail required and the breadth of components. When we talk about depth we are referring to what equipment components will be serialized. Some organizations may wish to keep track of lower levels than others; for example, some may want to serialize the mouse on every computer, whereas others will consider the mouse to be a part of the main asset, the computer. Even the specific computer may be considered only a part of the workstation; therefore, it will not even be serialized for the enterprise asset. This is one of the important decisions that need to be made in starting an ITAM program. The breadth of components refers to how much of the organization will be included into the program and what asset types will be considered. For obvious reasons it is hoped that the breadth becomes the whole of the organization and all of its assets, but in the first phases of the rollout it might be appropriate to limit the breadth to specific assets in one location or department and phase in the changes for managing IT assets over a period of time to make the transition to the whole organization more feasible.

Implementing an ITAM program must start from the top and have the full support and understanding of the benefits of the program in order to drive down through the organization to the implementation and assure the success of the program. Establishing an executive sponsor of the initiative is critical to implementing any new program to the organization, and beginning an ITAM program is no different.

The second challenge is establishing the team that will deliver and establishing the boundaries (breadth) of what the team is to work from. Attacking the entire organization, depending on the size, may be a daunting task, so having a strategic approach to discovering and implementation is important to assuring the success of the implementation. It is like the old adage: How do you eat an elephant? One bite at a time. It is important that the bites that you take are small enough to be digestible but big enough to make a visible difference.

It should be recognized that, depending on the organization and the state the current organization is in, the process of implementing ITAM could run from several months to several years. So it is important that you have a strategy that will support this long-term initiative and that you are able to see the tangible results that such implementation achieves and its intended benefits. A steady path toward maintenance excellence will not happen overnight, but the rewards of achieving it will definitely benefit the bottom line of the organization's financials.

To facilitate a further understanding of an approach to beginning an ITAM program the capabilities and tasks as outlined next are a possible roadmap for the enterprise to successfully launch an asset management program.

The idea of process maturity model is central to several prominent methodology structures that are heavily used in the IT industry. This concept is a useful way to show measurable progress in the various areas that make up the total life-cycle asset management (TLAM) model for ITAM.

"Begin with the goal in mind" is a central tenet of most modern quality management philosophies. If an organization has no clear goal or purpose, then it becomes impossible to define measurable key performance indicators (KPIs) and to articulate the processes that lead to success. Leveraging a maturity model is one method for driving toward a measurable means to reaching this goal that can be universally understood within the organization.

The maturity model defines "capable" in terms of the definition and communication of goals and "aware" in terms of understanding the targeted direction. The steps that must be taken to achieve a "capable" mission definition are as follows:

- Disseminate the goals of asset management to key stakeholders.
- Disseminate the goals of asset management throughout the agency.
- Prioritize the mission elements.
- Select the technical goals.
- Select the broad business process goals.

Whereas a mission statement can be painted in very broad brushstrokes, with goals that are very much "end-state" in their orientation, objectives are more process oriented. Objectives should begin to answer the question, "How do we achieve our goals?" In other words, objectives are the mile markers and road signs that tell us we are on the right route to achieve the mission.

The following are the steps that must be taken to achieve a "capable" state for the definition of objectives for implementation of a program:

- Gain consensus on specific business process objectives.
- Consider and identify the services and service levels that are needed to meet the business objectives.
- Establish measurable performance indicators.
- Set the timeline for near-term objectives to be achieved.

Once the asset management mission and objectives are established, then the real heavy lifting of establishing the asset management processes begins. These processes will provide the structure to enable members of the organization to support the business objectives that result in the successful achievement of the mission.

Establishing the asset management framework will address the governance issues that were identified as being a significant gap for the enterprise in terms of establishing a successful asset management program. Since the only governance framework is currently localized and fragmented, the creation of an enterprise-wide asset management framework will need to proceed through the "aware" stage first to arrive at the "capable" stage on the maturity model.

The following are the steps that must be taken to achieve an "aware" state for the asset management framework:

- Define the asset management framework, including process roles and standard operating procedures (SOPs).
- Define any necessary services and service-level agreements (SLAs) that are needed to meet the asset management objectives (services may be either internally or externally provided).
- Disseminate the asset management framework to your stakeholders.

Having taken the previous steps to achieve an aware state, the following steps must be taken to achieve a "capable" state for asset management framework:

- Use a feedback spiral to remove process failures from the asset management framework.
- Review and revise policies, procedures, and roles on a regular basis to reflect changing business conditions.
- Disseminate the asset management framework and processes agency-wide.
- Define conforming processes for each mission element and organization unit.

Once the management framework is established, then the true asset life-cycle management can begin. TLAM would apply nicely here and is a fundamental concept that should be included. As with the previous steps, it is necessary for you to progress through the "aware" state to reach a "capable" state on the maturity model.

Starting with the "aware" state, the following are the steps must be taken to achieve this for managing the asset life cycle:

- Identify asset tracking repositories.
- Consolidate asset tracking information.
- Nominate a final asset management repository (one source for the "truth").

Progressing to a "capable" state for managing the asset life cycle requires implementing the processes to support every phase of the asset management life cycle and implementing metrics to track costs, disposal, and recovery.

The verification and audit processes should be part of the asset management framework that was established. These processes use the asset data established in the asset tracking repositories and help to ensure that the data have a high degree of accuracy and reliability.

The verifying and auditing of asset information follows again within our maturity model with the following steps that must be taken to achieve an "aware" state:

- Update supplier contact and contract information annually.
- Track software license volume.
- Track procurement volume.
- Check software licenses as part of any IT asset change.
- Schedule and execute audits on major assets.
- Track recognized deficiencies.

Once the achievement of the "aware" state has again been reached, the following steps must be taken to achieve a "capable" state for verifying and auditing asset information:

- Track the asset life cycle, capturing asset class, asset source, and supplier.
- Track performance against service-level agreements.
- Track software licenses by usage and deployment.
- Track software licensing exceptions and compliance failure.
- Enforce license policies and communicate policy enforcement agency-wide.
- Establish agency-wide audit procedures that are codified and repeatable.
- Trigger automatic audits when unauthorized IT assets connect to the network.
- Work continuously to remove conflicts among asset management, agency mission, and legal requirements.

If all stakeholders have a high degree of confidence in the accuracy and reliability of asset data (because of the processes that are in place), then it becomes necessary to support the demand for reporting and analysis of capabilities.

The beginning of this is through the "aware" state to analyze and provide asset information. The following steps must be taken to achieve this state of "aware" for analyzing and providing asset information:

- Document and implement existing automated reporting.
- Implement repeatable analysis functions.
- Implement project-based reporting (history, storage, disposal, value, cost, licenses, assets by location, owned and leased assets, contract or license expiration, inventory variance).

Like all good steps it is expected that, once you move through the "aware" state, you will start to take the following steps to achieve a "capable" state for analyzing and providing asset information:

- Implement automated methods for most reporting.
- Expand analysis automation.
- Implement standardized asset reporting (similar categories as project-based reporting).

Once standardized asset management reporting and analysis tools are developed, then the processes of continuous improvement by management monitoring of performance metrics can take place.

To achieve an "aware" state for evaluating asset management performance, the organization must consolidate and document existing measurements and performance indicators to track asset management efforts.

The steps to be taken to achieve a "capable" state for evaluating asset management performance are to determine measurements and performance indicators that match decision support requirements and then report and disseminate measurements and performance indicators to key stakeholders.

16.9   ORGANIZATIONAL CLARITY AND CAPABILITY

Direction and control roles define how the organization establishes accountability for asset management. Reaching the "aware" state requires the departments to have an assigned owner of departmental assets, whereas achieving a "capable" state requires the departmental asset owner to be assigned to manage central IT assets.

Execution roles are used to define responsibility for asset management within the organization. The steps needed to reach an "aware" state are the following:

- Asset management roles and responsibilities have been discussed but may not be fully agreed to by all parties.
- Connections between actual asset management roles and mission and role descriptions may be vague.
- Measurements exist, but there may be few meaningful measurements.

In contrast, the steps to achieve a "capable" state are the following:

- Asset management execution roles and responsibilities are agreed to by all parties but may not yet be documented.
- Responsibilities generally tie to mission and role descriptions.
- General qualitative measurements exist.

The area of skills and desired behaviors refers to the skills and desired behaviors needed for an organization to perform at the various levels of asset management maturity. The "aware" state requires that the asset management skills are good but relate to only some of the data requirements and that the skills are primarily technical.

However, a "capable" state requires the following to be included:

- Determining technical and financial skills needed by asset management resources
- Determining the technical and financial skills needed by asset management resources to be effective in the function of asset management
- Documenting technical and financial skills of asset management resources
- Documenting the technical and financial skills needed by asset management resources to be effective in the function of asset management
- Comparing technical and financial skills needed with technical and financial skills that the asset management resources have, identify any gaps, and develop training to address any gaps
- Implementing technical and financial skill training of asset management resources

16.10   MEASURING AND IMPROVING THE PROCESS

16.10.1   MEASUREMENTS

Metrics provide the means to determine the performance of our processes, to set meaningful improvement goals, and to measure whether improvements to the process really made a difference. Discussions of product or service expectations with the process customer helps focus process performance measures on those characteristics the customer values. An understanding of the causes that affect process performance provides insight that aids in setting preventive measures and defining requirements in procedures. Since metrics drive performance by focusing attention on ways to improve the measure, care should be given not to suboptimize the system by applying flow or efficiency measures on overcapacity processes.

The organization should carefully monitor the actual metric data values, especially in the first few reporting periods after the metric has been established. It is important to validate the data gathered to ensure an accurate process assessment. It is also important to validate the consistency of measurement collection methods. Prior to taking actions to stabilize the process, the organization should determine the current "baseline" performance for each metric. The organization can then see the result of stabilizing actions by comparing the baseline performance with the stabilized performance.

The following steps must be taken to achieve an "aware" state in the measurements arena:

- Directorate and mission-based asset management performance are formally tracked.
- Begin the project to identify key asset management measurements.
- Meet with the process customer and establish the end-item product or service measures. Negotiate the minimum acceptable performance levels for each. *Minimum acceptable levels are not to be confused with goals.*
- Define and document each measure by clearly describing what is being measured and the calculation formula to be used.

The following steps must be taken to achieve a "capable" state in the measurements arena:

- The measurements feed the decision support loop for process mission and objectives.
- The measurement crosses directorate and process boundaries and addresses efficiency and performance.
- A repeatable and scheduled process gathers and aggregates project-level measurements.
- Measurements include business and operational data.
- Communicate with the "creators" and "collectors" of the measurement data to confirm the definitions and to agree to a method of data collection that ensures the consistent use of the definitions over time.

16.10.2 Understanding Customer Satisfaction with the Process

The actual tasks and actions taken to complete a process must be documented so they can be internalized across the organization. They must be repeatable to the point that anyone, given the proper skill and experience, can complete the process correctly and report customer satisfaction on a repeatable basis. Documenting process steps is the piece of the puzzle that is most often left out when defining process. Many organizations develop process maps but fail to document the steps in understanding customer satisfaction. The result is a process that does not accurately reflect the steps and tasks that really occur throughout the process.

The following steps must be taken to achieve an "aware" state in understanding customer satisfaction with the process:

- Gather historical customer satisfaction data.
- Gather and document baseline customer satisfaction information.
- Identify all contractual, regulatory, and program requirements that apply to the process and customer satisfaction.
- Identify any undocumented customer needs and expectations.

However, to achieve a "capable" state in understanding customer satisfaction with the process, the organization must take steps to implement repeatable customer satisfaction reporting and include the nonfinancial costs of performance.

16.10.3 IMPROVING THE PROCESS

An improvement plan should be documented, accessible, and communicated to all involved parties to ensure that progress can continue even if resources are reallocated. It is necessary to ensure a balanced set of metrics by reviewing customer measures, preventive process measures, and product control point measures. The goal is to create a balance to identify when you are improving one at the expense of another while focusing on the business strategy. Further, it is necessary to identify how each of the major cause elements will be managed. Controlling the right causes will enable an organization to prevent or predict errors before they occur. Items such as tool setting or employee skill level could be identified as a requirement in a procedure. This could result in other causes needing to be measured, inspected, or audited.

Process stability and process improvement are process characteristics that are calculated from performance data using easy-to-learn statistics. These are not a subjective assessment of a process. A stable process does not exhibit the unusual data values or patterns that indicate the presence of special causes. Once the process is stable, analyzing the variation of process data helps identify what to control to meet the requirements of the process. The stable process performance is compared with requirements and aids in the decision on whether to invest in improvements.

Once annual metrics have been set, then multiyear metrics or comparative thresholds for processes can be considered, and the organization can treat each potential area of improvement as a problem/issue to be carefully analyzed and ultimately improved.

To achieve an "aware" state in improving the process the organization needs to assure the process is addressed when incidents make process failures evident and that process improvement begins when technical or financial issues are apparent.

The following steps must be taken to achieve a "capable" state in improving the process:

- Complete routine process improvements.
- Establish measures for the asset management processes.
- Implement a repeatable process for trend, control, and Pareto reporting.
- Include personnel performance as a discrete analysis and is included in process improvement.
- Determine the process average for more than three reporting periods. This provides a representative summary of past performance that can be compared with performance after improvements are made. The resulting improvement in performance can be easily calculated using this baseline.
- Develop an improvement plan.
- Initiate the actions in the improvement plan to make the selected process changes.
- Communicate process changes to all stakeholders.
- Document lessons learned.

16.11 DELIVERY OF EXCELLENCE

Having described the work breakdown structure of tasks that are a necessary part of building to "capable" on the asset management maturity model, the following steps are needed to address the major "gaps" that were previously identified and move to a higher level on the maturity model.

16.11.1 STEP 1: GOVERNANCE

The steps to forming a good governance model consist of the following:

- Form a governance body with decision-making authority that can develop, prioritize, and shape asset management efforts to support agency-wide objectives.
- Gain consensus from key stakeholders on the structure and scope of the governance body.
- Form an agency-spanning committee for asset management efforts.
- Define the channels for communicating customer needs and policy acceptance to the committee and for garnering process and status information from the committee.
- Charge the governance body to develop, implement, and control asset management vision, strategy, and processes.

16.11.2 STEP 2: ORGANIZATION ASSET MANAGEMENT VISION

The governance body needs to do the following:

- Define the asset management vision.
- Define the asset management mission.
- Establish the asset management objectives.
- Gain stakeholder consensus on the asset management objectives.

16.11.3 STEP 3: ORGANIZATION ASSET MANAGEMENT STRATEGY

The organization asset management strategy will consist of the following:

- Define the asset management framework, including asset management policies, business process roles, and responsibilities.
- Define any necessary services and SLAs needed to meet the asset management objectives (services may be either internally or externally provided).
- Disseminate the asset management framework to stakeholders.
- Define high-level KPIs and process metrics to allow measurement and monitoring of asset management performance.

16.11.4   STEP 4: ORGANIZATION ASSET MANAGEMENT PROCESSES

The organization asset management processes will do the following:

- Establish a central repository for asset management data that becomes the "single source for the truth." (This will likely mean selection of a single integrated technology tool.)
- Define necessary SOPs and work instructions to support the agreed upon policies, roles, and technology tools.
- Use system engineering processes to consolidate process redundancies caused by systems with duplicate or conflicting roles.
- Define data standards and naming conventions.
- Map existing systems to data types, asset classes, access types, and locations.
- Perform initial baseline physical inventory.
- Populate the central data repository from "cleansed" asset data based on the physical inventory (this can include verified or reconciled network discovery).
- Implement business process controls for asset management processes to ensure complete and usable information.
- Implement the KPIs and process metrics that were defined in the asset management strategy, to include the reporting process as well as management roles and responsibilities for monitoring and corrective action.

16.11.5   STEP 5: RATIONALIZATION OF PROJECTS AND SYSTEMS

The individuals and divisions within the organization must recognize the challenges that exist within asset management. In an effort to address the asset management challenges, divisions have to initiate multiple asset management programs and systems. Unfortunately, the uncoordinated and nonintegrated nature of these efforts is agitating rather than resolving the organizational asset management challenge. After addressing governance, vision, strategy, and processes, the organization should be able to rationalize existing asset management projects and systems.

Large systems and collections of systems, such as asset management within the enterprise, are cocreated by teams from various disciplines and departments. System development is an emergent process. The lack of a central control in this process may result in conflicting systems and processes that need to be simplified and connected together to support the organization's objectives. Rationalization is the process of examining current efforts against a developed course of action and determining which efforts support the course of action, which efforts detract from the course of action, and what additional efforts need to be included in order to make the changes needed to gain the desired capabilities. From the rationalization analysis, an executable plan is developed to remove systems and projects that do not support the organization, to accelerate efforts that do support the organization, and to expand existing efforts or initiate new efforts to fill any gaps.

16.12 SUMMARY

IT assets are not just found in the big data centers or offices but have infiltrated into the very core of manufacturing and plant floor operations. Through statistical process controllers, SCADA networks, and handheld devices for controlling inventory, IT assets are fundamental to the success of all plants and businesses. Proper management and maintenance of these assets is essential to the success of the business. Having an effective ITAM solution in place that aligns with the TLAM principles will fulfil this need and support the needs of the business.

ITAM is only a subset of the overall integration of EAM and the TLAM for the enterprise with a specialized focus on a technology that is young and explosive in its impacts to the organization. ITAM and TLAM need to be part of every company's efforts in managing its assets and achieving maintenance excellence.

Through this rendition it is hoped that you now have an understanding of why information technology is part of the five asset classes any organization needs to consider when looking for the management of IT assets and driving toward maintenance excellence. IT has become and will continue to expand its role in the successful organizations of the future, and the importance of managing the maintenance of those IT assets through the life cycle of the asset is the foundational philosophy behind this book and TLAM for ITAM.

17 Achieving Asset Management Excellence

Don Barry
Original by John D. Campbell

CONTENTS

17.1 INTRODUCTION

The preceding sections described: the evolution from reactive to proactive maintenance, managing equipment reliability to reduce failure frequency, and optimizing equipment performance by streamlining maintenance for total life-cycle economics. In this chapter, we look at the specifics of implementing the concepts and methods of this process.

We take you through a three-step approach to put a maintenance management improvement program into place and achieve results:

- **Step 1: Discover:** learn where you are in a maintenance maturity profile, establish your vision and strategy based on research and benchmarking, and know your priorities, the size of the gap, and how much of it you want to close for now.
- **Step 2: Develop:** build the conceptual framework and detailed design, set your action plan and schedule to implement the design, obligate the financial resources, and commit the managers and skilled staff to execute the plan.
- **Step 3: Deploy:** document and delegate who is accountable and responsible, fix milestones, set performance measures and reporting, select pilot areas, and establish detailed specifications and policy for procurement, installation, and training.

Finally, we review the why's and how's of managing successful change in the work environment.

A friend who travels extensively in his work in the mining industry describes his most anxious moment abroad: "I was just back in San Paulo from the Amazon basin, driving and not paying attention, when it dawned on me that I hadn't a clue where I was. It was dark, I don't speak Portuguese, my rental was the best car in sight, and I was sure I was getting eye-balled by some of the locals." Luckily, he got back to familiar territory, but not without a lot of stress and wasted effort. Like this intrepid traveler, you need to keep your wits about you to successfully reach your destination. So, to achieve maintenance excellence, you must begin by first checking where you are.

The cornerstone of this described approach is the maintenance excellence pyramid (Figure 17.1). The most important principle embodied in this approach is that the successful implementation of the upper layers of the pyramid depends on a solid foundation having been laid at the lower levels. The pyramid is based on original material from the Coopers & Lybrand Consulting Library John Campbell's book *Uptime: Strategies for Excellence in Maintenance Management* and is further expanded in this book.

The process for achieving maintenance excellence is depicted in Figure 17.2.

Before you implement reliability improvement and maintenance optimization, you must set up an organization or team mandated to effect change. You will need an executive sponsor to fund the resources and a champion to spearhead the program. You will need a steering group to set and modify direction. Members are typically representatives from the affected areas, such as maintenance, operations, materials, information technology, human resources, and engineering. A facilitator is invaluable, particularly one who has been through this process before and understands the shortcuts and pitfalls. Last, but certainly not least, you need a team of dedicated workhorses to execute the initiatives.

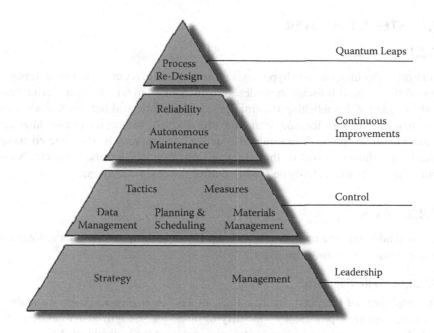

FIGURE 17.1   Maintenance excellence pyramid.

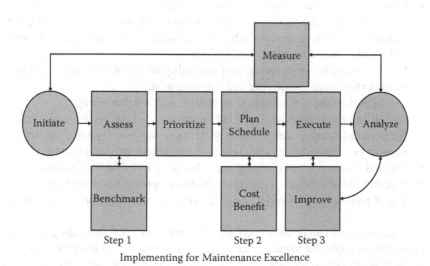

FIGURE 17.2   Process for achieving maintenance excellence.

17.2 STEP 1: DISCOVER

17.2.1 INITIATE

Initiating is defining an early hypothesis for what you are trying to improve. It is getting off the mark. It is establishing design principles from which the initiative must work or achieve. Establishing the final project structure, staking out a work area, conference room, or office, and setting the broad "draft" project charter are also part of this mobilizing phase. You would also determine who may be the affected stakeholders and should be part of the initiative. You then develop the "first-cut" work plan, which will undoubtedly be modified after the next step, *assess*.

17.2.2 ASSESS

This step addresses the question "Where am I?" and follows a strict methodology to ensure completeness and objectivity.

17.2.2.1 Self-Diagnostic

The employees of the plant, facility, fleet, or operation give feedback on maintenance management proficiency—identifying areas of strength and weakness at a fairly high level, based on a standard questionnaire. One is shown in Maintenance Strategy Assessment Questionnaire Appendix 24, designed as an initial improvement assessment of the following:

- Current maintenance strategy and acceptance level within the operations
- Maintenance organization structure
- Human resources and employee empowerment
- Use of maintenance tactics (e.g., preventive maintenance [PM], predictive maintenance [PdM], condition-based maintenance [CBM], run to failure [RTF], time-based maintenance [TBM])
- Use of reliability engineering and reliability-based approaches to equipment performance monitoring and improvement
- Use of performance monitoring, measures, and benchmarking
- Use of information technology and management systems with particular focus on integrating with existing systems or any new systems needed to support best practices (e.g., document management, project planning)
- Use and effectiveness of planning and scheduling and shutdown management
- Procurement and materials management that support maintenance operations
- Use of process analysis and redesign to optimize organizational effectiveness

The questionnaire can be developed with different emphasis, depending on the area of focus. For example, for maintenance optimization, research could be conducted on what the best practice companies are doing with expert systems, modeling techniques, equipment and component replacement decision making, and life-cycle economics. Questions are posed to reflect how these best practices are understood and adopted. This self-assessment exercise builds ownership of the need to change, improve, and close the gaps between current and best practices. The replies from the questionnaires are summarized, graphed, and analyzed and augment information from the next activity.

17.2.2.2 Data Collection and Analysis

While the questionnaire is being completed, gather operating performance data at the plant for the review. The data needed include the following:

- Published maintenance strategy, philosophy, goals, objectives, value, or other statements (design principles) that must be adhered to
- Organization charts and staffing levels for each division and its maintenance organization
- Maintenance budgets for the last year (showing actual costs compared with budgeted costs, noting any extraordinary items) and for current year
- Current maintenance specific policies, practices, and procedures (including collective agreement, if applicable)
- Sample maintenance reports that are currently in use (e.g., weekly, monthly)
- Current process or work flow diagrams or charts
- Descriptions or contracts concerning outsourced or shared services
- Descriptions of decision support tools
- Summaries of typical spreadsheets, databases, and maintenance information management systems
- Descriptions of models and special tools
- Position or job descriptions used for current maintenance positions, including planning, engineering, and other technical and administrative positions, as well as line functions

Although this information is usually sufficient, you sometimes discover additional needs once work starts at the plant. Compare against the background data, and reconcile any issues that come to light.

17.2.2.3 Site Visits and Interviews

If you're not familiar with the plant's layout or operation, you will need to spend time on a tour and learning safety procedures. A thorough tour that follows the production flow through the plant is best.

To facilitate the interview process on site, present the self-diagnostic results to management and other key personnel in a kickoff meeting. This will serve as an introduction to best practices—the model presented as a perfect score on the self-diagnostic. Introduce what you are doing; describe generic industrial best practices and what you will do on site. Then, conduct interviews of various plant personnel, using both the self-assessment and documentation collected earlier as question guides.

The composition of the interviews can be driven by the areas of weakness and strength initially revealed through the self-assessment and any other input received from the previously collected data. The interviews can be used as a tool for delving into specific problem areas and their causes. In particular, you may identify any organizational, systemic, or human factors that may be at the root of any problems or areas of high performance.

The interviewees will generally be the following:

- Plant manager, human resources/industrial relations manager
- Operations/production managers
- Information technology manager and systems administrators
- Purchasing manager, stores manager/supervisors
- Maintenance manager; superintendent, maintenance/plant engineers, planners, supervisors
- Several members of the maintenance work force (at least two from each major trade or area group)
- Representatives of any collective bargaining unit, or employee association

Conduct interviews in either a private office or the plant while the interviewee walks through his or her job and workplace. You may also want to observe how a planner, mechanic, or technician, for instance, spends the workday. This can reveal a lot about systemic and people issues as well as opportunities that can impact best practices.

17.2.2.4   Maintenance Process Mapping

A group session is one of the best ways to identify major processes and activities. The processes should be the actual ones practiced at the plants, which may not coincide with the process maps developed for an ISO 9000 certification. Treat existing charts or maps as "as-designed" drawings, not necessarily "as-is" ones. These are mapped graphically to illustrate how work, inventory, and other maintenance practices are managed and performed. Through the mapping process, draw out criticisms of the various steps to reveal weaknesses. This will not only help you understand what is happening now but also can be used for the as-is drawings. Or, it can provide a baseline for process redesign, along with best practices determined from benchmarking or expert advice.

Through process mapping, you will gain significant insight into the current degree of system integration and areas where it may help. Redesigning processes will be part of any implementation work that follows the diagnostic.

The key processes to examine in your interviews and mapping include the following:

- PM development and refinement
- Procurement (stores, nonstores, services)
- Demand/corrective maintenance
- Emergency maintenance
- Maintenance prevention
- Work order management
- Planning and scheduling of shutdowns
- Parts inventory management (receiving, stocking, issuing, distribution, review of inventory investment)
- Maintenance long-term planning and budgeting
- Preventive/predictive maintenance planning, scheduling, and execution

Through the interviews, identify and map other processes unique to your operations.

17.2.2.5 Report and Recommendations

At the end of the site visit, compile all the results into a single report showing overall strengths and weakness, the specific performance measures that identify and verify performance, and the most significant gaps between current practices and the vision. The report should contain an "opportunity map" that plots each recommendation on a grid, showing relative benefit compared with the level of difficulty to achieve it (see Section 14.4).

17.2.3 BENCHMARK

In many companies, benchmarking is more industrial tourism than an improvement strategy. Even the most earnest may answer benchmark questions based on how they want to perceive themselves or how they want to be perceived. After you have completed a diagnostic assessment, select the key factors for maintenance success specific to your circumstances. You need to focus sharply to gain useful information that can actually be implemented. Look within and outside your industry to discover who excels at those factors. Compare performance measures, the process they now use, and how they were able to achieve excellence. Recognize, however, that finding organizations that have excellent maintenance optimization can be difficult. There may only be a few dozen, for example, who have successfully implemented CBM optimization using stochastic techniques and software. Benchmarking can more often be a tool to help you understand the categories of ideas that are being deployed by leading organizations outside your targeted process rather than the specific degree to which it is being accurately executed and measured. To this end, benchmarking can be used to categorize what you wish to baseline measure in your own operation as a starting point for measured improvements.

A focus on "leading practices" may serve as a better external barometer than traditional benchmarking. With so many changes in asset and asset management technologies, picking the most effective asset management process or procedure at a given point in time would be like shooting at a moving target. Viewing how engineering, operations, and maintenance are working together to maintain functions, not assets, could help change the culture in your organization. Recognizing that more and more assets allow manual inspections to give way to onboard diagnostics and expert decision support systems that can automatically initiate a call in the work management process may be a process benchmark you could consider and adopt in your organization.

17.3 STEP 2: DEVELOP

17.3.1 PRIORITIZE

You will not be able to implement everything at once. In fact, limited resources and varying benefits will whittle down your "short list." One technique to quickly see the highest value initiatives versus those with limited benefit and expensive price tags is shown in Figure 17.3.

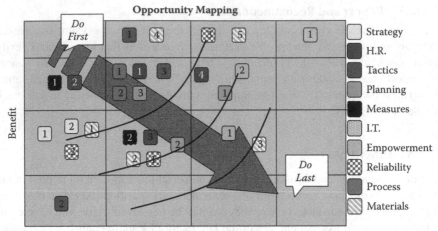

FIGURE 17.3 Maintenance opportunity prioritization map.

The benefit is often measured in a log 10 savings scale, for example, $10,000, $100,000, $1,000,000, or $10,000,000. The degree of difficulty could be shown on a similar scale, an implementation timeline such as one year, two years, three years, four years, or a factor related to the degree of change required in the organization for the initiative.

This technique will clearly identify the apparent "high benefit, low effort" opportunities (also often called "low-hanging fruit"). Often many of these identified ideas are executed for immediate savings and will help a project or major initiative declare early identified savings.

17.3.2 STRATEGIZE, PLAN, AND SCHEDULE

The new operating model will be developed from the prioritized requirements that surface from the prioritization step. The proposed "new operating model" will need to be flexible and robust. Inputs into its strategy development could include the following:

- Information about related initiatives already under way
- Current corporate strategy, mission/vision, values, desired behaviors, design principles, and design points
- Existing relevant maintenance and asset management policy documentation
- Relevant leadership or management training documents that would address the values and culture on "how things get done"
- Information available from existing systems to support the change
- Information about the regulatory environment and pressures that apply to the operating environment of the organization

Planning how you would implement the prioritized actions from the opportunity map may require some thought as to whether there are any dependencies to accomplish some of the originally described "easy-to-implement and high-benefit" improvements. At times, it may be determined that consolidating some of the early actions into a more substantial sequence of initiatives will make sense. Items such as "field training on numerous topics" or implementing a more rigorous integrated system or process may allow you to take some of the prioritized items and group them into subproject initiatives.

Using software such as MS Project, you will make the planning and scheduling process more rigorous. List the projects in priority (given in Section 14.4) and with credible timelines. No one will commend you for taking the fast track in the planning phase if the project fails several months along. Be sure to include people from other functions or processes related to maintenance, if they can help ensure project success. Typically, you'll look to procurement, production planning, human resources, finance and accounting, general engineering, information technology, contractors, vendors, or service providers.

17.3.3   COST/BENEFIT

The priority assessment given in Section 14.4 broadly groups the recommendations. If there is a significant capital or employee time investment, calculate the cost/benefit. Whereas costs are usually fairly straightforward—hardware, software, training, consulting, and time—estimating benefits can be a lot more nebulous. What are the benefits of implementing root-cause failure analysis, a reliability-centered maintenance (RCM) program, or the EXAKT software for CBM optimization? In Chapter 1, we described how to estimate the benefits of moving to more planned and preventive maintenance, which can be modified for specific projects targeting unplanned maintenance. Often, the equipment and hardware manufacturer or software/methodology designer can help, based on their experience with similar applications. Benchmarking partners or Web sites focused on maintenance management are other sources of useful information.

17.4   STEP 3: DEPLOY

17.4.1   EXECUTE

If possible, consider a pilot approach to improvement initiatives. This not only provides proof of concept, it more importantly acts as an excellent marketing tool for a full rollout.

What does experience teach us about program management? Program management is managing a group of projects with a common theme. Some of its key elements are as follows:

- Define the scope (one large enough to capture management's attention and get meaningful results).
- Follow a documented approach.
- Delineate roles and responsibilities.
- Put a lot of effort up front in the discover and develop phases (the measure-twice/cut-once approach).

- Assess the "do-ability" of the plan with all stakeholders.
- Take no shortcuts or fast tracking.
- Estimate the risk beyond the budget.
- Work to get the right champion who will be in it for the long run.

17.4.2 MEASURE

The more detailed the implementation plan, the easier to measure progress. Include scope, detailed activities, responsibilities, resources, budgets, timelines for expenditures and activities, and milestones. Ensure maximum visibility of these measures by posting them where everyone involved can see. Often forgotten are the review mechanisms: supervisor to subordinate; internal peer group or project team; management reviews; start of day huddle. They are all performance measures.

17.4.3 ANALYZE/IMPROVE

After the initial round of piloting and implementation comes the reality. Is the program actually delivering the expected results? How are the cost/benefit "actuals" measuring up? Despite best efforts, excellent planning, and execution, adjustments are often still needed to the original scope, work plan, staffing, or expected results. But if you have closely followed the discover and design stages and set up and managed performance measures, there will be few surprises at this stage.

17.4.4 MANAGING THE CHANGE

Finally, let's look at managing successful change during this three-stage process. There are eight critical elements:

- Ensure that everyone understands the compelling need to change the current order, to close the "gap" between what is done today and the vision, and that the justification for the new investment is clear.
- Build this vision of what the new order will be like so that it is shared and all accountable can buy in to it and that there are understood long-term goals and scope of change.
- Obtain visible and committed leadership so that the implementation has a high-level executive sponsor or sponsoring group, so that the executive committee shares the same goals as the front-line managers, and so that an effective project office team and the financial resources are assigned to get the job done.
- Promote broad-based stakeholder-wide participation toward a single program focus, with related activities effectively aligned and coordinated.
- Get buy-in from those most affected by the change through linking performance with rewards and recognition.
- Monitor performance and exercise leadership and control, especially when the implementation begins to drift off course.

- Establish a strong project management discipline with consistent milestones established, roles and responsibilities clearly defined and made visible, effective project goals in existence, an enterprise-wide culture change being considered, and skills available to implement the change.
- Communicate results at every step of the way, and at regular intervals. This is the single most talked about pitfall by organizations that have stumbled in achieving the results hoped for at the outset of an improvement project. Provide targeted, effective communications so that individual needs are met, so that there is consistency in the messages, so that effective two-way communications are in existence, so that successes are being leveraged, and so that enterprise-wide learning can take place.

Over and over, many organizations will implement process and technology change without considering the impact on the stakeholders. This often means that the user of the process is not fully engaged when the process change is executed, and this will often result in less than desired outcomes. Experience with managing people implications of major changes has provided us with a number of lessons learned:

- Early organization change management drives business ownership and accountability at the appropriate levels of the business and ultimately contributes to sustaining the changes that the project will introduce.
- Appropriate participation of stakeholders in the identification of business benefits and impacts will allow people to identify benefits and impacts for themselves and plan accordingly.
- Early identification of organization change management risk factors allows for the development of a robust organization change management strategy.
- Ensuring that "people" systems and structures are properly aligned with technology, process, and system changes increases the odds of realizing the intended benefits from the system implementation.
- Creating a change team and change network with clear roles, responsibilities, and competencies enhances the effectiveness of the change effort.

Managing change can be the "neverending quest." Recognizing your stakeholder groups and where they are in accepting and internalizing change is key to ensuring that you continue to promote the process and culture change or take other appropriate risk actions until it produces the full desired effect (Figure 17.4).

Every implementation to improve the way we manage physical assets will be different. This is because we are all individuals, operating in a unique company culture, implementing projects that are people driven. In this book, we have tried to impart our knowledge of what works best in most cases so you can apply it to your own particular circumstances.

Although you will strive to use the most cost-effective methods and the best tools, these alone will not guarantee success. What will is a committed sponsor with sufficient resources, an enthusiastic champion from each affected area to lead the way, a project manager to stay the course, well-executed training when required, and a motivated team to take action and set the example for the rest of the organization.

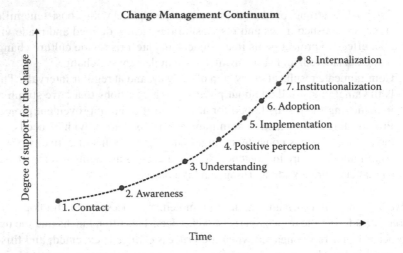

FIGURE 17.4   Change management continuum.

REFERENCES

1. Adapted from Campbell, J. D. Coopers and Lybrand Consultants Library, 1994.
2. Campbell, J. D. *Uptime: Strategies for Excellence in Maintenance Management.* Portland, OR, 1995.
3. IBM Global Business Consulting AMCoe, 2006.

18 The Future of Asset Management Solutions
Consolidation, Capability, Convergence

Joel McGlynn and Don Fenhagen

CONTENTS

18.1   INTRODUCTION

We are at an exciting point in the maturity stage of asset management. Asset management has been around as long as there has been infrastructure that needed to be repaired or maintained. However, developments in this area are accelerating at a rapid pace. Some of the key factors affecting this acceleration are newer technologies such as mobile asset management solutions; geographic information system (GIS) and other spatial solutions; radio-frequency identification (RFID) and sensors and actuators; the growing intelligence of assets; the convergence of asset classes that can include onboard information technology (IT) for assets and assets that have Internet Protocol (IP) addresses; and the convergence of asset management, product life-cycle management (PLM), and service life-cycle management (SLM). These new developments are adding new capability, flexibility, and new efficiencies but have changing organizational implications as well. It is helpful to review the past evolution of asset management solutions to better understand the future trends.

18.2   ASSET MANAGEMENT IN THE PAST

Asset management solutions and processes have progressed significantly over the last 25 years, from the early work order tracking and management solutions, which basically borrowed the material requirements planning (MRP) components (e.g., bill of materials [BOM]) as the asset hierarchy and routings as the work order, to computerized maintenance management systems (CMMS) and early enterprise asset management (EAM) solutions. Though these solutions progressed from a feature function perspective, they remain largely tracking systems—tracking the activities that were applied to equipment and reminding us of preventive activities that needed to be accomplished. The success or benefit that these solutions provided to organizations was largely dependent on data being correctly entered and updated in the system. This was not just dependent on the work order or activity data being entered but also the underlying structure that these data were being assimilated into, whether it was the asset hierarchy, work order types, failure codes, or something else. What made this even more challenging for companies was that each department or plant used its own approach to many of these structures. Finally, a further complication was introduced with multiple asset class solutions: one for fleet maintenance, another for the plant, another for facilities, and yet another for IT assets. In addition to the multiple asset class solutions, organizations now have to manage multiple asset classes across multiple organizations and sites that all may contain different processes, procedures, and structures.

18.3   ASSET MANAGEMENT SYSTEM CONSOLIDATION

Asset management system consolidation is abundantly apparent today. It is centered on the needs of IT organizations to become more efficient while improving the quality of services they deliver to their maintenance professional constituents. Asset management systems drive the operational need to standardize business processes across the business to increase operational efficiency. A key avenue in this pursuit is EAM system consolidation. Today most companies rely on many solutions from multiple vendors, each used to address the needs of separate departments and functions and each used for different types of assets. The challenge to IT organizations is worsened by widespread use of weeds like "pop-up" asset management applications built on MS Access or Excel by individuals in far recesses of the organization. These unmanaged and unaccounted-for pop-up applications, often used to manage critical asset-related information, represent operational risk to the organization and can undermine Sarbanes-Oxley compliance.

Replacing these various commercial and homegrown applications with one system brings clear value to IT organizations:

- Lower initial software and implementation costs, lower lifetime ownership costs
- Lower IT administration costs
- Lower hardware and software infrastructure costs
- Lower end-user training costs
- Lower operational risk
- Higher Sarbanes-Oxley compliance

Asset management software solutions allow customers to achieve these benefits today by offering a single "rational suite" of capabilities for managing all the critical assets in the enterprise, both operational and IT. These solutions share a common asset repository, are built on a common service-oriented architecture (SOA) platform, and support shared business processes across the asset classes. It is critical to have a strong integration framework that allows your asset management system to be interoperable and interconnected with your corporate ecosystem of software solutions. The obvious savings are the reducing long-term maintenance costs of these multiple disparate systems.

18.4 CAPABILITY: EMERGING TECHNOLOGY TRENDS IN ASSET MANAGEMENT

Technology trends in asset management include mobile asset management solutions, GIS and other spatial solutions, RFID, and sensors and actuators. All of these solutions are quickly becoming more intelligent and giving businesses the ability to conduct operational analytics to their assets to understand history of asset behavior and make predictions about the future. With some of the emerging trends in asset management, businesses are applying advanced algorithms to common asset occurrences to understand spatio-temporal (space and time) relationships that may not have been possible in past years. Asset management is driving toward a more intelligent, interconnected, and instrumented environment, which is leading to a smarter planet. (IBM)

18.4.1 MOBILIZING ASSET MANAGEMENT

People in asset-intense organizations are finding that mobile applications of asset management systems are more important in today's mobile world. Lack of access to asset data can mean mobile field technicians do not have the latest asset repair history, troubleshooting procedures, parts, and work location on hand—costing precious time while accurate, up-to-date information resides in the asset management system, instantly accessible from an appropriate mobile application and device. In many cases, technicians spend less than 50% of their time working on an asset. The rest is spent recording data to be manually fed back into the asset management system, or waiting for documentation, parts, tools, or other resources required to perform the work.

While mobile computing is now mainstream, many organizations are moving to extend the reach of their asset management investments to mobile workers. Companies in asset-intensive industries stand to realize significant benefits quickly by implementing an integrated, secure, and proven mobile asset management solution. Mobile asset management improves the efficiency and accuracy of asset reporting, helps maximize use of assets, and streamlines regulatory compliance while also enhancing the value of IT investments. Mobilizing your asset management solution automates the workflow and process needs of the repair or operations technician—enabling the real-time flow of vital information wherever and whenever the work is required.

18.4.2 BUSINESS BENEFITS

A mobile asset management solution empowers your maintenance organization to do the following:

- Create more "wrench time" and less paperwork time, allowing more completed work orders per day per person.
- Prevent production incidents.
- React faster to exception conditions with appropriate corrective measures.
- Maximize asset reliability and minimize downtime.
- Decrease travel time and truck rolls.
- Streamline compliance reporting.
- Reduce the time it takes to request, assign, complete, sign off on, and close out work orders.
- Improve asset inventory, accuracy of data collection, and timeliness of information flow.
- Manage warranty and maintenance contracts with outside providers.
- Improve reliable and clean enterprise data for better analytics and decision making.

By helping your business work smarter, faster, and safer, mobile asset management capabilities reduce operational cost and risk while improving service levels and enhancing corporate agility.

18.4.3 SPATIAL ASSET MANAGEMENT

GIS technology already play a major role in many asset management environments across utilities, energy, government, transportation, telecommunications, and many other asset-intensive industries by providing the capability to gather and summarize data about the diverse geographic locations and movements of strategic assets.

As GIS technology move from departmental, desktop-based solutions based on proprietary architectures to enterprise systems based on technology standards, organizations now have the opportunity to "spatially enable" a wide range of enterprise applications, including asset and service management solutions. Spatially enabled applications can support complex data analysis based on geographic location, such as representing data on maps in various spatial or geographic contexts and determining proximity, adjacency, and other location-based relationships among objects. One of the important things now occurring in the asset management world is convergence of the GIS business units and the asset management business units within organizations. As asset management systems are becoming integrated with GIS technology to spatially enable the asset management teams, GIS is becoming a part of asset management. GIS users are now asset management system users and vice versa. The footprint of GIS is increasing in all organizations, since companies now want to know not only what the status of their assets is but also where the assets are and how that relates to the world around them.

By combining GIS with asset and service management business processes in a modern SOA, a particularly powerful geospatial solution can be created—one that enables decision makers across the enterprise to make better informed decisions and helps organizations increase productivity and efficiency while improving service to customers.

18.4.4 RFID

RFID-enabled solutions help companies improve their customers' shopping experience, make warehouse management more efficient, cut down on labor cost, and increase the profit margin. The benefits are clear:

- An improved customer shopping experience
- The ability to ensure that stock in a warehouse is available and can meet customers' requirements successfully
- A decrease in broken or unusable goods
- An improved automation level and decrease in labor cost
- Improvement in the total operation efficiency of the warehouse

RFID has been used in many sectors, including warehouse management systems (WMSs), location awareness, asset tracking and EAM systems, transportation, health care, and banking.

18.4.5 SENSE AND RESPOND

As RFID and other sensor devices gain popularity, the terms *sensors* and *actuators* have become increasingly prevalent. All assets are moving toward being able to sense their current state and environment. To address the current challenges of increasingly global and dispersed assets, many companies are shifting toward sense-and-respond strategies for addressing asset management and maintenance of assets. As one example, trends are emerging to save organization labor costs on inspectuation tasks when asset components have expiry dates and need to be disassembled for a technician to physically read the date on the component. Sense-and-respond–based organizations can monitor, manage, and optimize business exceptions—anomalous events that occur within supply chains—with a limited need for human intervention. They can provide event assessment and optimize asset use and performance between planning and execution, based on real-time information. Ultimately, they allow businesses to remain nimble and responsive to shifting demand. And in a proactive business environment, sense-and-respond maintenance organizations can be used to influence market demand.

18.4.6 SMART DUST

As sensors and actuators and sense-and-respond technologies increase in popularity, reliability, and efficiency and decrease in cost we are going to see our asset management environment evolve into a world that has the ability to put sensors on virtually everything around us. An example of what is to come in the future is a new technology being tested around the world called "smart dust." The University of California (UC)–Berkeley and other state-of-the-art technology institutes are working closely with very small sensors, or "motes" that are decreasing in size exponentially every year. UC–Berkeley has been seen using smart dust to monitor temperature and humidity of the nearby redwoods in California. "In a couple of hours Wednesday afternoon, a team of graduate students using rock climbing equipment mounted a 120-foot-tall redwood and placed 10 sensors, each housed inside a tiny plastic cylinder."* With this technology in place, the researchers are working to analyze the temperature and humidity of the environment and create a wireless network of sensors that talk to one another and then share information with a data collection device, in this case a laptop.

These small "motes" are able to accurately and quickly monitor temperature, humidity, and location. Other sensors have other monitoring capabilities, depending on the situation. The smart dust technology is able to strategically place the sensors in many places, unnoticed by the human eye yet powerful enough to provide readers or databases with important asset information that will enable decision makers and investigators to act on information provided by the smart dust enabled asset. The name smart dust evolved from the thought that, at some point in the not-so-distant future, sensors will be microscopic, close to the size of dust but with the power of a laptop. This concept alone can make it apparent that maintenance organizations of the future will be able to make more accurate decisions on all types of assets and asset classes no matter what the size or value of the asset. As can be seen from the previous example, the application of the future technology is endless, exciting, and not so far away.

---

* Brand, W., *Oakland Tribune*, July 21, 2003.

18.4.7 THE AGE OF THE SMART ASSET

We are in the age of the smart asset. Many common assets today contain more computing power than your average laptop computer. Examples abound, such as aircraft and robotic welders in automobile factories and the electronics in over-the-road trucks and even newer-model cars. There is a wide array of sensors on, for example, pipelines, utility lines, tunnels, highways, tires, and buildings. The new Minneapolis Bridge over Interstate 35, which replaced the bridge that collapsed in August 2007 killing 13 and injuring 145, became operational on September 18, 2008. The new span is equipped with "Smart Bridge" technology designed to detect small problems before they become bigger problems. Within this bridge are 323 sensors that will detect stress and strain, loads, and vibrations in addition to traffic flow, speeds, accidents, stalls, and other disruptions. In addition, sensors will actuate anti-icing systems and identify intruders into unauthorized areas such as the hollow concrete girders. The data these sensors collect will be fed into networked computers at the University of Minnesota for analysis.

Assets are increasingly being built "smarter" or are made smarter by retrofitting. Some prototype smart refrigerators are built with RFID sensors that will monitor food going in and out, inside temperatures and humidity, product origination, how long the food has been in transit, expiration dates, and even tampering.

In an example of reeducating older assets, sensors are being attached to railcars identifying, for example, brake temperature, shock, and stress.

Most of these smart assets increasingly have IP addresses and are networked. Some of the maintenance decisions can be handled in the network itself such as diagnostic data and corrective routines. As assets become smarter, the implication for asset management is blurring the line between EAM and IT asset management creating an asset class convergence.

18.4.8 ASSET CLASS CONVERGENCE

Asset class convergence is an emerging trend in asset management. Equipment and IT assets are increasingly integrated into one system that together delivers services to the business. This trend is driving closer together the business processes for managing disparate types of assets. With increasing frequency production equipment, facilities and transportation assets are relying on embedded IT systems to improve their performance:

- Plant floor equipment incorporates operating system software and production software applications and is often networked with an IP address. Equipment makes increasing use of built-in condition monitoring, active RFID, and chipsets of various kinds. (Entek from Rockwell)
- Transportation vehicles, on road or rail, use onboard monitors as well as GIS.
- Building automation for climate control, security, and infrastructure management makes wide use of computer hardware and software. (EnNet from Gridlogix; Metasys from Johnson Controls)

- Intelligent utility network (IUN) helps boost the adoption of grid computing within the energy sector. IUN provides an information architecture that allows for the automated real-time monitoring of assets like meters, power lines, and customer usage to improve service and reliability. IUN includes a grid that will provide data, information, and analytics to help workers improve outage detection and restoration times along with ongoing operations.

The challenge, however, lies in managing the blurred distinction between the industrial asset and the associated IT. From the IT perspective, business processes historically used for managing IT hardware and software can no longer afford to distinguish between the IT asset and the operating asset it's integrated with. Handling degrading or failed services requires focus on restoration of service and root-cause analysis, both associated with service-level agreements. Understanding whether faults lie with the IT or the operating asset is more difficult, but it is becoming more important to know. Change and configuration management are processes dedicated to service management—and apply equally to a retrofit of a hydraulic system, a manufacturer recall of impeller assemblies, refreshing an operating system build, applying a patch to the software application, or upgrading memory. All of these processes need to manage the entire aggregated asset, of which IT is one element.

Figure 18.1 describes the type of questions you might ask if you are about to repair a motor/welder before the age of smart assets compared to a server you just bought.

18.5 ASSET CONVERGENCE—SCENARIOS

18.5.1 Energy and Utilities—Nuclear Power Generating Station

18.5.1.1 Description

Operating a nuclear powered electric generating station requires constant attention to detail, reliance upon standardized business processes, and a real-time understanding of the asset infrastructure. A nuclear power station is composed of numerous mission critical assets that range from heavy industrial (e.g., feedwater pumps, reactor vessel) to information technology (e.g., digital asset health monitoring, control room operations).

The business of operating a nuclear power station is highly regulated and highly process oriented; executives to engineering to maintenance personnel rely heavily upon technology to help them provide a safe, reliable, and profitable product to the market. This technology manifests itself as software systems, monitoring devices, logic controls, and other devices that help manage and monitor the operation of the plant and its individual assets.

18.5.1.2 Convergence Asset Examples

- Feed-water pump with numerous health monitors that analyze, record, and capture readings
- Reactor control rod drives (CRDs) that are controlled remotely from the control room via complex software programs
- Whole body count machines that take human dosage readings and store information for downstream analysis

18.5.1.3 Scenario A: The Corrective Action Process

Personnel working at a nuclear power station (employees and contractors) must be empowered to submit corrective action requests and incidents at any time. These requests, ranging from a general complaint to a comment about asset performance, must be brought into a repeatable process for analysis, tracking, and action. The corrective action process is a foundational element of the "safety culture" at a nuclear power station.

18.5.1.4 Scenario B: Surveillance Testing

Nuclear power stations are licensed by governmental organizations to operate the plants within certain parameters. These licenses spell out certain "surveillance" tests that a plant must perform to verify IT operations are within the parameters. These tests occur on a variety of assets, over a variety of time periods. However, all tests must be recorded, and the results must be available at any time to support regulatory inspections and other relevant business processes.

18.6 PRODUCT AND SERVICE LIFE-CYCLE MANAGEMENT CONVERGENCE: SERVICE MANAGEMENT AND ASSET MANAGEMENT

Organizations are always looking for ways to increase growth and introduce efficiencies while managing the risk of their operation to create a sustainable future. They do this by making decisions regarding capital investments and by directing operational spending. Capital investments lead to the acquisition of new technology and assets such as complex machinery, equipment, robots and other automation, servers, laptops, and software. And operational spending impacts the performance and useful life of assets once they are acquired. Traditionally, organizations have left the management of newly acquired technology and assets to the technologists who purchased them, resulting in a fragmented set of systems and processes for each asset or asset category. Historically this approach has met the needs of most organizations; however, organizational requirements for asset management are changing. They are changing because of the ongoing innovation of the assets themselves. In addition, there is more pressure on the business and increased influence from outside the organization for tighter governance and regulatory compliance. The three compelling reasons for changing the way an organization manages assets are (1) the increased interdependency of assets; (2) the need to understand the way individual assets affect service to the business; and (3) the requirement to provide visibility to properly address questions relating to risk management and compliance. The current fragmented approach to managing assets and services has been a "good enough" solution for most organizations. However, this fragmentation implicitly leads to organizational "blind spots." The upside of this situation is that it can also be considered an opportunity for organizations willing to innovate. Bringing asset and service management together will enable organizations to deliver the expected performance, to increase comprehension of service dependencies, and to manage the associated financial and operational risks. Looking at asset management in a different light.

Most executives have understood the need to invest in technology to foster improvements in productivity and to manage business complexity. However, in most cases the technology was left to the technologists and engineers to implement and manage. The result is an organization that lacks a standardized control framework to manage all assets from cradle to grave. Most organizations treat the asset management function as a departmental discipline. It is often tied specifically to the type of asset, and not part of an enterprise-level function or process. Each company, division, or department develops and uses its own processes and requirements and often makes separate investments for individual assets or groups of assets.

Appendix A: References, Facts, Figures, and Formulas

A.1 MEAN TIME TO FAILURE

The expected life, or the expected time during which a component will perform successfully, is defined as

$$E(t) = \int_0^\infty t f(t) dt$$

E(t) is also known as the mean time to failure (or MTTF).

Integrate

$$\int_0^\infty R(t) dt$$

by parts using

$$\int u \, dv = uv - \int v \, du$$

and letting $u = R(t)$ and $dv = dt$:

$$\frac{du}{dt} = \frac{dR(t)}{dt}$$

but by examining Figure 10.2, the probability density function, we see that

$$R(t) = 1 - F(t)$$

and

$$\frac{dF(t)}{dt} = f(t)$$

therefore,

$$\frac{du}{dt} = \frac{dR(t)}{dt} = -\frac{dF(t)}{dt} = -f(t)$$

Then, $du = -f(t) dt$

Substituting into

$$\int u\,dv = uv - \int v\,du$$

$$\int\limits_0^\infty R(t)dt = \Big[tR(t)\Big]_0^\infty + \int\limits_0^\infty tf(t)dt$$

but $\lim\limits_{t\to\infty} tR(t) = 0$ because the part will fail eventually. That is, the reliability at infinity is 0.

Therefore, the term

$$\Big[tR(t)\Big]_0^\infty$$

in the above equation is 0. Hence,

$$\int\limits_0^\infty R(t)dt = \int\limits_0^\infty tf(t)dt = E(t) = MTTF$$

A.2   MEDIAN RANKS

When only a few failure observations are available (say ≤ 20, use is made of the median rank tables:

	1	2	3	4	5	6	7	8	9	10	11	12
1	50	29.289	20.630	15.910	12.945	10.910	9.428	8.300	7.412	6.697	6.107	5.613
2		70.711	50.000	38.573	31.381	26.445	22.849	20.113	17.962	16.226	14.796	13.598
3			79.370	61.427	50.000	42.141	36.412	32.052	28.624	25.857	23.578	21.669
4				84.090	68.619	57.859	50.000	44.015	39.308	35.510	32.390	29.758
5					87.055	73.555	63.588	55.984	50.000	45.169	41.189	37.853
6						89.090	77.151	67.948	60.691	54.811	50.000	45.951

Example: Bearing failures times (in months): 2, 3, 3.5, 4, 6. From median rank tables:

							7	8	9	10	11	12
7							90.572	79.887	71.376	64.490	58.811	54.049
8								91.700	82.018	74.142	67.620	62.147
9									92.587	83.774	76.421	70.242
10										93.303	85.204	78.331
11											93.893	86.402
12												94.387

From a Weibull analysis, $\mu = 3.75$ months and $\sigma = 1.5$ months.

	Median Rank (%)	
1st failure time	13	2 months
2nd failure time	31.5	3 months
3rd failure time	50	3.5 months
4th failure time	68.8	4 months
5th failure time	87.1	6 months

Benard's formula is a convenient and reasonable estimate for the median ranks.

$$\boxed{\dfrac{Cumulative\ Probability\ Estimator}{Benard's\ Formula} = \dfrac{i-0.3}{N+0.4}}$$

A.3 CENSORED DATA

Hours	Event	Order	Modified Order	Median Rank
67	F	1	1	0.13
120	S	2		
130	F	3	2.5	0.41
220	F	4	3.75	0.64
290	F	5	5.625	0.99

A.3.1 PROCEDURE

$$I = \dfrac{(n+1)\cdot\left(previous\ order\ number\right)}{1+\left(number\ of\ items\ following\ suspended\ set\right)}$$

where I = increment.

The first order number remains unchanged. For the second, applying the equation for the increment I, we obtain

$$I = \dfrac{(5+1)\cdot\left(1\right)}{1+\left(3\right)} = 1.5$$

Adding 1.5 to the previous order number 1 gives the order number of 2.5 to the second failure:

$$\boxed{\dfrac{Cumulative\ Probability\ Estimator}{Benard's\ Formula} = \dfrac{i-0.3}{N+0.4}}$$

$$I = \frac{(5+1)\cdot\left(2.5\right)}{1+\left(3\right)} = 3.75$$

$$I = \frac{(5+1)\cdot\left(3.75\right)}{1+\left(3\right)} = 5.625$$

Applying Benard's Formula to estimate median ranks
For first failure:

$$\boxed{median\,rank = \frac{1-0.3}{5+0.4} = 0.13}$$

Similarly for the second, third, and fourth failures, respectively, we have $(2.5 - 0.3)/5.4 = 0.41$; $(3.75 - 0.3)/5.4 = 0.64$; and $(5.625 - 0.3)/5.4 = 0.99$.

A.4   THE 3-PARAMETER WEIBULL FUNCTION

Failure Number i	Time of Failure t_i	Median Ranks N = 20 $F(t_i)$
1	550	3.406
2	720	8.251
3	880	13.147
4	1020	18.055
5	1180	22.967
6	1330	27.880
7	1490	32.975
8	1610	37.710
9	1750	42.626
10	1920	47.542
11	2150	52.458
12	2325	57.374
13–20	Censored data	

Item: 3 Parameter Example

The curvature suggests that the location parameter is greater than 0.

Failure Number i	Time of Failure t_i	Median Ranks N = 20 $F(t_i)$
1	0	3.406
2	170	8.251
3	330	13.147
4	470	18.055
5	630	22.967
6	780	27.880
7	940	32.975
8	1060	37.710
9	1200	42.626
10	1370	47.542
11	1600	52.458
12	1775	57.374
13–20	Censored data	

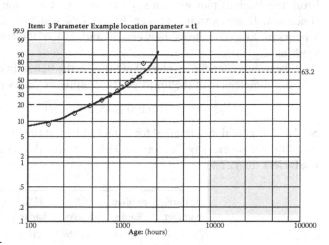

Now we get a line that is curved the other way, proving that the location parameter has a value between 0 and 550. $\gamma = 375$ yields a straight line as shown below.

Failure Number i	Time of Failure t_i	Median Ranks N = 20 $F(t_i)$
1	375	3.406
2	495	8.251
3	705	13.147
4	845	18.055
5	1005	22.967
6	1155	27.880
7	1315	32.975
8	1435	37.710
9	1575	42.626
10	1745	47.542
11	1975	52.458
12	2150	57.374
13–20	Censored data	

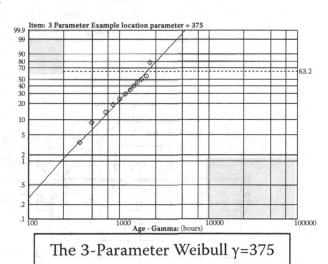

The 3-Parameter Weibull $\gamma=375$

A.5   CONFIDENCE INTERVALS

From the Weibull plot we can say that at time t = 7, the cumulative distribution function will have a value between 30 and 73% with 90% confidence. That is, after 7 hours in 90% of the tests, between 30 and 73% of the batteries will have stopped working.

If we want a confidence interval of 90% on the Reliability $R(t)$ at the time $t = 7$, we take the complement of the limits on the confidence interval for $F(t)$:

$$(100 - 73, 100 - 30)$$

So the 90% confidence interval for the reliability at time $wt = 7$ hours is between 27 and 70%. Or we can say that we are 95% sure that the reliability after 7 hours will not be less than 27%.

Failure Number	Median Ranks	5% Ranks	95% Ranks	t_i
1	5.613	0.426	22.092	1.25
2	13.598	3.046	33.868	2.40
3	21.669	7.187	43.811	3.20
4	29.758	12.285	52.733	4.50
5	37.853	18.102	60.914	5.00
6	45.941	24.530	68.476	6.50
7	54.049	31.524	75.470	7.00
8	62.147	39.086	81.898	8.25
9–12	Still operating			

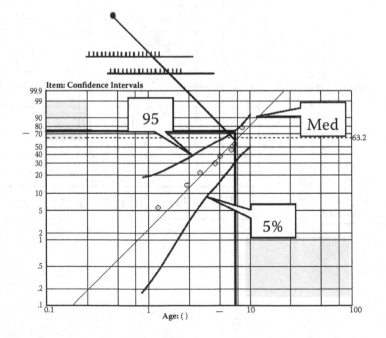

5% Ranks

	1	2	3	4	5	6	7	8	9	10	11	12
1	5.00	2.53	1.70	1.27	1.02	0.85	0.71	0.64	0.57	0.51	0.47	0.43
2		22.36	13.054	9.76	7.64	6.28	5.34	4.62	4.10	3.68	3.33	3.05
3			36.84	24.86	18.92	15.31	12.88	11.11	9.78	8.73	7.88	7.19
4				47.24	34.26	27.13	22.53	19.29	16.88	15.00	13.51	12.29
5					54.93	41.82	34.13	28.92	25.14	22.24	19.96	18.10
6						60.70	47.91	40.03	34.49	30.35	27.12	24.53
7							65.18	52.9	45.04	39.34	34.98	31.52
8								68.77	57.09	49.31	43.56	39.09
9									71.69	60.58	52.99	47.27
10										74.11	63.56	56.19
11											76.16	66.13
12												77.91

95% Ranks

	1	2	3	4	5	6	7	8	9	10	11	12
1	95	77.64	63.16	52.71	45.07	39.30	34.82	31.23	28.31	25.89	23.84	22.09
2		97.47	86.46	75.14	65.74	58.18	52.07	47.07	42.91	39.42	36.44	33.87
3			98.31	90.24	81.08	72.87	65.87	59.97	54.96	50.69	47.01	43.81
4				98.73	92.36	84.68	77.47	71.08	65.51	60.66	56.44	52.73
5					98.98	93.72	87.12	80.71	74.86	69.65	65.02	60.90
6						99.15	94.66	88.89	83.13	77.76	72.88	68.48
7							99.27	95.36	90.23	85.00	80.04	75.47
8								99.36	95.90	91.27	86.49	81.90
9									99.43	96.32	92.12	87.22
10										99.49	96.67	92.81
11											99.54	96.95
12												99.57

A.6   ESTIMATING (ALSO CALLED FITTING) THE DISTRIBUTION

A.6.1   MAXIMUM LIKELIHOOD ESTIMATE (MLE) METHOD

$$\frac{\sum_{i=1}^{n} x_i^{\hat{\beta}} \ln x_i}{\sum_{i=1}^{n} x_i^{\hat{\beta}}} - \frac{1}{n}\sum_{i=1}^{n} \ln x - \frac{1}{\hat{\beta}} = 0 \qquad \hat{\eta} = \frac{\sum_{i=1}^{n} x_i^{\hat{\beta}\frac{1}{\hat{\beta}}}}{n}$$

A.6.2 LEAST SQUARES ESTIMATE METHOD

$$\hat{\beta} = \frac{\sum\limits_{i=1}^{n} x_i^2 - \dfrac{\left(\sum\limits_{i=1}^{n} x_i\right)^2}{n}}{\sum\limits_{i=1}^{n} x_i y_i - \dfrac{\sum\limits_{i=1}^{n} x_i \sum\limits_{i=1}^{n} y_i}{n}} \qquad \hat{\eta} = e^{\hat{A}}$$

where $y_i = \ln(t_i)$ and $x_i = \ln(\ln(1\text{-Median Rank of } y_i))$,

$$\hat{A} = \frac{\sum y_i}{n} - \frac{1}{\beta}\frac{\sum x_i}{n}$$

The MLE method needs an iterative solution for the Beta estimate. Statisticians prefer maximum likelihood estimates to all other methods because MLE has excellent statistical properties. They recommend MLE as the primary method. In contrast, most engineers recommend the method of least squares estimate. In general, both methods should be used because each has advantages and disadvantages in different situations. MLE is more precise. On the other hand, for small samples, it will be more biased than rank regression estimates (from Reference 2 in Chapter 10).

A.7 KOLMOGOROV–SMIRNOV GOODNESS-OF-FIT TEST

A.7.1 STEPS

1. Determine the distribution to which you want to fit the data. Then determine the parameters of the chosen distribution.
2. Determine the significance level of the test (α usually at 1.5% or 10%). It is the probability of rejecting the hypothesis that the data follows the chosen distribution assuming the hypothesis is true.
3. Determine $F(t_i)$ using the parameters assumed in Step 1. $F(t_i)$ is the value of the theoretical distribution for failure number i.
4. From the failure data, compute $\hat{F}(t_i)$ using the Median Ranks if appropriate.
5. Determine the maximum value of

$$\left\{ \begin{array}{c} \left| F(t_i) - \hat{F}(t_i) \right| \\ \left| F(t_i) - \hat{F}(t_{i-1}) \right| \end{array} \right\} = d$$

If $d > d_\alpha$, we reject the hypothesis that the data can be adjusted to the distribution chosen in Step 1. (d_α is obtained from the K-S statistic table.)

We have tested five items to failure. Here are the failure times: $t_i = 1, 5, 6, 8,$ 10 hours. We assume that the data follow a normal distribution and will check this assumption with a K-S Goodness-of-Fit test.

A.7.2 SOLUTION

The solution is as follows:

Estimate the parameters of the chosen distribution: Estimate of $\mu = \Sigma t_i/n = 6$ and the estimate of $\sigma^2 = \Sigma(t_i - t)^2/(n - 1) = s^2$

t_i	$F(t_i)$	$\|\hat{F}(t_i)\|$	$\|F(t_i) - \hat{F}(t_i)\|$	$\|F(t_i) - \hat{F}(t_{i-1})\|$	d
1	0.070	0.129	0.059		0.059
5	0.390	0.314	0.076	0.261 d_{max}	0.261
6	0.500	0.500	0.0	0.186	0.186
8	0.720	0.686	0.034	0.220	0.220
10	0.880	0.871	0.009	0.194	0.194

The values of $F(t_i)$ are obtained from the normal distribution table.

Sample Size	K-S Level of Significance (d_α)				
n	0.20	0.15	0.10	0.05	0.01
1	0.900	0.925	0.950	0.975	0.995
2	0.684	0.726	0.776	0.842	0.929
3	0.565	0.597	0.642	0.708	0.828
4	0.494	0.525	0.564	0.624	0.783
5	0.446	0.474	0.510	0.565	0.669
6	0.410	0.436	0.470	0.521	0.618
7	0.381	0.405	0.438	0.486	0.577
8	0.358	0.381	0.411	0.457	0.543
9	0.339	0.360	0.388	0.432	0.514
10	0.322	0.342	0.368	0.410	0.490
11	0.307	0.326	0.352	0.391	0.468
12	0.285	0.313	0.338	0.375	0.450
13	0.284	0.302	0.325	0.361	0.433
14	0.274	0.292	0.314	0.349	0.418
15	0.266	0.283	0.304	0.338	0.404
16	0.258	0.274	0.295	0.328	0.392
17	0.250	0.266	0.286	0.318	0.381
18	0.244	0.259	0.278	0.309	0.371
19	0.237	0.252	0.272	0.301	0.363
20	0.231	0.246	0.264	0.294	0.356
25	0.21	0.22	0.24	0.27	0.32
30	0.19	0.20	0.22	0.24	0.29
35	0.18	0.19	0.21	0.023	0.27
Over 35	$1.07/\sqrt{n}$	$1.14/\sqrt{n}$	$1.22/\sqrt{n}$	$1.36/\sqrt{n}$	$1.63/\sqrt{n}$

Here we are engaging in Hypothesis Testing. A significance level is applied by some authority or standard governing the situation: either 0.01, 0.05, 0.10, 0.15, or 0.20.

We apply our K-S statistic, 0.261, to the row for a sample size of $n = 5$. Assuming we wish to conform to a significance level of 0.20, we note that 0.261 is not greater than 0.446. That means that if we were to reject the model, there will be a high probability (20%) that we are wrong (in rejecting a good model). Hence we say that the model is not rejected based on a 20% significance level. Frequently, the less stringent 5% significance level is applied.

A.8   PRESENT VALUE

A.8.1   PRESENT VALUE FORMULAE

To introduce the present value criterion (or present discounted criterion), consider the following. If a sum of money, say $1,000, is deposited in a bank where the compound interest rate on such deposits is 10% per annum, payable annually, then after 1 year there will be $1,100 in the account. If this $1,100 is left in the account for a further year, there will then be $1,210 in the account.

In symbol notation we are saying that if $L is invested and the relevant interest rate is $i\%$ per annum, payable annually, then after n years the sum $S resulting from the initial investment is

$$s = \$L\left(1 + \frac{i}{100}\right)^n \tag{A8.1}$$

Thus, if $L = $1,000, $i = 10\%$, and $n = 2$ years, then

$$\$S = 1,000\left(1 + 0.1\right)^2 = \$1,210$$

The present-day value of a sum of money to be spent or received in the future is obtained by doing the reverse calculation of that above. Namely, if $S is to be spent or received n years in the future, and $i\%$ is the relevant interest rate, then the present value of $S is

$$PV = \$S\left(\frac{1}{1 + \frac{i}{100}}\right)^n \tag{A8.2}$$

where

$$\left(\frac{1}{1 + \frac{i}{100}}\right) = r$$

is the discount factor.

Thus the present-day value of $1,210 to be received 2 years from now is

$$PV = \$1,210 \left(\frac{1}{1+0.1}\right)^2 = \$1,000$$

That is, $1,000 today is "equivalent" to $1,210 2 years from now when $i = 10\%$.

It has been assumed that the interest rate is paid once per year. Interest rates may, in fact, be paid weekly, monthly, quarterly, semi-annually, etc.; and when this is the case, the Equations A8.1 and A8.2 must be modified.

In practice, with replacement problems, it is usual to assume that interest rates are payable once per year and so Equation A8.2 is used in the present value calculations.

It is usual to assume that the interest rate i is given as a decimal, and not in percentage terms. Equation A8.1 is then written as

$$PV = \$S \left(\frac{1}{1+i}\right)^n \tag{A8.3}$$

An illustration of the sort of problems where the present value criterion is used is the following. If a series of payments S0, S1, S2, ..., Sn, illustrated in Figure A.8.13.1, are to be made annually over a period of n years then the present value of such a series is

$$PV = S_o + S_1 \left(\frac{1}{1+i}\right)^1 + S_2 \left(\frac{1}{1+i}\right)^2 + ... + S_n \left(\frac{1}{1+i}\right)^n \tag{A8.4}$$

If the payments S_j, where $j = 0, 1, 2, ..., n$, are equal, then the series of payments is termed an annuity and Equation (A8.4) becomes

$$PV = S + S \left(\frac{1}{1+i}\right) + S \left(\frac{1}{1+i}\right)^2 + ... + S \left(\frac{1}{1+i}\right)^n \tag{A8.5}$$

which is a geometric progression, and the sum of $n + 1$ terms of a geometric progression gives

$$PV = S \left[\frac{1-\left(\frac{1}{1+i}\right)^{n+1}}{1-\left(\frac{1}{1+i}\right)}\right] = S \left(\frac{1-r^{n+1}}{1-r}\right) \tag{A8.6}$$

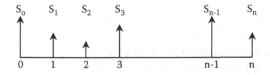

FIGURE A.8.13.1

If the series of payments of Equation (A8.5) is assumed to continue over an infinite period of time (i.e., n → ∞, then from the sum to infinity of a geometric progression, we obtain

$$PV = \frac{S}{1-r} \tag{A8.7}$$

In all of the above formulae, we have assumed that i remains constant over time. If this is not a reasonable assumption, then equation (A8.4) should be modified slightly; for example, we might let i take the values $i_1, i_2, ..., i_n$ in the different periods.

A.8.2 EXAMPLE: ONE-SHOT DECISION

To illustrate the application of the present value criterion in order to decide on the best of a set of alternative investment opportunities, we consider the following problem.

A subcontractor obtains a contract to maintain specialized equipment for a period of 3 years, with no possibility of an extension of this period. To cope with the work, the contractor must purchase a speciainpurpose machine tool. Given the costs and salvage values of the table for three equally effective machine tools (A, B, C), which should the contractor purchase? We assume that the appropriate discount factor is 0.9 and that operating costs are paid at the end of the year in which they are incurred.

Machine Tool	Purchase Price $	Installation Cost $	Operating Cost $			Salvage Value
			Year 1	Year 2	Year 3	
A	50,000	1000	1000	1000	1000	30,000
B	30,000	1000	2000	3000	4000	15,000
C	60,000	1000	500	800	1000	35,000

For machine tool A:

Present value = 50,000 + 1000 + 1000 (0.9) + 1000(0.9)² +
           1000(0.9)³ − 30,000(0.9)³
      = \$31,570

For machine tool B:

Present value = 30,000 + 1000 + 2000 (0.9) +
           3000(0.9)² + 4000(0.9)³ − 15,000(0.9)³
      = \$27,210

For machine tool C:

Present value = 60,000 + 1000 + 500 (0.9) +
           800(0.9)² + 1000(0.9)³ − 35,000(0.9)³
      = \$37,310

Thus, equipment *B* should be purchased because it gives the minimum present value of the costs.

A.8.3 FURTHER COMMENTS

In the above machine tool purchasing example, note that the same decision on the tool to purchase would not have been reached if no account had been taken of the time value of money. Note also that many of the figures used in such an analysis will be estimates of future costs or returns. Where there is uncertainty about any such estimates, or where the present value calculation indicates several equally acceptable alternatives (because their present values are more or less the same), then a sensitivity analysis of some of the estimates may provide information to enable making an "obvious" decision. If this is not the case, then we may impute other factors such as "knowledge of supplier," "spares availability," etc. to assist us in coming to a decision. Of course, when estimating future costs and returns, account should be taken of possible increases in materials costs, wages, etc. (i.e., inflationary effects).

When dealing with capital investment decisions, a criterion other than present value is sometimes used. For a discussion of such criteria (e.g., "pay-back period" and "internal rate of return"), the reader is referred to the engineering economic literature.

A.9 COST OF CAPITAL REQUIRED FOR ECONOMIC LIFE CALCULATIONS

Assume that the cost associated with borrowing money is an interest rate charge of about 20% per annum (p.a.). The reason for this high interest rate is due, in part, to inflation. When economic calculations are made, it is acceptable either to work in terms of "nominal" dollars (i.e., dollars having the value of the year in which they are spent [or received]), or in "real" dollars, (i.e., dollars having present-day value). Provided the correct cost of capital is used in the analysis, the same total discounted cost is obtained whether nominal or real dollars are used. (This does require that inflation proceeds at a constant annual rate.)

To illustrate the method for obtaining the cost of capital to use if all calculations are done in present-day (i.e., real) dollars, consider the following:

- *Assuming no inflation:* I put $100 in the bank today, leave it for 1 year, and bank will pay me 5% interest for doing so. Thus, at end of 1 year, I can still buy $100 worth of goods plus an item costing $5.00. Thus, my return for doing without my $100 for 1 year is to buy an item costing $5 (and still have goods costing $100 if I wish).
- *Assume inflation at 10% p.a.:* I put $100 in the bank today. To obtain a "real" return of 5% by foregoing the use of this $100 today requires that I can still buy $100 worth of goods—which in 1 year would cost $110 since inflation occurs at 10%, plus the item originally costing $5.00—which

1 year later would cost $5.00 + 10% of $5.00, which is $5.50. Thus, at end of 1 year, I need to have $100 + $10 + $5 + $0.50 + $115.50. Thus, the interest required on my $100 investment is

$$\$10 + \$5 + \$0.50$$
$$= \theta + i + i\theta$$
$$= 10\% + 5\% + 0.5\%$$
$$= 15.5\%$$

where θ = inflation rate, i = interest rate alternatively—if today the interest rate for discounting is 15.5%, and inflation is occurring at 10%, then the "real" interest rate is

$$i = \left(\frac{1+0.155}{1+0.10}\right) - 1$$

$$= \left(\frac{1.155}{1.10}\right) - 1$$

$$= 1.05 - 1$$

$$= 0.05 \text{ (or 5\%)}$$

Formally, the appropriate cost of capital when working in present-day dollars is

$$i = \frac{1 + \text{Cost of capital (as a decimal and with inflation)}}{1 + \text{Inflation rate (as a decimal)}}$$

Example: Illustrate the use of "real" and "nominal" dollars. Assume the following:

- We are in the year 2000
- Cost of truck in 2000 = $75,000
- Maintenance cost for a new truck in 2000 = $5,000
- Maintenance cost for a 1-year-old truck in 2000 = $10,000
- Maintenance cost for a 2-year-old truck in 2000 = $15,000

Assuming that the real cost of capital is equal to 15%, then the total discounted cost of the above series of cash flow is

$$\$75,000 + \$5,000 + \$10,000\left(\frac{1}{1+0.15}\right)^1 + 15,000\left(\frac{1}{1+0.15}\right)^2$$

$$= \$75,000 + \$5,000 + \$10,000\,(0.8696) + \$15,000\,(0.8696)^2$$

$$= \$75,000 + \$5,000 + \$8,696 + \$11,342$$

$$= \$100,038$$

Assuming inflation now occurs at an average rate of 10% per annum, then the cost of capital would be

$$0.15 + 0.10 + 0.10(0.15) = 0.265$$

The cost data would now be

Cost of truck in 2000 = $75,000
Maintenance cost of new truck for 1 year in 2000 = $5,000
Maintenance cost of a 1-year-old truck for 1 year in 2001 = $10, 000
  $(1 + 0.1) = \$11,000$
Maintenance cost of a 2-year-old truck for 1 year in 2002 = $15,000
  $(1 + .1)^2 = \$18,150$

Using a cost of capital of 26.5%, then the total discounted cost would now be

$$\$75,000+\$5,000+\$11,000\left(\frac{1}{1+0.265}\right)^1+18,150\left(\frac{1}{1+0.265}\right)^2$$

$$= \$75,000 + \$5,000 + \$11,000 \,(0.7905) + \$18,150 \,(0.6249)$$
$$= \$75,000 + \$5,000 + \$8,696 + \$11,343$$
$$= \$100,038$$

which is identical to that obtained when present-day dollars were used along with an "uninflationary" cost of capital.

A.9.1 EQUIVALENT ANNUAL COST

The economic life model gives a dollar cost that is a consequence of an infinite chain of replacements, or by modification (see following) for the first N cycles. To ease interpretation of this total discounted cost, it is useful to convert the total discounted costs associated with the economic life to an equivalent annual cost (EAC).

In the calculations performed above, the total discounted cost over the 3 years was $100,038 to an equivalent annual cost. The formula for calculating the capital recovery factor (CRF) is

$$CRF = \frac{i\left(1+i\right)^n}{\left(1+i\right)^n -1}$$

Rather than use the above formula, most books dealing with financial analysis include CRF tables for a variety of interest rates. In such tables, one would find for $i = 15\%$ and $n = 3$ years, that

$$CRF = 0.4380$$

$$EAC = \$100,038 \,(0.4380) = \$43,816.644 \text{ p.a.}$$

This means that a constant payment of \$43,816.644 p.a. would result in a total discounted cost of \$100,038 if the cost of capital is taken at 15%. To check this:

$$\text{T.D.C.} = \$43,816.644 \left(\frac{1}{1+0.15}\right)^1 + 43,816.644 \left(\frac{1}{1+0.15}\right)^2$$

$$+ \$43,816.644 \left(\frac{1}{1+0.15}\right)^3$$

$$= \$38,101.43 + \$33,131.678 + \$28,810.152$$

$$= \$100,043 \text{ (given rounding errors, this is equal to \$100,038)}$$

A.10 OPTIMAL NUMBER OF WORKSHOP MACHINES TO MEET A FLUCTUATING WORKLOAD

A.10.1 STATEMENT OF PROBLEM

From time to time, jobs requiring the use of workshop machines (e.g., lathes) are sent from various production facilities within an organization to the maintenance workshop. Depending on the workload of the workshop, these jobs will be returned to production after some time has elapsed. The problem is to determine the optimal number of machines that minimizes the total cost of the system. This cost has two components: the cost of the workshop facilities and the cost of downtime incurred due to jobs waiting in the workshop queue and then being repaired.

A.10.2 CONSTRUCTION OF MODEL

1. The arrival rate of jobs to the workshop requiring work on a lathe is Poisson distributed with arrival rate λ.
2. The service time a job requires on a lathe is negative exponentially distributed with mean $1/\mu$.
3. The downtime cost per unit time for a job waiting in the system (i.e., being served or in the queue) is C_d.
4. The cost of operation per unit time for one lathe (either operating or idle) is C_i.
5. The objective is to determine the optimal number of lathes n to minimize the total cost per unit time $C(n)$ of the system.

   $C(n)$ = Cost per unit time of the lathes + Downtime cost per unit time due to jobs being in the system
   Cost per unit time of the lathes = Number of lathes × Cost per unit time per lathe = nC_l

Downtime cost per unit time of jobs being in the system

   = Average wait in the system per job
   × Arrival rate of jobs in the system per unit time

×Downtime cost per unit time/job
$$= W_s \lambda C_d$$

where W_s = mean wait of a job in the system. Hence

$$C(n) = nC_l + W_s \lambda C_d$$

This is a model of the problem relating the number of machines n to total cost $C(n)$.

A.11   OPTIMAL SIZE OF A MAINTENANCE WORKFORCE TO MEET A FLUCTUATING WORKLOAD, TAKING INTO ACCOUNT SUBCONTRACTING OPPORTUNITIES

A.11.1   STATEMENT OF PROBLEM

The workload for the maintenance crew is specified at the beginning of a period, say a week. By the end of the week, all the workload must be completed. The size of the workforce is fixed; thus there is a fixed number of man-hours available per week. If demand at the beginning of the week requires fewer man-hours than the fixed capacity, then no subcontracting takes place. If, however, the demand is greater than the capacity, then the excess workload is subcontracted and returned from the subcontractor by the end of the week.

Two types of costs are incurred:

1. Fixed cost depending on the size of the workforce
2. Variable cost depending on the mix of internal/external workload

Because the fixed cost is increases through increasing the size of the workforce, there is less chance that subcontracting will be necessary. However, there may frequently be occasions when fixed costs will be incurred yet demand may be low, that is, considerable under-utilization of the workforce. The problem is to determine the optimal size of the workforce to meet a fluctuating demand to minimize expected total cost per unit time.

A.11.2   CONSTRUCTION OF MODEL

1. The demand per unit time is distributed according to a probability density function $f(r)$, where r is the number of jobs.
2. The average number of jobs processed per man per unit time is m.
3. The total capacity of the workforce per unit time is mn, where n is the number of men in the workforce.
4. The average cost of processing one job by the workforce is C_w.
5. The average cost of processing one job by the subcontractor is C_s.
6. The fixed cost per man per unit time is C_f.

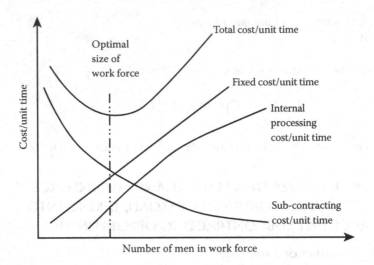

FIGURE A.11.1

The basic conflicts of this problem are illustrated in Figure A.11.1 from which it is seen that the expected total cost per unit time $C(n)$ is

> $C(n)$ = Fixed cost per unit time + Variable internal processing per unit time + Variable subcontracting processing cost per unit time
> Fixed cost per unit time = Size of workforce × Fixed cost per man = nC_f
> Variable internal processing cost per unit time = Average number of jobs processed internally per unit time × Cost per job

Now, the number of jobs processed internally per unit time will be

(1) Equal to the capacity when demand is greater than capacity
(2) Equal to demand when demand is less than, or equal to, capacity

$$\text{Probability of (1)} = \int_{nm}^{\infty} f(r)\,dr$$

$$\text{Probability of (2)} = \int_{-\infty}^{nm} f(r)\,dr = 1 - \int_{nm}^{\infty} f(r)\,dr$$

When (2) occurs, the average demand will be

$$\frac{\int_{0}^{nm} rf(r)}{\int_{0}^{nm} f(r)}$$

Therefore, the *variable internal processing cost per unit time* is

$$\left(nm \int_{nm}^{\infty} f(r)dr + \frac{\int_{0}^{nm} rf(r)}{\int_{0}^{nm} r(r)} \int_{0}^{nm} f(r) \right) C_w$$

Variable subcontracting processing cost per unit time = Average number of jobs processed externally per unit time × Cost per job

Now, the number of jobs processed externally will be

(1) Zero when the demand is less than the workforce capacity
(2) Equal to the difference between demand and capacity when demand is greater than capacity

$$\text{Probability of (1)} = \int_{0}^{nm} f(r)dr$$

$$\text{Probability of (2)} = \int_{nm}^{\infty} f(r)dr = 1 - \int_{0}^{nm} f(r)dr$$

When (2) occurs, the average number of jobs subcontracted is

$$\int_{nm}^{\infty} (r-nm)f(r)dr \bigg| \int_{nm}^{\infty} f(r)dr$$

Therefore, the *variable subcontracting processing cost per unit time* is

$$\left(0 \times \int_{0}^{nm} f(r)dr + \frac{\int_{nm}^{\infty} (r-nm)f(r)dr}{\int_{nm}^{\infty} f(r)dr} \int_{nm}^{\infty} f(r)dr \right) C_s$$

and

$$c(n) = nC_f + \left(nm \int_{nm}^{\infty} f(r)dr + \int_{0}^{nm} rf(r)dr \right) C_w$$

$$+ \left(\int_{nm}^{\infty} (r-nm)f(r)dr \right) C_s$$

This is a model of the problem relating workforce size n to total cost per unit time $C(n)$.

A.12 OPTIMAL INTERVAL BETWEEN PREVENTIVE REPLACEMENTS OF EQUIPMENT SUBJECT TO BREAKDOWN

A.12.1 STATEMENT OF PROBLEM

Equipment is subject to sudden failure and when failure occurs, the equipment must be replaced. Because failure is unexpected, it is not unreasonable to assume that a failure replacement is more costly than a preventive replacement. For example, a preventive replacement is planned and arrangements are made to perform it without unnecessary delays, or perhaps a failure may cause damage to other equipment. To reduce the number of failures, preventive replacements can be scheduled to occur at specified intervals. However, a balance is required between the amount spent on the preventive replacements and their resulting benefits, that is, reduced failure replacements.

In this appendix it is assumed, not unreasonably, that we are dealing with a long period of time over which the equipment is to be operated and the intervals between the preventive replacements are relatively short. When this is the case, we need consider only one cycle of operations and develop a model for the cycle. If the interval between the preventive replacements was "long," it would be necessary to use the discounting approach and the series of cycles would have to be included in the model.

The replacement policy is one where preventive replacements occur at fixed intervals of time, and failure replacements occur when necessary, and we want to determine the optimal interval between the preventive replacements to minimize the total expected cost of replacing the equipment per unit time.

A.12.2 CONSTRUCTION MODEL

1. C_p is the cost of a preventive replacement.
2. C_f is the cost of a failure replacement.
3. $f(t)$ is the probability density function of the equipment's failure times.
4. The replacement policy is to perform preventive replacements at constant intervals of length t_p, irrespective of the age of the equipment, and failure replacements occur as many times as required in interval $(0, t_p)$. The policy is illustrated in the figure below.
5. The objective is to determine the optimal interval between preventive replacements to minimize the total expected replacement cost per unit time.

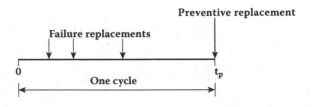

The total expected cost per unit time, for preventive replacement at time t_p, denoted $C(t_p)$ is

$$C\left(t_p\right) = \frac{\text{Total expected cost in interval } \left(0, t_p\right)}{\text{Length of interval}}$$

Total expected cost in interval $(0, t_p)$ = Cost of a preventive replacement +

Expected cost of failure replacements = $C_p + C_f H\left(t_p\right)$

where $H(t_p)$ is the expected number of failures in interval $(0, t_p)$.

Length of interval $= t_p$

Therefore,

$$C\left(t_p\right) = \frac{C_p + C_f H\left(t_p\right)}{t_p}$$

This is a model of the problem relating replacement interval t_p to total cost $C(t_p)$.

A.13   OPTIMAL PREVENTIVE REPLACEMENT AGE OF EQUIPMENT SUBJECT TO BREAKDOWN

A.13.1   STATEMENT OF PROBLEM

This problem is similar to that of Section A.12 except that instead of making preventive replacements at fixed intervals, thus incurring the possibility of performing a preventive replacement shortly after a failure replacement, the time at which the preventive replacement occurs depends on the age of the equipment. When failures occur failure replacements are made.

Again, the problem is to balance the cost of the preventive replacements against their benefits, and we do this by determining the optimal preventive replacement age for the equipment to minimize the total expected cost of replacements per unit time.

A.13.2   CONSTRUCTION OF MODEL

1. C_p is the cost of a preventive replacement.
2. C_f is the cost of a failure replacement.
3. $f(t)$ is the probability density function of the failure times of the equipment.
4. The replacement policy is to perform preventive replacements once the equipment has reached a specified age t_p, plus failure replacements when necessary. The policy is illustrated in Figure A.13.1.
5. The objective is to determine the optimal replacement age of the equipment to minimize the total expected replacement cost per unit time.

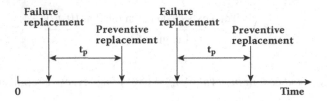

FIGURE A.13.1

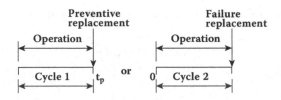

FIGURE A.13.2

In this problem, there are two possible cycles of operation: one cycle being determined by the equipment reaching its planned replacement age t_p, the other being determined by the equipment ceasing to operate due to a failure occurring before the planned replacement time. These two possible cycles are illustrated in Figure A.13.2.

The total expected replacement cost per unit time $C(t_p)$ is

$$C(t_p) = \frac{\text{Total expected replacement cost per cycle}}{\text{Expected cycle length}}$$

Total expected replacement cost per cycle

= Cost of a preventive cycle ×

Probability of a preventive cycle +

Cost of a failure cycle ×

Probability of a failure cycle

$$= C_p R(t_p) + C_f \left[1 - R(t_p)\right]$$

Remember: If $f(t)$ is as illustrated in Figure A.13.3, then the probability of a preventive cycle equals the probability of failure occurring after time t_p; that is, it is equivalent to the shaded area, which is denoted $R(t_p)$.

The probability of a failure cycle is the probability of a failure occurring before time t_p, which is the unshaded area of Figure A.13.3. Because the area under the curve equals unity, then the unshaded area is $[1 - R(t_p)]$.

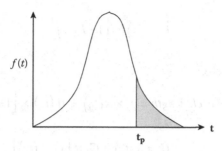

FIGURE A.13.3

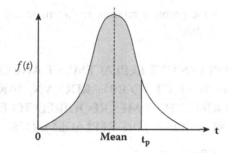

FIGURE A.13.4

Expected cycle length

= Length of a preventive cycle ×
Probability of a preventive cycle +
Expected length of a failure cycle ×
Probability of a failure cycle

$$= t_p \times R(t_p) + (\text{Expected length of a failure cycle}) \times \left[1 - R(t_p)\right]$$

To determine the expected length of a failure cycle, consider Figure A.13.4. The mean time to failure of the complete distribution is

$$\int_{-\infty}^{\infty} t f(t) dt$$

which for the normal distribution equals the mode (peak) of the distribution. If a preventive replacement occurs at time t_p, then the mean time to failure is the mean of the shaded portion of Figure above because the unshaded area is an impossible region for failures. The mean of the shaded area is

$$\int_{-\infty}^{t_p} tf(t)dt / \left[1 - R(t_p)\right]$$

denoted $M(t_p)$. Therefore,

$$Expected\ cycle\ length = t_p \times R(t_p) + M(t_p) \times \left[1 - R(t_p)\right]$$

$$C(t_p) = \frac{C_p \times R(t_p) + C_f \times \left[1 - R(t_p)\right]}{t_p \times R(t_p) + M(t_p) \times \left[1 - R(t_p)\right]}$$

This is now a model of the problem relating replacement age t_p to total expected replacement cost per unit time.

A.14 OPTIMAL PREVENTIVE REPLACEMENT AGE OF EQUIPMENT SUBJECT TO BREAKDOWN, TAKING INTO ACCOUNT THE TIMES REQUIRED TO EFFECT FAILURE AND PREVENTIVE REPLACEMENTS

A.14.1 STATEMENT OF PROBLEM

The problem definition is identical to that of Section A.13 except that, instead of assuming that the failure and preventive replacements are made instantaneously, the time required to make these replacements is taken into account.

The optimal preventive replacement age of the equipment is again taken as that age which minimizes the total expected cost of replacements per unit time.

A.14.2 CONSTRUCTION OF MODEL

1. C_p is the cost of a preventive replacement.
2. C_f is the cost of a failure replacement.
3. T_p is the time required to make a preventive replacement.
4. T_f is the time required to make a failure replacement.
5. $f(t)$ is the probability density function of the failure times of the equipment.
6. $M(t_p)$ is the mean time to failure when preventive replacement occurs at time t_p.
7. The replacement policy is to perform a preventive replacement once the equipment has reached a specified age t_p, plus failure replacements when necessary. This policy is illustrated in Figure A.14.1.

8. The objective is to determine the optimal preventive replacement age of the equipment to minimize the total expected replacement cost per unit time.

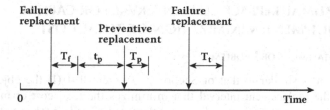

FIGURE A.14.1

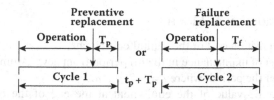

FIGURE A.14.2

As was the case for the problem of Section A.13, there are two possible cycles of operation and they are illustrated in Figure A.14.2.

The total expected replacement cost per unit time, denoted $C(t_p)$, is

$$C(t_p) = \frac{\text{Total expected replacement cost per cycle}}{\text{Expected cycle length}}$$

Total expected replacement cost per cycle

$$= C_p \times R(t_p) + C_f \left[1 - R(t_p)\right]$$

Expected cycle length

= Length of a preventive cycle ×
Probability of a preventive cycle +
Expected length of a failure cycle ×
Probability of a failure cycle

$$= (t_p + T_p) R(t_p) + \left[M(t_p) + T_f\right]\left[1 - R(t_p)\right]$$

$$C(t_p) = \frac{C_p R(t_p) + C_f \left[1 - R(t_p)\right]}{(t_p + T_p) R(t_p) + \left[M(t_p) + T_f\right]\left[1 - R(t_p)\right]}$$

This is a model of the problem relating preventive replacement age t_p to the total expected replacement cost per unit time.

A.15 OPTIMAL REPLACEMENT INTERVAL FOR CAPITAL EQUIPMENT: MINIMIZATION OF TOTAL COST

A.15.1 STATEMENT OF PROBLEM

This problem is similar to that of Section A.9 except that (1) the objective is to determine the replacement interval that minimizes the total cost of maintenance and replacement over a long period; and (2) the trend in costs is taken to be discrete, rather than continuous.

A.15.2 CONSTRUCTION OF MODEL

1. A is the acquisition cost of the capital equipment.
2. C_i is the cost of maintenance in the ith period from new, assumed to be paid at the end of the period, where $i = 1, 2, \ldots, n$.
3. S_i is the resale value of the equipment at the end of the nth period of operation.
4. r is the discount rate.
5. n is the age in periods of the equipment when replaced.
6. $C(n)$ is the total discounted cost of maintaining and replacing the equipment (with identical equipment) over a long period of time with replacements occurring at intervals of n periods.
7. The objective is to determine the optimal interval between replacements to minimize total discounted costs, $C(n)$.

The replacement policy is illustrated in Figure A.15.1.

Consider the first cycle of operation. The total cost over first cycle of operation, with equipment already installed is

$$C_1\left(n\right) = C_1 r + C_2 r^2 + C_3 r^3 + \ldots + C_n r^n + A r^n - S_n r^n$$

$$= \sum_{i=1}^{n} C_i r^i + r^n \left(A - S_n\right)$$

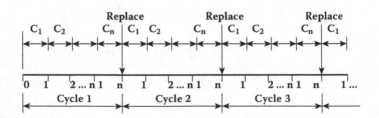

FIGURE A.15.1

For the second cycle, the total cost discounted to the start of the second cycle is

$$C_2(n) = \sum_{i=1}^{n} C_i r^i + r^n (A - S_n)$$

Similarly, the total costs of the third, fourth, etc. cycle, discounted back to the start of their cycle, can be obtained.

The total discounted costs, when discounting is taken to the start of the operation (i.e., at time 0) is

$$C(n) = C_1(n) + C_2(n) r^n + C_3(n) r^{2n} + \ldots + C_n(n) r^{(n-1)n} + \ldots$$

Because $C_1(n) = C_2(n) = C_3(n) = \ldots = C_n(n) = \ldots$, we have a geometric progression that gives, over an infinite period,

$$C(n) = \frac{C_1(n)}{1 - r^n} = \frac{\sum_{i=1}^{n} C_i r^i + r^n (A - S_n)}{1 - r^n}$$

This is a model of the problem relating replacement interval n to total costs.

A.16 THE ECONOMIC LIFE MODEL USED IN PERDEC

The economic life model used in PERDEC is

$$EAC(n) = \left[\frac{A + \sum_{i=0}^{n} C_i r^i - S_n r^n}{1 - r^n} \right] * i$$

where
    A = acquisition cost
    C_i = O&M costs of a bus in its ith year of life, assuming payable at the start of year, $i = 1, 2, \ldots n$
    r = discount factor
    S_n = resale value of a bus of age n years
    N = replacement age
    $C(n)$ = total discounted cost for a chain of replacements every n years

A.17   ECONOMIC LIFE OF PASSENGER BUSES

The purpose of this appendix is to demonstrate how the standard economic life model for equipment replacement can be modified slightly to enable the economic life of buses to be determined, taking into account declining utilization of a bus over its lifetime. Specifically, new buses are mostlyutilized to meet base load demand while older buses are used to meet peak demands, such as "rush hours."

The case study describes a fleet of 2,000 buses whose annual fleet demand is 80 million kilometers. The recommendations resulting from the study were implemented and substantial savings were reported.

A.17.1   INTRODUCTION

The study took place in Montreal, Canada, where Montreal's Transit Commission is responsible for providing bus services to approximately 2 million people. To meet the bus schedules requires a fleet of 2,000 buses undertaking approximately 80,000,000 kilometers per year. In terms of North American bus operators, Montreal has the third largest fleet, falling in behind New York and Chicago. The bulk of the buses used in Montreal are standard 42-seat GMC buses.

The objective of the study was to analyze bus operations and maintenance costs to determine the economic life of a bus and to identify a steady-state replacement policy, that is, one where a fixed proportion of the fleet would be replaced on an annual basis.

A.17.2   THE ECONOMIC LIFE MODEL

Figure A.17.1 illustrates the standard conflicts associated with capital equipment replacement problems.

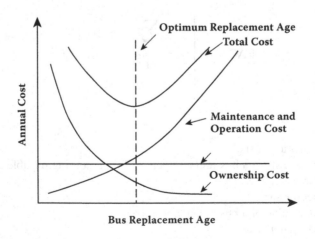

FIGURE A.17.1

The standard economic life model is

$$C(n) = \frac{A + C_1 + \sum_{i=2}^{n} C_i r^{i-1} - S_n r^n}{1 - r^n}$$

where

A = acquisition cost

C_1 = operations and maintenance (O&M) cost of a bus in its first year of life

C_i = O&M cost of a bus in its ith year of life, assuming payable at the start of the year, $i = 2, 3, \ldots n$

r = discount factor

S_n = resale value of a bus of age n years

n = Replacement age

$C(n)$ = total discounted cost for a chain of replacements every n years

The total disocunted cost can be converted to an equivalent annual cost (EAC) by the Capital Recovery Factor, which in this case is i, the interest rate appropriate for discounting. *Note*: $r = 1/(1 + i)$.

A.17.3 DATA ACQUISITION

- *Acquisition cost:* In terms of 1980 dollars, the acquisition cost of a bus was $96,330.
- *Resale value:* The policy being implemented by Montreal Transit was the replacement of a bus on a 20-year cycle, at which age the value of the bus was $1,000. There was considerable uncertainty in the resale value of a bus and, for purposes of the study, two "extremes" were evaluated: one being termed a "high" trend in resale value and the other a "low" trend in resale value.

"High" Trend in Resale Value		"Low" Trend in Resale Value	
Replacement Age (Years)	Resale Value ($)	Replacement Age (Years)	Resale Value ($)
1	77,000	1	2,000
2	65,000	2	2,000
3	59,000	3	2,000
4	54,000	4	2,000
5	50,000	5	2,000
6	46,000	6	2,000
7	42,000	7	2,000
8	38,000	8	2,000
9	34,000	9	2,000
10	31,000	10	2,000
11	28,000	11	2,000

(continued on next page)

"High" Trend in Resale Value		"Low" Trend in Resale Value	
Replacement Age (Years)	Resale Value ($)	Replacement Age (Years)	Resale Value ($)
12	25,000	12	2,000
13	22,000	13	2,000
14	19,000	14	2,000
15	16,000	15	2,000
16	13,000	16	2,000
17	10,000	17	2,000
18	7,000	18	2,000
19	4,000	19	2,000
20	1,000	20	1,000

A.17.4 INTEREST RATE

The interest rate appropriate for discount was "uncertain." In the study, a range was used (from 0% to 20%) to check the sensitivity of the economic life to variations in interest rate. The major conclusions of the study were based on an "inflation fee" interest rate of 6%.

A.17.5 OPERATIONS AND MAINTENANCE COSTS

O&M costs are influenced by both the age of a bus and its cumulative utilization. O&M data were obtained for six cost categories: fuel, lubrication, tires, oil, parts, and labor. Analysis of the costs resulted in the following trend being identified:

$$c(k) = 0.302 + 0.723 \left(\frac{k}{10^6} \right)^2$$

where
k = cumulative kilometers traveled by the bus since new
$c(k)$ = trend in O&M costs (in $/km) for a bus of age k kilometers

A.17.6 BUS UTILIZATION

Figure A.17.2 shows the trend line that was fitted to the relationship between bus utilization (km/yr) and bus age (newest to oldest). The reason for this relationship is that new buses are highly utilized to meet base load requirements, with the older buses being used to meet peak demands.

In the analysis that was undertaken, it was assumed that the relationship identified in Figure A17.2 would be independent of the replacement age of the bus. For example, using the present policy of replacing buses on a 20-year cycle then, in a steady state, 2000/20 = 100 buses would be replaced annually. The total work done by the newest 100 buses in their first year of life would then be:

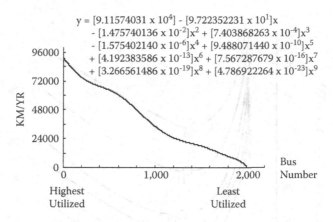

$$y = [9.11574031 \times 10^4] - [9.722352231 \times 10^1]x$$
$$- [1.475740136 \times 10^{-2}]x^2 + [7.403868263 \times 10^{-4}]x^3$$
$$- [1.575402140 \times 10^{-6}]x^4 + [9.488071440 \times 10^{-10}]x^5$$
$$+ [4.192383586 \times 10^{-13}]x^6 + [7.567287679 \times 10^{-16}]x^7$$
$$+ [3.266561486 \times 10^{-19}]x^8 + [4.786922264 \times 10^{-23}]x^9$$

FIGURE A.17.2

$$\int_0^{86361} \left[0.302 + 0.723 \left(\frac{k}{10^6} \right)^2 \right] dk$$

Similarly, the O&M costs for the remaining 19 years can be calculated and inserted into the economic life model, $C(n)$, to enable $C(20)$ to be calculated.

The entire process is then repeated for other possible replacement ages to enable the optimal n to be identified.

Figure A.17.3 shows the results when i ranges from 0% to 20% and the "low" trend in resale value is used.

A.17.7 CONCLUSIONS

Montreal Transit decided to implement a 3-year bus purchase policy on the basis of a 16-year replacement age. The resultant savings were approximately $4 million per year. The class of problem discussed in this appendix is is typical of those found in many transport operations. For example:

- Haulage fleets undertaking both long-distance and local deliveries: When new vehicles are used on long-haul routes initially, and then as they age they are relegated to local delivery work.
- Stores with their own fleets of delivery vehicles, where there are peak demands around Christmas time: The older vehicles in the fleet are retained to meet these predictable demands. Because of this unequal utilization, it is necessary to evaluate the economic life of the vehicle by viewing the fleet as a whole, rather than focusing on the individual vehicle.

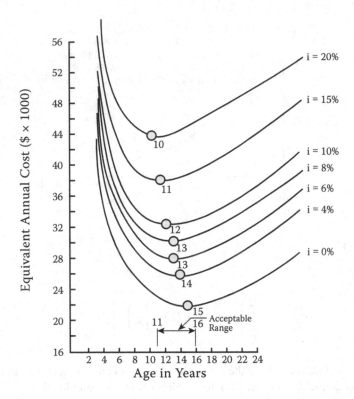

FIGURE A.17.3

REFERENCES

Simms, B.W., Lamarre, B.G., Jardine, A.K.S. and Boudreau, A., Optimal buy, operate and sell policies for fleets of vehicles, *European Journal of Operational Research,* 15(2), 183–195February 1984. (This paper discusses a dynamic programming/linear programming approach to the problem that enabled alternative utilization policies to be incorporated into the model).

AGE/CON User's Manual, OMI, Specialized Software Division, Suite 704, 3455 Drummond Street, Montreal, Quebec, Canada H3G 2R6. (AGE/CON is a software package used for vehicle fleet replacement decisions. It includes the routines for handling problems of the class described in this appendix).

A.18   OPTIMAL REPLACEMENT AGE OF AN ASSET TAKING INTO ACCOUNT TAX CONSIDERATIONS

The formula is

$$\text{NPV}(k) = A\left[1 - \frac{td}{i+d}\left(\frac{1+i/2}{1+i}\right)\right] - \frac{it\,A\left(1-\frac{d}{2}\right)\left(1-d\right)^{k-1}}{\left(i+d\right)\left(1+i\right)^{k}} - \frac{S_k\left(1-t\right)}{\left(1+i\right)^{k}} + \sum_{j=1}^{k}C_j\frac{\left(1-t\right)}{\left(1+i\right)^{j-1}}$$

where

i = Interest rate

d = Capital cost allowance rate

t = Corporation tax rate

A = Acquisition cost

S_k = Resale price

C_j = Operations and maintenance cost in the jth year

NPV(k) = Net present value in the kth year

Remark: $A(1-\frac{d}{2})(1-d)^{k-1}$ is the nondepreciated capital cost.

The equivalent annual cost (EAC) is then obtained by multiplying the NPV(k) by the capital recovery factor

$$\frac{i(1+i)^k}{(1+i)^k - 1}$$

A.19 OPTIMAL REPLACEMENT POLICY FOR CAPITAL EQUIPMENT TAKING INTO ACCOUNT TECHNOLOGICAL IMPROVEMENT: INFINITE PLANNING HORIZON

A.19.1 STATEMENT OF PROBLEM

For this replacement problem it is assumed that once the decision has been made to replace the current asset with the technologically improved equipment, then this equipment will continue to be used and a replacement policy (periodic) will be required for it. It will be assumed that replacement will continue to be made with the technologically improved equipment. Again, we wish to determine the policy that minimizes total discounted costs of maintenance and replacement.

A.19.2 CONSTRUCTION OF MODEL

1. n is the economic life of the technologically improved equipment.
2. $C_{p,t}$ is the maintenance cost of the present equipment in the ith period from now, payable at time i, where $i = 1, 2, ..., n$.
3. $S_{p,t}$ is the resale value of the present equipment at the end of the ith period from now, where $i = 0, 1, 2, ..., n$.
4. A is the acquisition cost of the technologically improved equipment.
5. $C_{i,j}$ is the maintenance cost of the technologically improved equipment in the jth period after its installation and payable at time j, where $j = 1, 2, ..., n$.
6. $S_{i,j}$ is the resale value of the technologically improved equipment at the end of its jth period of operation ($j = 0, 1, 2, ..., n; j = 0$ is included so that we

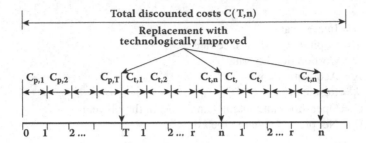

FIGURE A.19.1

can then define $S_{t,o} = A$. This then enables Ar^n in the model to be cancelled if no change is made.

Note that it is assumed that if a replacement is to be made at all, then it is with the technologically improved equipment. This is not unreasonable as it may be that the equipment currently in use is no longer on the market.

7. r is the discount factor.
8. The replacement policy is illustrated in Figure A.19.1.

The total discounted cost over a long period of time with replacement of the present equipment at the end of T period of operation, followed by replacements of the technologically improved equipment at intervals of n, is

$$C(T,n) = \text{Costs over interval } (0,T) + \text{Future costs}$$

$$\text{Costs over interval } (0,T) = \sum_{i=1}^{n} C_{p,t} r^i - S_{p,T} r^T + Ar^T$$

Future costs, discounted to time T, can be obtained by the method described in Section A.16 where the economic life of equipment is calculated. We replace C_i by $C_{t,j}$ to obtain

$$C(n) = \frac{\sum_{i=1}^{n} C_{t,j} r^i + r^n \left(A - S_n \right)}{1 - r^n}$$

Therefore, $C(n)$ discounted to time zero is $C(n)r^T$ and

$$C(T,n) = \sum_{i=1}^{T} C_{p,t} r^i - S_{p,T} r^T + Ar^T + \left(\frac{\sum_{j=1}^{n} C_{t,j} r^i + r^n \left(A - S_n \right)}{1 - r^n} \right) r^T$$

This is a model of the problem relating changeover time to technologically improved equipment, T, and economic life of new equipment, n, to total discounted costs $C(T,n)$.

A.20 OPTIMAL REPLACEMENT POLICY FOR CAPITAL EQUIPMENT TAKING INTO ACCOUNT TECHNOLOGICAL IMPROVEMENT: FINITE PLANNING HORIZON

A.20.1 STATEMENT OF PROBLEM

When determining a replacement policy, there may be equipment on the market that is, in some way, a technological improvement on the equipment currently used. For example, maintenance and operating costs may be lower, throughput may be greater, quality of output may be better, etc. The problem discussed in this section is how to determine when, if at all, to take advantage of the technologically improved equipment.

It will be assumed that there is a fixed period of time from now during which equipment will be required and, if replacement is with the new equipment, then this equipment will remain in use until the end of the fixed period. The objective will be to determine when to make the replacement, if at all, to minimize total discounted costs of maintenance and replacement.

A.20.2 CONSTRUCTION OF MODEL

1. n is the number of operating periods during which equipment will be required.
2. The objective is to determine that value of T, at which replacement should take place, with the new equipment, $T = 0, 1, 2, ..., n$. The policy is illustrated in the figure below.

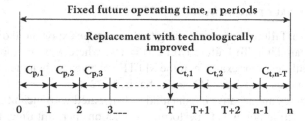

The total discounted cost over n periods, with replacement occurring at the end of the Tth period is

$C(T)$ = Discounted maintenance costs for present equipment over period $(0,T)$

+ Discounted maintenance costs for technologically improved equipment over period (T,n)

+ Discounted acquisition cost of new equipment

− Discounted resale value of present equipment at end of Tth period

− Discounted resale value of technologically improved equipment at end of nth period

$$= \left(C_{p,1} r^1 + C_{p,2} r^2 + C_{p,3} r^3 + \ldots + C_{p,T} r^T \right)$$

$$+ \left(C_{t,1} r^{T+1} + C_{t,2} r^{T+2} + \ldots + C_{t,n-T} r^n \right) + A r^T$$

$$- \left(S_{p,T} r^T + S_{t,n-T} r^n \right)$$

Therefore,

$$C(T) = \sum_{i=1}^{n} C_{p,i} r^i + \sum_{j=1}^{n-T} C_{t,j} r^{T+j} + A r^T - \left(S_{p,T} r^T + S_{t,n-T} r^n \right).$$

This is a model of the problem relating replacement time T to total discounted cost $C(T)$.

A.21 OPTIMAL INSPECTION FREQUENCY: MINIMIZATION OF DOWNTIME

A.21.1 STATEMENT OF PROBLEM

The problem of this section assumes that equipment breaks down from time to time and, to reduce the breakdowns, inspections and consequent minor modifications can be made. The problem is to determine the inspection policy that minimizes the total downtime per unit time incurred due to breakdowns and inspections.

A.21.2 CONSTRUCTION OF MODEL

1. Equipment failures occur according to the negative exponential distribution with Mean Time To Failure (MTTF) = $1/\lambda$, where λ is the mean arrival rate of failures. (For example, if the MTTF = 0.5 years, then the mean number of failures per year = $1/0.5 = 2$, that is, $\lambda = 2$.).
2. Repair times are negative exponentially distributed with mean time $1/\mu$.
3. The inspection policy is to perform n inspections per unit time. Inspection times are negative exponentially distributed with mean time $1/i$.

The objective is to choose n to minimize total downtime per unit time. The total downtime per unit will be a function of the inspection frequency n, denoted $D(n)$. Therefore,

$$D(n) = \text{Downtime incurred due to repairs per unit time}$$
$$+ \text{Downtime incurred due to inspection per unit time}$$

$$= \frac{\lambda(n)}{\mu} + \frac{n}{i}$$

The above equation is a model of the problem relating inspection frequency n to total downtime $D(n)$.

A.22 MAINTENANCE STRATEGIC ASSESSMENT (MSA) QUESTIONNAIRE

This questionnaire should be filled out by the following personnel:

Department	Positions	How Many?
Production	Managers	All
	Superintendents	All
	Supervisors	1 for each area (min.)
	Operators	1 for each area (min.)
Maintenance	Managers	All
	Superintendents	All
	Supervisors	1 for each area (min.)
	Trades	1 for each trade (min.)
Others	Managers	All

The results of this assessment will not be disclosed without the written permission of the firm being evaluated.

MSA Questionnaire

INSTRUCTIONS:
Assign a score to each of the statements in the following questionnaire based on how well you think your maintenance organisation adheres to the statement. The following rating scale must be used:

	Score
Strongly agree	4
Mostly agree	3
Partially agree	2
Totally disagree	1
Do not understand	0

It is not necessary to add the scores up; that will be done when we enter them into our database. The final results will be presented on a scale of 100.

First, please tell us about yourself:

Name: _____   Job Title: _____

Plant/Site: _____   Division: _____

Primary Responsibility	Department
_____Management	_____ Maintenance
_____Supervision	_____ Operations/Production
_____Trades/Hourly	_____ Purchasing
_____Administrative	_____ Tech/IT Support
_____Other	_____ Other

MSA QUESTIONNAIRE

1. MAINTENANCE STRATEGY

Statement	Score (4,3,2,1,0)
The maintenance department has a defined mission, mandate, and a set of objectives that are well documented and understood by all personnel concerned.	
The maintenance mission statement and objectives clearly support a published statement of the company's objectives and goals, and the role of maintenance in achieving the company's objectives is understood.	
We have a long-term plan or strategy to guide maintenance improvement efforts that supports, and is linked to, the overall corporate strategy.	
We have a set of policies or guiding principles for maintenance. Maintenance is seen as a process, not a function.	
Our approach to maintenance is proactive. We do our best to prevent breakdowns; and when something breaks, we fix it immediately.	
Annual maintenance budget is prepared based on a long-term improvement plan, scheduled overhaul strategy, and history of equipment performance. Maintenance budget is related to expected performance, and indications are provided as to the likely outcome if work is to be deferred.	
The maintenance budget has an allowance for any project work being done by the maintenance department. If not, project work is budgeted separately and accounted for outside of maintenance.	
Total (max. 28)	

Note: **Strongly Agree (4), Mostly Agree (3), Partially Agree (2), Totally Disagree (1), Do Not Understand (0)**

2. ORGANIZATION/HUMAN RESOURCES

Statement	Score (4,3,2,1,0)
Maintenance staffing level is adequate, highly capable, and experienced.	
Functions covering plant needs are fully defined, our employees understand what is/is not expected of them, and organizational charts are current.	
The maintenance organization is mostly decentralized and organized by area or product line.	
First line supervisors are responsible for at least 12 to 15 maintenance workers.	
Adequate support staff are available to allow supervisors to spend more than 75% of their time in direct support of their people.	
Overtime represents less than 5% of the total annual maintenance man-hours. Overtime is not concentrated in one trade group or area, but is well distributed.	
Regular technical training is provided to all employees and is more than 5 days per year per employee. Maintenance supervisors have also received formal supervisory training.	
A formal established apprenticeship program is employed to address the maintenance department's needs for qualified trades. Clear standards are set for completing the apprenticeship programs.	
Part of the pay is based on demonstrated skills and knowledge and/or results and productivity.	
Contractors are used to augment plant staff during shutdowns or for specific projects or specialized jobs. Their cost/benefit is periodically reviewed.	
Total (max. 40)	

Note: **Strongly Agree (4), Mostly Agree (3), Partially Agree (2), Totally Disagree (1), Do Not Understand (0)**

3. EMPLOYEE EMPOWERMENT

Statement	Score (4,3,2,1,0)
We don't have a "Command and Control" organization with highly disciplined procedures.	
Multi-skilled tradespeople (e.g., electricians doing minor mechanical work, mechanics doing minor electrical work, etc.) are a key feature of the organization.	
Operators understand the equipment they run, perform minor maintenance activities like cleaning, lubricating, minor adjustments, inspections, and minor repairs (not generally requiring the use of tools).	
Supervisors regularly discuss performance and costs with their work teams.	
Continuous improvement teams are in place and active.	
Much of the work is performed by self-directed work teams of operators, maintainers, and engineers.	
Maintenance is a part of the team involved during design and commissioning of equipment modifications or capital additions to the plant.	
Trades usually respond to call-outs after hours. Operations can get needed support from maintenance trades quickly and with a minimum of effort.	
Call-outs are performed by an on-shift maintainer who decides what support is needed without reference to a supervisor for guidance. Operations do not decide who will be called.	
Partnerships have been established with key suppliers and contractors. Risk-sharing is a feature of these arrangements.	
Total (max. 40)	

Note: **Strongly Agree (4), Mostly Agree (3), Partially Agree (2), Totally Disagree (1), Do Not Understand (0)**

4. MAINTENANCE TACTICS

Statement	Score (4,3,2,1,0)
Less than 5% of the total maintenance work man-hours is devoted to emergencies (e.g., unscheduled shutdowns).	
Condition-based maintenance is favored over time- or cycle-based maintenance.	
Use of condition-based maintenance techniques such as vibration analysis, oil sampling, non-destructive testing (NDT), and performance monitoring is widespread.	
Preventive and/or predictive maintenance represents 60% or more of the total maintenance man-hours.	

Statement	Score (4,3,2,1,0)
Compliance with the PM program is high: 95% or more of the PM work is completed as scheduled.	
Results from PM inspections and failure history data are used to continually refine and improve effectiveness of the PM program.	
For new equipment, we review the manufacturer's maintenance recommendations and revise them as appropriate for our specific operating environment and demands.	
We used a formal reliability-based program for determining the correct PM routines to perform. That program is still used for continuously fine-tuning and improving our PM performance.	
Total (max. 40)	

Note: **Strongly Agree (4), Mostly Agree (3), Partially Agree (2), Totally Disagree (1), Do Not Understand (0)**

5. RELIABILITY ANALYSIS

Statement	Score (4,3,2,1,0)
Equipment history is maintained for all key pieces of equipment, showing cause of failure and repair work completed.	
Equipment failures are analyzed to determine root-cause and prescribe preventive measures.	
Our failure prevention efforts are mostly successful. We can usually eliminate the problems we focus on without creating new problems.	
Equipment Mean Time Between Failures (MTBF) and process or mechanical availability are logged/calculated/forecasted.	
Value-risk studies have been conducted to optimize maintenance programs.	
All equipment has been classified based on its importance to plant operations and safety. The classification is used to help to determine work order priorities and to direct engineering resources. We work first on the most critical equipment's problems.	
Reliability statistics are maintained even though our employees have a good feel for the best and worst equipment.	
Reliability-centered maintenance or other formal analysis is used to determine the optimum maintenance routines to perform on our equipment.	
Total (max. 32)	

Note: **Strongly Agree (4), Mostly Agree (3), Partially Agree (2), Totally Disagree (1), Do Not Understand (0)**

6. PERFORMANCE MEASURES/BENCHMARKING

Statement	Score (4,3,2,1,0)
Labor and material costs are accumulated and reported against key systems and equipment.	
Downtime records including causes are kept on key equipment and systems. These records are periodically analyzed to generate continuous improvement actions.	
The maintenance department has a set of performance indicators that are routinely measured and tracked to monitor results relative to the maintenance strategy and improvement process.	
All maintenance staff has been trained in or taught the significance of the measures we use. Most of us can read the measures and trends, and can determine whether or not we are improving our overall performance.	
All maintenance trades/areas can see and understand the relationship between their work and results of the department overall. If a particular trade/area is weak, they can see it and work to correct it.	
Performance measures are published or posted regularly and kept available/visible for all department staff and trades to see and read.	
Internal and industry norms are used for comparison.	
Maintenance performance of "best-in-class" organizations has been benchmarked and used to set targets for performance indicators.	
Total (max. 32)	

Note: **Strongly Agree (4), Mostly Agree (3), Partially Agree (2), Totally Disagree (1), Do Not Understand (0)**

7. INFORMATION TECHNOLOGY

Statement	Score (4,3,2,1,0)
A fully functional maintenance management system exists, which is linked to the plant financial and material management systems.	
Our maintenance and materials management information is considered a valuable asset and is used regularly. The system is not just a "black hole" for information or a burden to use that produces no benefit.	
Our maintenance management system is easy to use. Most of the maintenance department, especially supervisors and trades, has been trained on it, can use it, and do use it.	
Our planners/schedulers use the maintenance management system to plan jobs and to select and reserve spare parts and materials.	
Parts information is easily accessible and linked to equipment records. Finding parts for specific equipment is easy to do and the stock records are usually accurate.	

Statement	Score (4,3,2,1,0)
Scheduling for major shutdowns is done using a project management system that determines critical paths and required levels of resources.	
Condition-based maintenance techniques are supported by automated programs for data analysis and forecasting.	
Expert systems are used in areas where complex diagnostics are required.	
Total (max. 32)	

Note: **Strongly Agree (4), Mostly Agree (3), Partially Agree (2), Totally Disagree (1), Do Not Understand (0)**

8. PLANNING AND SCHEDULING

Statement	Score (4,3,2,1,0)
A plant equipment register exists, which lists all equipment in the plant that requires some form of maintenance or engineering support during its life.	
Over 90% of maintenance work is covered by a standard written work order, standing work order, PM work order, or a PM checklist or routine.	
Over 80% of maintenance work (preventive, predictive, and corrective) are formally planned by a planner, supervisor, or other person at least 24 hours or more before being assigned to the trades.	
Non-emergency work requests are screened, estimated, and planned (with tasks, materials, and tools identified and planned) by a dedicated planner.	
Realistic assessments of jobs are used to set standard times for repetitive tasks and to help schedule resources.	
A priority system is in use for all work requests/orders. Priorities are set using predefined criteria, which are not abused to circumvent the system.	
Work for the week is scheduled in consultation with production and is based on balancing work priorities set by production with the net capacity of each trade, taking into account emergency work and PM work.	
All shutdowns are scheduled using either critical path or other graphical methods to show jobs, resources, time frames, and sequences.	
Work backlog (ready to be scheduled) is measured and forecasted for each trade and is managed at less than 3 weeks per trade.	
Long-term plans (1 to 5 years) are used to forecast major shutdowns and maintenance work and are used to prepare the maintenance budget.	
Total (max. 40)	

Note: **Strongly Agree (4), Mostly Agree (3), Partially Agree (2), Totally Disagree (1), Do Not Understand (0)**

9. MATERIALS MANAGEMENT

Statement	Score (4,3,2,1,0)
Service levels are measured and are usually high. Stockouts represent less than 3% of orders placed at the storeroom.	
Parts and materials are readily available for use where and when needed.	
Distributed (satellite) stores are used throughout the plant for commonly used items (e.g., fasteners, fittings, common electrical parts).	
Parts and materials are restocked automatically before the inventory on hand runs out and without prompting by the maintenance crews.	
A central tool crib is used for special tools.	
Inventory is reviewed on a regular basis to delete obsolete or very infrequently used items. An ABC analysis is performed monthly.	
Purchasing/Stores is able to source and acquire rush emergency parts that are not stocked quickly and with sufficient time to avoid plant downtime.	
Average inventory turnovers are greater than 1.5 times.	
Order points and quantities are based on lead-time, safety stock, and economic order quantities.	
Inventory is controlled using a computerized system that is fully integrated with the maintenance management/planning system.	
Total (max. 40)	

Note: **Strongly Agree (4), Mostly Agree (3), Partially Agree (2), Totally Disagree (1), Do Not Understand (0)**

10. MAINTENANCE PROCESS REENGINEERING

Statement	Score (4,3,2,1,0)
Key maintenance processes (e.g., planning, corrective maintenance) have been identified, and "as-is" processes are mapped. Those maps are accurate reflections of the processes that are actually followed.	
Key maintenance processes are redesigned to reduce or eliminate non-value-added activities.	
The CMMS or other management systems are used to automate workflow processes.	
Process mapping and redesign have been extended to administration and technical support processes.	
Costs of quality and time for maintenance processes are routinely measured and monitored. Activity costs are known.	
Total (max. 20)	

Note: **Strongly Agree (4), Mostly Agree (3), Partially Agree (2), Totally Disagree (1), Do Not Understand (0)**

Appendix B: RFID Updates

Jordan Olivero, Taylor Teal, and Casey Hidaka

B.1 INTRODUCTION

Radio frequency identification (RFID) is an automatic identification technology that enables an organization's ability to track, monitor, and manage assets. While still in the adoption phase, it is clearly poised for widespread implementation across industries. Its price is dropping, its technology is advancing, and it is becoming a standard in industries that stand to benefit. Every so often, an enabling technology will disrupt an industry. RFID is on the verge of doing this, which is why we wanted to include an appendix to briefly explain RFID. Here, we describe the technology's components and highlight industries that have applied RFID to asset management.

B.2 AUTOMATIC IDENTIFICATION

RFID is an example of automatic identification and data capture (AIDC) technology by which a physical object can be identified automatically. Types of AIDC include UPC bar codes, radio frequency devices, smart cards, magnetic stripes, and vision systems. For our use, UPC bar codes are a familiar point of comparison to RFID. First, we take a look at the advantages of each.

B.2.1 Advantages

UPC BAR CODE	RFID

Advantages include:

- Lower cost
- Comparable accuracy rate
- Unaffected by material type
- Absence of international restrictions
- No social issues
- Mature technology with large installed base

B.2.2 UPC Bar Code

Advantages of the UPC bar code include the following:

- Support for nonstatistical data
- No need for line of sight
- Longer read range
- Larger data capacity

- Multiple simultaneous reads
- Sustainability/durability
- Intelligent behavior

On June 26, 1974, Marsh's Supermarket introduced the first bar code. The first product was a pack of Wrigley's Juicy Fruit gum. Some thought that the new technology would never pay off. Today, it is difficult to purchase a product without a bar code.

Previously, the bar code inventory manager would manually count and record items on a ledger, often resulting in an inaccurate count. Thus, bar codes were created to encode unique data about the product for fast and accurate readability. Although expensive to implement in the adoption phase, the bar code has proved its value because it automated repetitive processes, reduced counting errors, and thus reduced expenses associated with these costly inefficiencies. Its effectiveness drew entire supply chains to embrace the technology.

While certainly a game changer, the bar code has its limits. The bar code cannot store large amounts of data and is only able to store enough data to allow for identification at a stock keeping unit (SKU) level. The non-unique SKU describes general characteristics about a product but not specifics, such as when it expires or the condition of the asset it represents.

B.2.3   RFID

"Radio frequency identification (RFID) technology uses radio waves to automatically identify physical objects (either living beings or inanimate items)."[2]

How does it work? A radio device called a *tag* is attached to the physical object that needs to be identified. Unique identification data about this object is stored on the tag. When such a tagged object comes within range of an RFID reader, the tag transmits these data to the reader, which captures the data and forwards them over suitable communication channels, such as a network or a serial connection, to a software application.

B.2.3.1   RFID Tag[3]

Transmission range and data richness distinguish among the types of RFID tags: passive, active, and semi-active (also known as semi-passive).

- A passive tag does not have an on-board power source and uses the RF (radio frequency) signal emitted from the reader to energize itself and transmit its stored data to the reader.
- An active tag has an on-board power source (a battery or a source of power, such as solar energy) and electronics for performing specialized tasks. An active tag uses its on-board power supply to transmit its data—either autonomously or under system control—to a reader, typically over larger distances than passive tags. It does not utilize the reader's emitted RF power for data transmission. The on-board power supply enables (1) several readers to determine a tag's specific spatial location, and (2) functionality of microprocessors, sensors, and input/output ports.

- A semi-active tag has an on-board power source and electronics for performing specialized tasks. The on-board power supply provides energy to the tag for its operation. However, for transmitting its data, a semi-active tag uses the reader's emitted RF signal.

B.3   RFID BENEFITS

Imagine walking into a supermarket, filling up your shopping cart, and walking directly out of the store without stopping to check out. The cost of your groceries would automatically be debited from an electronic account. Imagine receiving reminders from medicine bottles to take prescribed medications. Imagine a business owner forecasting items' expiration in real-time. Imagine your house automatically setting a room's mood based on an individual's preferences. In fact, these situations are not far from reality. The truth is: Businesses are already innovating their processes with RFID because they (and users) understand the benefits.

RFID is not the solution in every case. Its value is realized only after human interaction has reached the efficiency limit and process discipline cannot be advanced with another AIDC technology. In summary, an RFID implementation can enable lower operating expenses and maximized profitability by

- Reducing inventory
- Heightening security
- Lowering store and warehouse labor expenses (i.e., maintenance, insurance, etc.)
- Preventing out-of-stock items
- Reducing damage, pilferage, and shrinkage

Increased supply chain visibility, by knowing the location and the condition of the asset, has enabled decision makers to decrease lead-times. With this new visibility, they can adjust orders days or weeks ahead of the traditional arrive-time to notice a damaged item or note that the container's security had been compromised. Once on the shelf, auditors can also catch obsolete inventory items faster. We mention only a few of the broad and varied benefits before taking a closer look at how RFID impacts Asset Management.

B.4   ASSET MANAGEMENT

The following direct excerpt is used with permission from the author's publisher.[4]
What characterizes an RFID-enabled Asset Management application?

- The need to manage an asset by knowing its inventory characteristics (manufacturer, model, serial number, description, configuration level (if appropriate), and storage or installation location)
- Attaching a means of identification that contains a unique identifier associated with a background database of information containing specific asset information; this tag may be a bar code, RFID tag, or combination tag

- Detecting the location and other properties and states of this asset in real time by attempting to read the tag data on a periodic as well as an on-demand basis

Asset management applications can tie the unique identity of an asset to its location. This can be accomplished using RFID tags; *passive* RFID tags will provide information about location based on when the tag was last read (zoned location or proximity), and *active* RFID tags can determine more refined location in real-time because tags can be pinpointed using triangulation techniques (also called real-time location systems (RTLS)).

As an example, the ANSI (American National Standards Institute) INCITS 371 standard, developed by the *International Committee for Information Technology Standards,* enables users to locate, manage, and optimize mobile assets throughout the supply chain. Stationary active RFID readers read the asset tags as they pass through zoned locations in a facility or yard. This data and the reader's location information are then transferred into an asset management system. Both local and global/wide-area monitoring is possible. Global asset monitoring is utilizing satellite communication networks to link RFID systems at remote sites.

B.4.1 FLEET MANAGEMENT

Used as a fleet management tool, RFID tags are mounted on transportation items such as power units, trailers, containers, dollies, and vehicles. These tags contain pertinent data about the item by which it can be identified and managed. Readers, both stationary and mobile, are placed at locations through which these tagged items move (for example, access-controlled gates, fuel pumps, dock doors, and maintenance areas). These readers automatically read the data from the tags and transmit it to distributed or centralized data centers as well as an asset-management system. This system can then allow or deny a vehicle access to a gate, fuel, maintenance facilities and so on. Thus, using the data from the tagged items and vehicles, an asset-management system can locate, control, and manage resources to optimize utilization on a continuous, real-time basis. The data captured from the tagged items can be timely and accurate, resulting in elimination of manual entry methods, which, in turn, reduces wait times in lanes and dwell times for drivers and equipment.

An extension of this would be the collection of vehicle diagnostic data and combination with the vehicle's unique ID to improve fleet life-cycle management. This would be an example of fleet monitoring and management.

B.4.2 BENEFITS OF ASSET MANAGEMENT

The benefits of asset management include

- Better use of assets. The ability to locate, control, and use an asset when needed allows fleet asset optimization.
- Improved operations. Accurate and automatic data capture coupled with intelligent control leads to better security of controlled areas, provides pro-active vehicle maintenance, and enhances fleet life.

- Improved communication. Real-time, accurate data provides better communication to customers, management, and operation personnel.

B.4.3 CAVEATS

Initial investment may be required for hardware and infrastructure. Cost increases with the fleet size, the number of data capture points, and the amount of custom implementation services required. In addition, for geographically dispersed operations, wide-area wireless communications such as satellite (or GPS]) communication may be needed, thus increasing [...] cost.

B.4.4 IMPLEMENTATION NOTES

Semi-active, read-only, and read-write tags with specialized on-board electronics (for example, to indicate the status of a data transaction) are generally used. Most importantly, such a tag can be integrated with a vehicle's on-board sensors to relay critical vehicle information such as fuel level, oil pressure, and temperature to a reader. The fleet-management system uses this data to determine proactive maintenance on vehicles, resulting in a longer fleet life.

B.5 APPLICATIONS

The applications of RFID are widespread; this section addresses those that pertain to asset management. Some of the more active entities implementing RFID technology are Wal-Mart and the Department of Defense (DoD). Other recognizable companies are: Target, Albertsons, Best Buy, Tesco, K2, DOW Chemical, Metro, and UPS.

B.5.1 RETAIL

Retail and consumer packaged goods companies have been among the early adopters of RFID technologies, both internally and within their shared supply chains. Wal-Mart rolled-out passive RFID at the pallet and case level in many of its distribution centers and retail locations when testing began in January 2005. The company reengineered its supply-chain management process with RFID technology and set strong expectations for its suppliers to adopt it as well.

B.5.2 GOVERNMENT: DEFENSE AND PHARMACEUTICAL

Municipal, state, and federal government agencies employ RFID to reduce costs and offer better services. Among the first to implement RFID in the supply chain, the United States Department of Defense (DoD) issued an RFID policy in 2004, affecting many of its 43,000 suppliers, with the aim of streamlining supply chain operations, rooting out inefficiencies, and meeting its strategic imperatives. Today, most containers include an RFID tag to provide greater visibility and flexibility in meeting the logistical challenges of managing war efforts 7,000 miles away, in Iraq and Afghanistan.

Another adopter in the government is the Food and Drug Administration (FDA), which has encouraged its suppliers to use RFID. Growers, distributors, and producers of food and food products use it to track location and monitor the temperature of food as it moves through the supply chain. RFID has also been helpful in preventing counterfeit drugs from infiltrating the pharmaceutical supply chain and controlling the theft of drug shipments. Some companies now use it to collect pedigree data, a useful tool to fight counterfeiting and ensure drug safety. The technology shows a lot of promise for tracking the pedigree of drugs and may be an extremely useful tool to fight counterfeiting and ensure drug safety.

B.5.3 AUTOMOTIVE

The automotive industry uses RFID to improve visibility and optimize just-in-time inventory.

Forecasting is a particularly challenging issue for most auto companies, stemming from the large number of automobile and feature configurations available. By using RFID tags, auto companies can reduce the time-to-market of a particular configuration. RFID is aiding the manufacturing process by tracking reusable containers, work-in-process, and finished inventory. Faster than UPC bar code technology, RFID-tagged parts enable manufacturers to locate needed parts, know when a part's quantity is low, and reduce inventory costs.

B.5.4 PERSONNEL EMERGENCY LOCATION AND SECURITY

Imagine a fire in a building of 600 employees. How would rescuers know if everyone had escaped? If RFID is able to determine that three people are still inside, how would they be located?

Active RFID tags combined with location awareness and safety software can provide location identification to a high degree of accuracy (using specialized technologies such as ultra wideband). Imagine the power of knowing where each employee is located and being able to grant or deny access to authorized users. In the event of an emergency, RFID may even save lives.

B.5.5 CONTAINER SHIPPING

RFID is revolutionizing the global supply chain by enabling status updates on cargo containers as well as near-real-time wireless global access to their content's status and location.

This "moving virtual warehouse" has the potential to update the status of its contents as soon as it arrives at its destination or at any node desired along the supply chain. As a result, supply chain stakeholders can receive and respond to product data. The benefits spread across stakeholders from consignees to operation managers to Customs' authorities: lowered costs to market, improved control of just-in-time delivery of components for assembly, and reduced warehousing overheads.

RFID enables the ability to monitor, maintain, and verify the required conditions from source to market. The sharing of this diverse information on containers and their contents is changing the nature of the global transportation industry.

B.6   CHALLENGES

As an emerging technology, RFID's largest challengers are information privacy and read accuracy. Still, industry experts say there is no reason to stall RFID implementation projects.

For example, read accuracy is improving as the industry develops best practices. Rather than reflect radio waves, liquids, and metals absorb them, making it generally more difficult to read tags close to liquids or metals. Innovative manufacturers are countering this limitation by extensive research to design an easy-to-read tag in these particular environments.

Privacy for the consumer and the corporation is one of the most frequently discussed topics, and developers are working on ways to secure sensitive information. To enhance consumer privacy, the Clipped Tag was introduced by IBM researchers Paul Moskowitz and Guenter Karjoth in 2005. Applied in the retail environment, a consumer can tear off a portion of the tag after the point of sale. The Clipped Tag is in use today and enables retailers to tag individual consumer items. Despite some challenges, the technology in its current form today remains widely successful.

B.7   CONCLUSION

In this appendix, we looked at the components of RFID and its merits in asset management. RFID stands, as UPC bar codes did in the early 1970s, on the brink of mass implementation. It will impact and transform business by providing greater asset visibility and enabling a dynamic infrastructure. As improvements are made and the technology drops in price, the market will innovate ways to integrate RFID into many cross-industry applications. Marketplace leaders are already embracing RFID—the technology is at the foundation of making the planet's assets more interconnected, instrumented, and intelligent. We stand at the beginning of the "Internet of Things."

REFERENCES

1. Lahiri, Sandip, *RFID Sourcebook,* ©2006. Reprinted by permission of Pearson Education, Inc., Upper Saddle River, NJ: p.128.
2. Lahiri, Sandip, *RFID Sourcebook,* ©2006. Reprinted by permission of Pearson Education, Inc., Upper Saddle River, NJ: p.1
3. Lahiri, Sandip, *RFID Sourcebook,* ©2006. Reprinted by permission of Pearson Education, Inc., Upper Saddle River, NJ: p. 9–17
4. Lahiri, Sandip, *RFID Sourcebook,* ©2006. Reprinted by permission of Pearson Education, Inc., Upper Saddle River, NJ: p. 74–77.

Appendix C: PAS 55—An Emerging Standard for Asset Management in the Industry

Don Barry and Jeffrey Kurkowski

C.1 INTRODUCTION

This appendix documents, at a high level, another approach to assessing and managing assets that is now formalized in a standard (PAS 55). It originated in the United Kingdom and is now being globally recognized to some degree. We will describe the definition, background, and objectives of the PAS 55 standard and how our Asset Management Center of Excellence models and currently available enterprise asset management (EAM) systems can enable, leverage, and support the implementation of this standard.

The Publicly Available Specification (PAS) 55-1: 2008: Asset Management Standard was first published in 2004. The Institute of Asset Management and the British Standards Institute (BSI) worked together to develop strategies to help reduce risks to business-critical assets. This project resulted in PAS 55. This new standard is the culmination of their latest thinking in terms of leading practices in asset management systems. PAS 55 is becoming internationally accepted as an industry standard for quality asset management. This standard was first applied in the electrical power generation, transmission, and distribution sector in the United Kingdom and has since evolved to be applied in other business sectors and geographies. The standard can act as a valuable guideline for asset life-cycle management, quality control, and compliance.

Significant amounts of money and time are spent on managing business-critical assets each year, yet there is still confusion over terminology, and a wide variety of management approaches are in use. In many cases, these approaches serve asset management needs well, but often they do not, sometimes resulting in high-profile failures. To the Institute of Asset Management (United Kingdom), it was apparent that there was a crucial need to provide a consistent framework for asset management systems.

In this appendix, one will find references to the terms "leading" and "best" practices. It is a contention that one will be presented with what some refer to as "best practices" when looking to improve a certain area or set of processes. It is universally recognized that a "best practice" or set of "best practices" that may be "best" for one enterprise or organization, may not be the "best" for another. Therefore, the concept promoted in here takes the approach that an enterprise or organization will review what is the most common, or "leading practices," when looking to improve an area or set of processes. They will adopt the ones they believe are a good fit for them and apply them as necessary. When they have completed the exercise, they will have defined their "best practices."

C.2 THE PAS 55 ASSET MANAGEMENT STANDARD

The PAS 55 asset management standard gives guidance toward leading practices on asset management and is typically relevant for all asset-intensive industries. PAS 55 defines asset management as "systematic and coordinated activities and practices through which an organization optimally and sustainably manages its assets and asset systems, their associated performance, risks and expenditures over their life-cycles for the purpose of achieving its organizational strategic plan." [1]

There are different levels at which asset units (or classes) can be identified and managed—ranging from discrete assets to more complex functional asset systems, networks, sites, or diverse portfolios. Figure C.1 shows examples of priorities that might be evident at the different levels of asset integration and management.

The standard can be focused on all types of assets, varying from critical or strategic *physical assets* to *human assets*. The physical assets are positioned in the following five asset classes (Figure C.2):

- Real Estate and facilities (offices, schools, hospitals)
- Plant and Production (oil, gas, chemicals, pharmaceuticals, food, electronics, power generation)
- Mobile Assets (military, airlines, trucking, shipping, rail)
- Infrastructure (railways, highways, telecommunications, water and wastewater, electric and gas distribution)
- Information Technology (computers, routers, networks, software, auto discovery, service desk)

PAS 55 states that the definition of asset management represents a significantly greater scope than just the maintenance or care of physical assets.

C.3 THE SCOPE OF PAS 55

To be successful, it is vital that this standard be implemented as an integral part of the overall business environment of an organization. Data that should already be available on condition, performance, activities, costs, and opportunities is needed for the foundation of a successful implementation. It is also important that intangible assets are taken into account regarding reputation, image, and social impact. From a financial perspective, information about life-cycle costs, capital investment criteria, and operating cost is essential. The human-asset perspective is necessary to get a good view of motivation, expertise, and roles and responsibilities of the people and

Typical Priorities and Concerns

FIGURE C.1   The typical priorities and concerns evident when integrating and managing assets and asset systems.

Real Estate and Facilities   Plant and Production   Mobile Assets   Infrastructure   Information Technology

FIGURE C.2   The Asset Class Model defines unique asset tracking requirements for each of the classes in both manual and automated systems. *Source*: IBM Asset Management Center of Excellence (AMcoe) Models.

leadership teams involved within the organization. PAS 55 promotes the integrated approach as a key principle with attributes that include a holistic view, a systemic and systematic approach that validates risk, strives for optimization, and is sustainable (Figure C.3). These key principles are key to driving strategic transformation and are key change design points.

C.4 PAS 55 AND COMPLIANCE

PAS 55 is designed to help organizations display asset management competency by meeting a particular set of requirements. Requirements address "leading practices" rather than "best practices" in each area.

All applied processes must be effective and must require evidence of what is being done and why. The standard is nonprescriptive—as in standards like the International Organization for Standardization (ISO) 9001, ISO 14001, or Occupational Health and Safety Assessment Series (OHSAS) 18001. All elements of the standard framework need to be covered in the process.

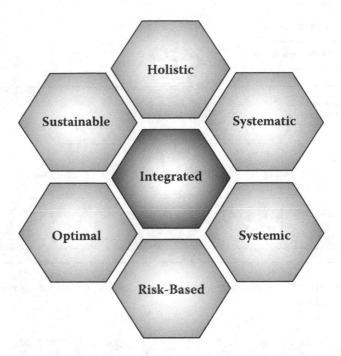

FIGURE C.3   This PAS 55 model highlights key principles of successful asset management that are grouped around an integrated coordinated approach.

The PAS 55 standard is independent of an asset distribution or asset ownership structure and is based on the concept of the PDCA cycle (Plan-Do-Check-Act), meaning that measurable continual improvement is an integral part of the approach. This makes the PAS 55 standard an ideal complement to certified management systems that may already be in place. Using the standard provides assurance to the organization and to its external stakeholders that physical infrastructure assets are managed in an optimal way as a result of an independent third-party audit.

However, it should be recognized that "compliance" still depends on the enterprise, how it uses the PAS 55 guidelines, and the auditing processes in place to confirm its overall "compliance" requirements.

C.5 THE BENEFITS OF PAS 55

In today's economy, factors that drive the need for good asset management are becoming more apparent. Asset risks are appearing more often on the boardroom agenda and there is more of a focus on regulatory compliance from governmental and industry institutions. Organizations are placing a clear emphasis on cost containment, price management, return on investment, and increased overall asset value. The bar is being raised by a worldwide interest in lean principles, asset management, and asset performance. Many sectors are seeing increased expectations from consumers about quality and service delivery, as well as green initiatives. On top of this, there is an increased complexity of assets, tools, and equipment as assets become more interconnected, instrumented, and intelligent.

The PAS 55 standard can benefit companies not only from the regulatory point of view, but also to help them gain competitive advantage by ensuring that they are effectively managing their assets. Using this standard methodology for comprehensive asset management can drive cost savings and service improvement.

Overall, using the PAS 55 standard encourages companies to do the following:

- Achieve asset management leading practices.
- Start processes to map the entire asset base and create the information strategy in accordance with the companyís overall strategy.
- Organize around true life-cycle asset management processes.
- Challenge and reduce current time-based work and replace with a ìrisk-basedî management approach.
- Position asset management-specific accountability from the ìshop floor to the top floorî and create motivational performance management.
- Focus on building the asset management knowledge base.
- Understand and target the tools, and engage the entire organization.
- Adopt a truly holistic approach by continuously challenging good or best practices.

C.6 PAS 55: FUTURE CHALLENGES AND DIRECTIONS

More companies are realizing the benefits of the PAS 55 and the asset life-cycle approach. The drivers for adaptation include the increasing requirements of different regulators, the influence of financial and insurance companies, and the desire to improve the overall image of the organization to the market.

Sector-specific application guideline projects are being launched, including guidelines for property asset management. Additionally, EAM vendors are preparing for alignment with the PAS 55 standard.

Specific challenges for asset management include the following:

- Integrating asset management into companiesí long-term strategies by creating a Chief Asset Officer (CAO) role on the board
- Connecting and integrating asset management with financial and asset management strategies and processes
- Developing a competency-building framework of asset management educational tools
- Assuring environmental, regulatory, and legal compliance to meet sustainable manufacturing requirements

C.7 PAS 55: MANAGEMENT SYSTEM STRUCTURE

Figure C.4 illustrates the elements of asset management as a continual cycle of activity.

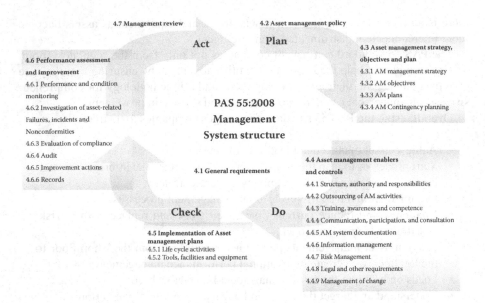

FIGURE C.4 PAS 55 defines the elements of asset management as a continual cycle of activity.

C.7.1 How Asset Management Models Compare
and Support the Benefits of PAS 55

The original edition of this book—and arguably the original version of the Maintenance Excellence Pyramid that has been in place since the early 1990s—have displayed a complete approach to asset management that supports much, if not all, of the assertions and attributes of a PAS 55 approach, and more. Since that time, IBM's Asset Management Center of Excellence has embraced the concepts, enhanced them, and defined a set of three comprehensive models and approaches to cover all aspects of maintenance operations. The models cover the differences in managing five classes of assets, one's overall organization, and an organization's total life-cycle strategy. We would not suggest that a detailed comparison be proposed here. Because PAS 55 is an emerging standard, we mention the two approaches as a point of reference and to drive and confirm the thought process within an enterprise. We also submit that simply applying a full maintenance assessment and executing the key opportunities would significantly improve one's maintenance operation. This would also contribute to an organization that elected to embrace the PAS 55 standard as a globally recognized means of provoking improvements in their management of assets. This approach can be complimentary to organizations that have embraced, or have contemplated embracing, PAS 55 requirements but are also looking for the prescriptive details behind the holistic approach to way they do Asset Management.

In Figure C.5, either approach creates provocation for improvement and change. It is how one identifies the need for improvement and change and aligns them with PAS 55 requirements is the suggested approach to assessing one's asset management complete requirements while addressing PAS 55 compliance.

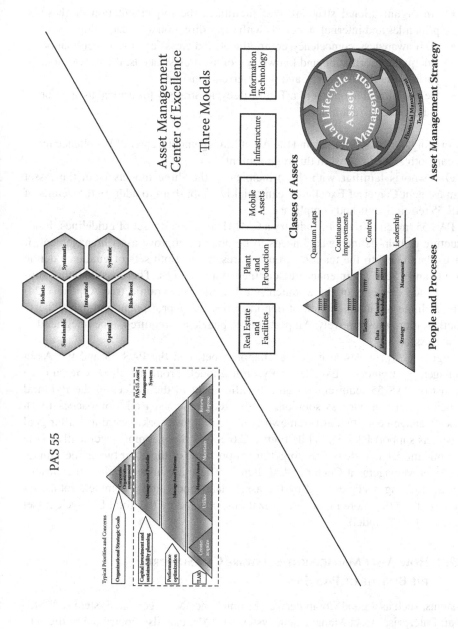

FIGURE C.5 A set of essential base principle models that should be considered for successful asset management execution.

A base principles model(s) for asset management is considered essential in any successful execution. The following principles are considered essential to the successful implementation of an asset management transformation:

- An organizational structure that facilitates the implementation of these principles and internal standards with clear direction and leadership
- Staff awareness, competency, commitment, and cross-functional coordination
- Adequate information and knowledge of asset class needs, their condition, performance, risks, costs, and the interrelationships
- Definition and adoption of a Total Lifecycle Strategy for critical assets (for mature organizations)

Leveraging the PAS 55 model and the Asset Management Center of Excellence models can both serve to assist in this requirement.

One who is familiar with the principles of the three models from the Asset Management Center of Excellence can quickly adopt them to address the details of PAS 55 requirements and compliance

PAS 55, much like the IT industry's CMMI standard is a set of guidelines. It has structure to help define "what" needs to be done to improve asset management. It leaves the "how" to the team or group addressing it. Both sets of guidance define "good" and "best" maintenance disciplines and practices. The questions that support the ten elements of the Maintenance Excellence Pyramid (see Appendix 24) will facilitate the identification of opportunities for improvement and drive "best" practices more prescriptively. All parts of the puzzle are required to be integrated to be successful.

Figure C.6 provides a high-level mapping between the PAS 55 and the Asset Management Center of Excellence Pyramid model. From the chart, one can see the list of PAS 55 requirements and the alignment to the indices in the Pyramid model, which can address solutions to the requirements. Both approaches to an Asset Management Standard framework are looking to work toward a similar goal as well as support PAS 55. While Figure C.6 covers the mapping specifically to the Pyramid model, one should not forget the importance of the other two models under the Asset Management Center or Excellence. These are understanding the various unique tracking needs defined by the asset class model and a complete total lifecycle strategy (for more mature organizations) covered by the Total Lifecycle Asset Model (TLAM model).

C.7.2 How Asset Management Software Can Support the Benefits of PAS 55

Systems, such as a good Computerized Maintenance Management System (CMMS) or an Enterprise Asset Management system (EAM), can also contribute to the support of PAS 55 requirements.

PS 55: 2008 Sections	How To Support Section (from Pyramid)
4.1 General requirements	Maintenance Strategy
4.2 Asset management policy	Maintenance Strategy
4.3 Asset management strategy, objectives and plans	Maintenance Strategy
4.3.1 AM management strategy	Maintenance Strategy
4.3.2 AM objectives	Maintenance Strategy, Measures
4.3.3 AM plans	Maintenance Strategy
4.3.4 Contingency planning	Maintenance Strategy
4.4 Asset management enablers and controls	Maintenance Strategy
4.4.1 Structure, authority and responsibilities	Maintenance Strategy
4.4.2 Outsourcing of AM objectives	P&S, Maintenance Strategy
4.4.3 Training, awareness and competence	Management
4.4.4 Communication, participation, and consultation	Maintenance Strategy, Management
4.4.5 AM system documentation	Data
4.4.6 Information management	Data
4.4.7 Risk Management	Data, RCM, Maintenance Strategy
4.4.8 Legal and other requirements	Maintenance Strategy
4.4.9 Management of change	Maintenance Strategy, Management, P&S
4.5 Implementation of Asset management plans	Maintenance Strategy
4.5.1 Life cycle activities	RCM, P&S, Life-cycle
4.5.2 Tools, facilities and equipment	PCM, P&S, Data
4.6 Performance assessment and improvement	Maintenance Strategy
4.6.1 Performance and condition monitoring	Tactics
4.6.2 Investigation of asset-related Failures, incidents and nonconformities	Tactics, RCm
4.6.3 Evaluation of compliance	P&S, Measures
4.6.4 Audit	Measures, Maintenance Strategy
4.6.5 Improvement actions	BPR
4.6.6 Records	Data
	Maintenance Strategy, Management

FIGURE C.6 How Proven Methods from the Pyramid Model Can Help Drive a PAS 55 Requirement Compliant Solution.

A good system can provide the following:

- A single version of the truth for all asset-related information
- A place to define performance criteria, and associated key performance indicator (KPI) measures the identification of performance failure
- Tracking and management of incidents, problems, and change
- Support for corrective and preventive action(s)

The ability of a good system workflow engine to model and monitor processes and procedures provides a communication mechanism for the user community and a secure data repository for all asset-related information. For example, IBM Maximo Asset Management provides capabilities that support the benefits of PAS 55, providing a communication mechanism for the user community and a secure data repository for all asset-related information.

C.8 SUMMARY

Today's economy is driving the need for smarter asset management with increased expectations from companies, regulators, and shareholders at a time when assets are becoming much more interconnected, instrumented, and intelligent. Although formally first published in 2004, the PAS 55 standard is growing in acceptance and adoption by companies in the United Kingdom and beyond. While the PAS 55 standard works to ensure consistency across the growing asset management requirements, it is primarily intended to be aligned with other key processes within an organization.

Good execution of the principles of the three Asset Center of Excellence Models will significantly contribute to an organization that has started out to align with PAS 55 or is at least considering it. Also,, a good CMMS/EAM system can enable a natural alignment with PAS 55 by providing capabilities and functionalities that allow capital asset-intensive industries to leverage the implementation of this standard.

Like the three models that have been leveraged by multiple enterprises over many years, PAS 55 can be leveraged as a maintenance strategy transformation tool. A strategy and framework "tool" like the three models and PAS 55 is often required to help organizations facilitate change. If an organization is more likely to embrace an emerging new standard (PAS 55) as part of their transformation, it should be supported by all means. The three models and PAS 55 can act as the change agent. Both can also improve customer service, increase return on assets, enable greater compliance, improve asset performance, and reduce risk. Both work to accelerate this transition time to benefit, and both can be deployed, even together, to complement the transformation within a given organization.

REFERENCES

IAM – BSI - PAS55-1 2008. Asset Management. Part 1. Specification for the optimized management of physical assets.

IBM White Paper: Enabling the benefits of PAS 55: The new standard for asset management
in the industry – 2009.

John Woodhouse presentation. Process integration whole-life planning and optimization in
asset management. 2009. Managing Director of TWPL (UK), Chairman, Development
& Standards Committee, IAM.

Index

Printed in the United States
by Baker & Taylor Publisher Services